Engineering Mec

Volume 1

S

Engineering Mechanics

Volume 1

Statics

PER-ÅKE JANSSON AND RAGNAR GRAHN

DIVISION OF MECHANICS
CHALMERS UNIVERSITY OF TECHNOLOGY
GÖTEBORG, SWEDEN

PRENTICE HALL

New York London Toronto Sydney Tokyo Singapore

First published 1995 by
Prentice Hall International (UK) Limited
Campus 400, Maylands Avenue
Hemel Hempstead
Hertfordshire, HP2 7EZ
A division of
Simon & Schuster International Group

Typeset in 9.5/12 pt Palacio
by Mathematical Composition Setters Ltd, Salisbury, Wiltshire

Printed and bound in Great Britain by
T.J. Press (Padstow) Ltd

Library of Congress Cataloging-in-Publication Data

Jansson, Per-Åke.
Engineering mechanics / by Per-Åke Jansson and Ragnar Grahn.
p. cm.
Includes bibliographical references and index.
Contents: V. 1. Statics – v. 2. Dynamics.
ISBN 0-13-220146-1 (v. 1). – ISBN 0-13-309949-0 (v. 2)
1. Mechanics, Applied. I. Grahn, Ragnar. II. Title.
TA350.J33 1995
620.1–dc20 94-10526
CIP

British Library Cataloguing in Publication Data

A catalogue record for this book is available from the British Library

ISBN 0-13-220146-1 (pbk)

1 2 3 4 5 99 98 97 96 95

Contents

Preface

The present book, which consists of two volumes, *Statics* and *Dynamics,* is based on a textbook in Swedish that has been used for some years at Chalmers University of Technology in Göteborg, Sweden, for several undergraduate courses in mechanics.

The main purpose when the project started was to provide Swedish students with a book that was better adapted to their background and their abilities than existing books. Our aim has been to write a book at an adequate mathematical level with a treatment of the subject that is rigorous enough to be accepted for a university course. On the other hand, it is a well-known fact that a book of this kind must not be too wordy if students are to read it. Based on our experience from a number of courses we believe that the result of this process is a book that is not unnecessarily talkative but still sufficiently detailed. Our expectations are that the present English edition should be useful to our colleagues outside Sweden and their students.

The book should be suitable for elementary courses in mechanics at university level. It is also possible to select parts of the book for courses at a lower level. In fact, it has been used successfully in courses within a two-year engineering education programme in Sweden.

We have not incorporated topics normally covered in courses on strength of materials, dynamics of machines, etc. However, examples from such areas have been used to illustrate the applications of the fundamental principles. We assume that the student has a basic knowledge of calculus (differentiation and integration of simple functions, solution of elementary differential equations) and vector algebra. For the student who does not have the adequate background, or needs to refresh his or her memory, a summary of basic mathematics is given in Appendix 1.

Every lecturer in the field is aware of the fact that a large number of students have tremendous difficulties in grasping the meaning of the most fundamental concept of all in mechanics, namely *force*. The main reason is probably because the concept of force, as it appears in Newton's laws of motion, is not consistent with the everyday experiences that have formed the student's ideas from early childhood. During the last decades extensive pedagogical research has been carried out concerning the conceptions of physical phenomena among students. Since we have been personally involved in a research project where the understanding of the concept of force has been studied extensively, our experience has affected the approach considerably, especially in the *Dynamics* volume.

Elementary courses in mechanics are normally problem oriented. Hence, we have included several worked examples carefully chosen to illustrate the fundamental principles. In addition, a large number of problems suitable for solving in the classroom or for homework are provided. Within each section these have been arranged primarily according to the contents. The problems are classified with regard to the degree of difficulty. Thus a number of more elementary, relatively uncomplicated, problems are marked with the symbol '○'. These are all of an introductory character and are designed to help the student get started with problem solving. Normally, they are examples of straightforward applications of the theory. The more demanding problems are marked with the symbol ◎. These may involve lengthy calculations, algebraic or numerical, or require that some clever trick is used.

PER-ÅKE JANSSON
RAGNAR GRAHN

Notation

Latin letters

$\boldsymbol{A}$	area
$\boldsymbol{a}$ a	acceleration
arcsin ...	inverse function of sin ...
arsinh ...	inverse function of sinh ...
c	speed of light: 299 792 458 m/s
c	damping coefficient
$\mathrm{d}\boldsymbol{r}$ $\mathrm{d}s$	element of path
E	energy
$\mathbf{e}$	unit vector
e	base of natural logarithms
$\exp x$	exponential of x: e^x
e	coefficient of restitution
$\boldsymbol{F}$ F	force
f	frequency: $1/\tau$
G	gravitational constant: $6.6726 \cdot 10^{-11}\ \mathrm{m^3/kg/s^2}$
g	acceleration of free fall
I I_{xx}	moment of inertia
I_{xy}	product of inertia
$\mathbf{I}$	inertia matrix
i	imaginary unit: $\mathrm{i}^2 = -1$
k	spring constant, stiffness
k	radius of gyration
$\boldsymbol{L}$	angular momentum
L	length
$\ln x$	natural logarithm of x
$\boldsymbol{M}$ M	couple, moment of force, bending moment
m	mass
N	axial force
N	normal force
P	power
$\boldsymbol{p}$	(linear) momentum: $m\boldsymbol{v}$
q	load per unit length
$\mathbf{R}$	rotation matrix
$\boldsymbol{r}$	position vector
r R	radius
r	(r, φ) plane polar coordinates
S	cable tension
s	arc length, path variable
$\dot{s}$	tangential velocity
T_j^k	homogeneous transformation matrix from $(x_k y_k z_k)$ to $(x_j y_j z_j)$
T	kinetic energy
t	time
V	potential energy
V	shear force
V	volume
$\boldsymbol{v}$	velocity
v	speed, velocity component
W	work

Greek letters

$\boldsymbol{\alpha}$ α	angular acceleration
Δx	increment of x
ζ	(ξ, η, ζ), principal axes of inertia
ζ	damping factor: $x(t) = A \exp(-\zeta\omega t) \sin \omega_d t$
η	efficiency
η	(ξ, η, ζ), principal axes of inertia
θ	(φ, θ, ψ), Euler angles
λ	mass per unit length
μ_k μ_s	coefficient of kinetic, static friction
ξ	(ξ, η, ζ), principal axes of inertia
$\boldsymbol{\rho}$	relative position vector
ρ	(mass) density
ρ	radius of curvature
Σ	summation
ΣF	sum of external forces
ΣM	sum of moment of external forces
σ	mass per unit area
τ	period, periodic time: $1/f$
φ	(r, φ), plane polar coordinates
φ	(φ, θ, ψ), Euler angles
ψ	(φ, θ, ψ), Euler angles
$\boldsymbol{\Omega}$ Ω	angular velocity
$\boldsymbol{\omega}$ ω	angular velocity
ω	angular frequency: $2\pi f$
ω_d	angular frequency of damped oscillator

Other symbols

∞	infinity
$\dot{}$	time derivative: $\dot{x} = dx/dt$
$\bar{}$	refers to the centre of mass (CM); $\bar{v}$, $\bar{a}$ (velocity, acceleration of CM), $\bar{L}$ (angular momentum with respect to CM)
$\frown$ $\frown$	moment about A: $\overset{\frown}{A}$ clockwise, $\overset{\frown}{A}$ counterclockwise
$\to$	approaches
$\nearrow$ $\uparrow$	projection directions
$\overrightarrow{}$	position vector from A to B: $\overrightarrow{AB}$
^{T}	transpose of matrix $\mathbf{A}$: $\mathbf{A}^T$
$^{-1}$	inverse of matrix $\mathbf{A}$: $\mathbf{A}^{-1}$
$\nearrow$	vector notation for $\boldsymbol{M}$, $\boldsymbol{L}$ and $\boldsymbol{\omega}$
$\bullet$	product of scalars, $a \cdot b$; scalar product of vectors, $\boldsymbol{a} \cdot \boldsymbol{b}$
$\times$	vector product, $\boldsymbol{a} \times \boldsymbol{b}$

Introduction

I.1 Mechanics – a presentation

Mechanics is one of the main areas of physics and, at the same time, one of the fundamental disciplines within all branches of engineering education. The concepts, models and laws building up the subject have remained fairly unchanged for a very long time. Furthermore, the view on the structure and contents of the subject is very similar all over the world.

Mechanics is the study of *material systems* at rest or in motion and the corresponding *forces* acting. Mechanics is usually divided into areas according to Figure I.1.

The term *dynamics* refers to the study of systems in motion. It thus includes *kinematics* and *kinetics*. Courses and textbooks in mechanics are often divided into *statics* and *dynamics*.

The proportions between these areas in a certain education programme are, naturally, related to the field of study: in a civil engineering course more attention is paid to statics, while dynamics is the predominant part of a mechanical engineering course.

In this context the term 'mechanics' usually refers to what is called *classical mechanics*, a field of study that was founded in the seventeenth century by Galileo and, above all, Newton. Classical mechanics may be applied to systems of enormous variation in size, from the stars and other celestial bodies down to objects approaching a molecule in size. When we proceed further down and reach the atomic level, however, *quantum physics*[1] must be applied.

When the velocity v of a body is of the same order of magnitude as the speed of light c it is necessary to use *relativistic mechanics*[1]. The range of validity of classical mechanics thus does not include quantum physical or relativistic effects. However, this does not prevent classical mechanics from being a completely sufficient tool when the predominant parts of engineering problems are to be solved.

In mechanics we consider fundamental physical concepts and laws. We attach great importance to their background and mutual connections as well as to their applications.

1 *Quantum physics* and the *theory of relativity* were developed during the first part of the twentieth century. Important achievements were made by Planck, Einstein and Bohr among others.

Systems	Without regard to causes (forces)	With regard to causes (forces)
At rest	(Geometry)	Statics
In motion	Kinematics	Kinetics

Figure I.1

Generally, you cannot say that mechanics is an experimental *or* a theoretical discipline; in mechanics we study the real world but in doing so we utilize mathematics as a tool to a large extent. The experimental basis for the subject is obvious. In the laboratory arranged experiments may be carried out, but we also make observations of phenomena occurring around us, in the cosmos as well as in our close neighbourhood.

In fact, humans carry out mechanical experiments from early childhood with their own bodies, with toys and other objects. They study the surrounding world, the streaming water, the wind, the flight of the birds and the falling leaves. Through such experiences we have all developed concepts and ideas of the world from a mechanical point of view.

There are numerous examples of such concepts. In many cases they are rather diffuse. If we name them we may use terms like 'balance', 'force', 'inertia', 'pressure', 'power', etc. We often have a good practical experience of the laws governing mechanical phenomena, for instance the motion of a projectile, a pendulum or a bouncing body.

Later, at school, when we are confronted with the subject from a scientific point of view, we bring our own ideas of concepts, laws and designations. There are enormous advantages but also risks and difficulties in doing so. This is often painfully obvious in dynamics when concepts like 'inertia' and 'centrifugal force' are to be defined in a precise manner which in many cases is in conflict with the student's own opinion.

In an engineering education programme at university level mechanics is established as a fundamental subject of a pronounced mathematical character. Basic concepts and laws are normally described using mathematical quantities and models. Technical applications are essential and natural ingredients of the courses. Many subjects that are studied later are more or less based on courses in mechanics.

In mechanics, in a general sense, all types of material systems may be studied, like *particles, rigid bodies, deformable bodies* and *fluids*. These systems may be considered in various courses, e.g. *mechanics, solid mechanics, strength of materials, hydraulics* and *fluid mechanics*.

In engineering education at university level it is customary to study what is called the rigid body within the subject of mechanics while deformable bodies are treated in courses on solid mechanics or strength of materials. This convention will also be followed here. In dynamics the study of rigid bodies will be preceded by a treatment of particles and more general systems of particles. By a particle we mean a body that is so small that its extension in space may be neglected, i.e. its position is specified by a point.

Newton's laws

Classical mechanics is based on Newton's three laws. These will be considered in detail in Section 5.1 in the *Dynamics* volume but a short presentation, slightly adapted to modern terminology, follows here:

1. The law of inertia
   A body remains in its state of rest or motion along a straight line at a constant velocity unless it is forced by acting forces to change this state.
2. The law of motion
   The rate of change of the linear momentum of a body is proportional to the acting force and is in the direction of this force.
3. The law of action and reaction
   To every action there is always an opposed and equal reaction; or the mutual forces of two bodies upon each other are always equal in magnitude and opposite in direction.

The first and third laws constitute the foundation of statics and the treatment of forces exerted by bodies on each other.

I.2 Fundamental concepts

Matter and *material systems* are fundamental concepts within classical mechanics. It is assumed that to every material system we associate a certain physical quantity called *mass*, normally denoted by m.

A three-dimensional *space* and the *position of a point* in space are assumed to be familiar concepts. Physical quantities that are linked to these concepts are, for example, *length* and *volume*. The position of a point may then be determined by coordinates measured with respect to a given set of coordinate axes.

The concept of *time* plays a fundamental part in dynamics and may be given a clearer meaning in connection with events studied there.

It is assumed that the student has a sufficiently good understanding of these fundamental concepts through earlier studies and experiences. We thus refrain from attempts to define them in a more rigorous way.

In classical mechanics we assume that matter is *continuously distributed*. Hence, we can define a density function denoted by ρ according to

$$\rho(\mathrm{P}) = \lim_{\Delta V \to 0} \frac{\Delta m}{\Delta V}$$

where Δm is the mass of a material system of volume ΔV including point P, i.e. $\rho(\mathrm{P})$ denotes the value of the density at point P.

The fundamental concept of *force* is somewhat difficult to define in classical mechanics. A more extensive discussion is given below in Section I.4.

I.3 Units and dimensions

I.3.1 Fundamental quantities, systems of units

In order to measure physical quantities we need reference objects in terms of *units*. If we want to express the length l of a certain distance, for instance, we use a *numerical value* and a *unit*, e.g. $l = 23.7$ m, where 23.7 is the numerical value and m is the symbol of the unit. In this case, the unit is 1 metre (1 m). Generally, we can express this as follows: physical quantity = (numerical value) × (unit).

The units that are used for measurements of physical quantities are usually organized in a systematic way to form *systems of units*. There are a number of such systems that have been proposed and used throughout the years. Some are no longer in practical use, some have survived for certain specialized applications and some are still widely used. At this point we shall present two systems that are commonly used by practising engineers, the *SI system* and the *US customary system*. In the remainder of this book, however, we shall only use the SI system with a few minor exceptions.

Conversion factors between the SI system and the US customary system are given in Table I.1.

SI units

The SI system (Système International d'Unités; international system of units, ratified 1960) contains units of seven fundamental quantities, namely *length, mass, time, electric current, thermodynamic temperature, amount of substance* and *luminous intensity*.

The corresponding base units are: 1 metre (1 m), 1 kilogram (1 kg), 1 second (1 s), 1 ampere (1 A), 1 kelvin (1 K), 1 mole (1 mol) and 1 candela (1 cd).

The first three of these are of special interest in mechanics. Their definitions are:

- 1 m is the distance travelled by light in a vacuum during a time interval of 1/299 792 458 s.
- 1 kg is equal to the mass of the international kilogram prototype.
- 1 s is the duration of 9 192 631 770 periods of the radiation corresponding to the transition between the two hyperfine levels in the ground state of the caesium-133 atom.

The base units of a system must be defined independently of each other. From these it should be possible to derive all other units. Let us give a few examples.

The velocity of a body is the quotient of a length and a time interval. Consequently, the SI unit of velocity is 1 metre per second (1 m/s). The acceleration of a body is the quotient of a change in velocity and a time interval; the unit of acceleration is 1 m/s^2. According to Newton's second law a force is equal to the product of a mass and an acceleration, $F = ma$. The SI unit of force is called 1 newton (1 N) and is defined as $1\,N = 1\,kg\,m/s^2$. An angle is the ratio of the length of a circular arc and the radius of the circle. The corresponding SI unit is 1 radian (1 rad), where 1 rad = 1 (m/m). We observe that the SI units of various physical quantities can be derived from other units by using definitions or physical laws, e.g. $F = ma$.

Table I.1 **Conversion factors between the SI and US customary systems**

Quantity	US customary to SI	SI to US customary
Length	1 inch = 1 in = 25.40 mm	1 metre = 1 m = 39.37 in
	1 foot = 1 ft = 0.3048 m	1 m = 3.281 ft
	1 mile = 1 mi = 1.609 km	1 km = 0.6214 mi
Area	1 in^2 = 645.2 mm^2	1 m^2 = 1550 in^2
	1 ft^2 = 0.092 90 m^2	1 m^2 = 10.76 ft^2
Volume	1 in^3 = $16.39 \cdot 10^3$ mm^3	1 mm^3 = $61.02 \cdot 10^{-6}$ in^3
	1 ft^3 = 0.028 32 m^3	1 m^3 = 35.31 ft^3
Mass	1 slug = 14.59 kg	1 kilogram = 1 kg = 0.068 52 slug
Force	1 pound = 1 lb = 4.448 N	1 newton = 1 N = 0.2248 lb
Pressure, stress	1 psi = 1 lb/in^2 = 6.895 kPa	1 pascal = 1 Pa = $0.1450 \cdot 10^{-3}$ psi
Work, energy	1 lb ft = 1.356 J	1 joule = 1 J = 0.7376 lb ft
Power	1 lb ft/s = 1.356 W	1 watt = 1 W = 0.7376 lb ft/s

For practical and historical reasons we also use certain units that are not included in the SI system. Some typical examples are degree (°), hour (h), litre (l) and tonne (t), which are all frequently used in everyday language. Although we will try to keep to the SI system, we will allow ourselves to make use of such units in certain applications.

US customary units, gravitational units

Even though the SI system is the dominant system within science, another system of units, the US customary system, is still very common among practising engineers, at least in English-speaking countries. This system differs from the SI system in that the fundamental quantities are *length*, *time* and *force*. The corresponding base units are 1 foot (ft), 1 second (s) and 1 pound (lb), respectively.

The unit of time is the same as in the SI system, the foot is defined as 0.3048 m and the pound is defined as the *weight* of a platinum standard, the mass of which is 0.453 592 43 kg. This *standard pound* is kept at the National Bureau of Standards in Washington, DC. The weight of a body is, by definition, the gravitational force exerted by the earth on the body. Since the gravitational attraction varies with location, it has been decided that 1 pound is the weight of the standard pound at a latitude of 45° and at sea level. This means that 1 lb = 4.448 N. As the definition of the US customary units depends upon gravitation, the units are sometimes called *gravitational units*, and the unit system is said to be a *gravitational system*, contrary to the SI system, which is an *absolute system*.

In the US customary system the unit of mass is a derived unit called 1 slug. It is defined as the mass of a body which receives an acceleration of 1 ft/s^2 when acted upon by a force of 1 lb. From Newton's second law it follows that 1 slug = 1 lb s^2/ft = (4.448/0.3048) N s^2/m = 14.59 kg.

Unfortunately, the word 'pound' is used as a unit of mass in commercial transactions in English-speaking countries. In this case 1 pound is the mass of a body whose weight is 1 pound, e.g. the standard pound. Obviously, it is necessary to make a distinction between the pound as a unit of force and the pound as a unit of mass. For this reason the units are sometimes denoted by 1 lbf and 1 lbm, respectively. However, the reader is strongly advised *not* to use the pound as a unit of mass for scientific or engineering purposes.

It should be mentioned that US customary units are also referred to as British or English gravitational units, British engineering units, etc.

I.3.2 Dimensions

In this context the concept of *dimension* is useful. Every physical quantity has a certain dimension; if r is the radius of a circle, we say that the dimension of r is *length*. This can be expressed as $\dim r = \mathrm{L}$. The corresponding denotations for *mass* and *time* are M and T, respectively. The dimensions of all other quantities in mechanics can be expressed in terms of M, L and T. If a quantity is equal to the product of two other quantities, its dimension is equal to the product of the dimensions of the factors. Similar relations apply to quotients and powers of quantities.

If v is a velocity, then $\dim v = \mathrm{LT}^{-1}$, as a velocity is the quotient of a length and a time. In the same way the dimensions of an acceleration a and a force F are $\dim a = \mathrm{LT}^{-2}$ and $\dim F = \mathrm{MLT}^{-2}$, respectively. The dimension of an angle φ, which is the quotient of two lengths, is $\dim \varphi = \mathrm{LL}^{-1} = 1$. Such quantities are often, somewhat improperly, referred to as dimensionless.

Quantities that are equal to each other must have the same dimensions. This is also true for quantities that are added or subtracted. For this reason it is possible to use the concept of dimensions to check the validity of certain calculations.

EXAMPLE

A particle of mass m is released from a height h above the ground. The velocity of the particle when it hits the ground is v. From an elementary application of the concept of energy conservation, which is assumed to be familiar to the reader, we obtain $mv^2/2 = mgh$, i.e.

$$v = \sqrt{2gh} \qquad \text{(I.1)}$$

It is now possible to check that the right member $(2gh)^{1/2}$ has the correct dimension:

$\dim g = \mathrm{LT}^{-2}$ (as g is the acceleration due to gravity)

$\dim h = \mathrm{L}$

$\dim gh = \mathrm{L}^2\mathrm{T}^{-2}$ (as $\dim gh = \dim g \cdot \dim h$)

$\dim 2gh = \mathrm{L}^2\mathrm{T}^{-2}$ (multiplication by the dimensionless factor 2 does not change the dimension)

$\dim \sqrt{2gh} = \sqrt{\mathrm{L}^2\mathrm{T}^{-2}} = \mathrm{LT}^{-1}$

But $\mathrm{LT}^{-1} = \dim v$, meaning that both members of $v = (2gh)^{1/2}$ have the same dimensions. Equation (I.1) is said to be *homogeneous*. This is a *necessary condition* for the equation to be correct, but it is certainly *not* sufficient.

In physical applications expressions like exp x, sin x, etc., occur frequently. It is easy to show that exponents as well as arguments of trigonometric functions have to be dimensionless.

It is not possible to check the dimension of a certain expression unless the quantities involved are denoted by algebraic symbols. If numerical values are substituted into the expression at an earlier stage of the calculations, this opportunity will be lost.

I.4 The concept of force

I.4.1 The fundamental forces of physics

The physical quantity *force* is of fundamental importance in mechanics. At this stage it is difficult to define the concept of force in classical mechanics in a stringent manner. In general physics all forces are traced back to the four fundamental forces, which originate from *gravitation, electromagnetism* and the *weak* and *strong interaction*.

The *gravitational force* is an attractive force exerted by a body on another body in its vicinity. Two particles of masses m and M, respectively, attract each other with forces F, Figure I.2, according to Newton's law of gravitation:

$$F = G\,\frac{mM}{r^2} \tag{I.2}$$

Here r is the distance between the particles, and G is the *constant of gravitation*, which is a universal constant. Its value is approximately $G \approx 6.673 \cdot 10^{-11}\,\mathrm{N\,m^2/kg^2}$. From the law of gravitation, Equation (I.2), it is possible to derive expressions for the forces exerted on each other by more general material systems.

Two particles at rest with electric charges Q_1 and Q_2 exert forces on each other according to *Coulomb's law*:

$$F = \frac{Q_1 Q_2}{4\pi\epsilon_0 r^2} \tag{I.3}$$

Here r is the distance between the particles, and $\epsilon_0 = 8.854 \cdot 10^{-12}\ \mathrm{F/m^1}$ is called the permittivity of free space. If the charges are of equal sign, the particles will repel each other. Otherwise the forces will be attractive.

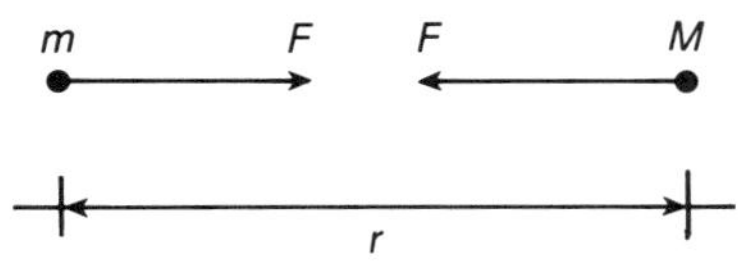

Figure I.2

In a more general case a particle of electric charge Q may travel with a velocity v in an electromagnetic field of electric field strength $\boldsymbol{E}$ and magnetic flux density $\boldsymbol{B}$. The force exerted on the particle is

$$\boldsymbol{F}=Q(\boldsymbol{E}+\boldsymbol{v}\times\boldsymbol{B}) \tag{I.4}$$

The contribution $Q(\boldsymbol{v}\times\boldsymbol{B})$ is called the *Lorentz force*. Some of the symbols above have been written in bold type to indicate that they correspond to vector quantities. Quantities like forces and velocities not only have numerical values but also directions. Consequently, it may be advantageous to describe them as vectors. We shall return to this matter later on. At this point, the matter of importance is that an electrical charge (Q) in motion ($\boldsymbol{v}$) in an electromagnetic field ($\boldsymbol{E}$, $\boldsymbol{B}$) will be subjected to a force ($\boldsymbol{F}$) which depends on the direction of motion and the directions of the field quantities. This is easily accomplished by using vector notation.

Electrical charges in motion generate electromagnetic fields. Consequently, electrically charged particles in motion will exert forces on each other according to Equation (I.4). Thus, if two bodies are in close contact with each other, the atoms in their respective surface layers will affect each other in a highly complicated manner. The overall effect is that bodies in close contact will exert what are called *contact forces* on each other. These forces have an electromagnetic origin, but it is practically impossible to derive them from the laws of electromagnetism.

The forces originating from the *weak* and *strong interaction* have a very short range of action. Actually, they can only be detected on the level of atomic nuclei. They are important when the fundamental structure of matter is considered, but in classical mechanics, on a *macroscopic level*, they play no part at all.

I.4.2 The concept of force in classical mechanics

In classical mechanics *force* is a quantity that describes the influence on a material system by its surroundings, e.g. by other material systems. This means that to every force there is always a source or a cause. Ultimately, every force can be derived from the four fundamental forces, but in classical mechanics this relation is in most cases disregarded. Consequently, if two bodies, A and B, are in contact, the influence of B upon A is described by a contact force. This contact force is the fundamental quantity from the classical mechanics point of view. The source of the force is body B, which by the contact exerts a force on body A. In classical mechanics we need not know that the contact force can be traced back to the electromagnetic interaction of the atoms of the two bodies.

It is not meaningful to use concepts like *rest* and *motion*, unless we have defined a frame of reference. We shall assume that all forces have the same values regardless of what frame of reference is used. A quantity, e.g. a force, that is unaffected by the choice of frame of reference is called an *objective quantity*.

Let us give a specific example. If a train passenger A is walking forward on the train, the passenger's velocity will depend on who is measuring it. Another passenger B, who is not moving with respect to the train, will measure one value, e.g. 1 m/s;

a person C standing on the ground watching the train passing by will measure a completely different value, e.g. 32 m/s. This is because the two observers B and C are fixed with respect to different frames of reference – the moving train and the surrounding landscape, respectively. However, the forces acting on passenger A are the same in all frames of reference. It does not matter how the train is moving (with a constant speed or accelerating, straight ahead or in a curve); the same forces act in both frames of reference.

Some quantities that are referred to as 'forces' in everyday language do *not* fulfil the requirement of being objective quantities. The well-known 'centrifugal force' is maybe the most typical example. If the train is travelling through a curve, the centrifugal force is a meaningful quantity to observer B but not to observer C. It is also impossible to define the source of the centrifugal force in the manner mentioned above. A similar example is the 'force' that seems to push the passengers forward, when the train comes to a sudden stop. These 'forces' are sometimes called fictitious forces. In dynamics it is extremely important to distinguish between real forces and fictitious forces. We shall discuss this question at length in Section 5.1. in the *Dynamics* volume. Hopefully, this distinction will be obvious to the reader when we get there.

Some of the more important forces of classical mechanics are presented here.

Gravitational forces

A particle is attracted by other particles in its vicinity according to the law of gravitation, Equation (I.2). We assume that the earth may be regarded as a material system consisting of a large number of particles. Each of these will attract a particle P of mass m located in the neighbourhood of the earth. By adding all these contributions from all particles belonging to the earth we obtain an expression for the gravitational force exerted by the earth on particle P.

We now assume that the earth is spherically symmetrical, i.e. the density at a certain point depends only on the distance to the centre of the earth. In this case it may be shown that the force exerted by the earth is identical to the force exerted by a particle of the same mass as the earth located at the centre of the earth; cf. Problem 5.10 in the *Dynamics* volume. Hence, the force is $F=GmM/r^2$, where M is the mass of the earth and r is the distance from particle P to the centre of the earth. If the particle is located in the immediate vicinity of the surface of the earth, r is equal to the radius of the earth (R). The attracting force is then equal to what we usually call the *weight* of particle P, $F=mg$. Here, $g=9.81\ \mathrm{m/s^2}$ is the acceleration due to gravity. Thus, $GM=gR^2$. It should be noted that this relationship is not exact owing to the rotation of the earth and to the fact that the earth is not exactly spherical.

Elastic forces

If a body is in contact with an elastic medium, it will be subjected to a force that depends on the deformation of the elastic medium. The simplest example is a body which is attached to a perfectly elastic spring, Figure I.3. The unstretched length of the

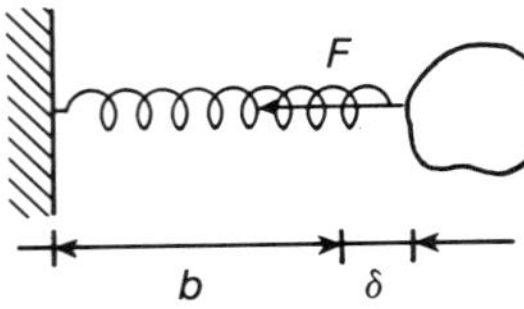

Figure I.3

spring is b. If the spring is stretched a distance δ, it will strive to regain its unstretched length.

The body will then be subjected to a spring force F according to Figure I.3. Under certain circumstances, for instance that δ is not too large, F will be directly proportional to δ. Hence,

$$F = k\delta \tag{I.5}$$

The constant of proportionality k is called the *spring constant* or the *stiffness*. If the spring is compressed, so that its length is smaller than b, it will try to expand. The force F will then be directed in the opposite sense. A spring of this type can in many cases be used to model more complex elastic systems.

Constraining forces

Two bodies that are in contact will exert contact forces on each other. Elastic forces like the spring force of the previous paragraph are typical examples. The body attached to the spring is free to move, and once the location of the body is known, it is possible to calculate the spring force. In another case a body A in contact with another body B may be completely prevented from moving by the presence of body B. Body A is then subject to a *constraint* and the corresponding contact force is referred to as a *constraining force*. Constraining forces are not determined solely by the location of the body. More information is needed; for instance, it is necessary to know what other forces are acting.

A typical example of a constraining force is given in Figure I.4. A body suspended by an inextensible cord is prevented from moving downwards by the cord. This constraint corresponds to a constraining force S directed upwards exerted by the cord on the suspended body. The direction of the constraining force is opposite to the direction in which the body is prevented from moving.

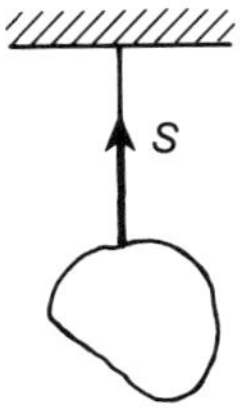

Figure I.4

A more thorough exposition of constraining forces will be given in Sections 2.2.2 and 2.3.2.

I.4.3 Distributed forces – concentrated forces

From earlier studies the reader is probably used to regarding forces as acting at a certain point, the point of application. This is normally an idealization. Let us assume that a material system occupies a certain volume V and that a number of forces are acting at various points of V. In a special case the number of forces may be very large and even approach infinity in such a way that the points of application completely 'fill up' the volume. The force is then said to be continuously distributed over volume V. The gravitational force is a typical example. Since each small part of a material system is subjected to a gravitational force, the overall effect will be a force that is distributed over the entire volume occupied by the system.

In a similar way a force may be distributed over a surface. As an example we may consider a vessel containing a gas under high pressure. The gas will exert compressive forces on the interior walls of the vessel. These forces are then distributed over a surface.

In reality *all* forces are distributed over volumes or surfaces. In some cases the region may be so small that it may be approximated by a point. Hence, the force can be regarded as concentrated; there is a point of application. Force S of Figure I.4 is distributed over the cross-section of the cord, but if the diameter of the cross-section is small compared with other lengths of interest, it is reasonable to consider S as applied at a point.

As we shall see in Chapter 1 it is always possible to replace more complicated systems of forces by simpler, equivalent systems, in many cases consisting of one single force with a certain point of application. This will be used later on several occasions, for instance in Chapter 3. At this stage we merely observe that studying *concentrated forces* should be of value for future purposes.

1

Force systems

1.1 Vector properties, fundamental operations

The concept of force was introduced in Section I.4 above. We noted that even if all real forces are distributed over surfaces or volumes, there are reasons to study what are called concentrated forces. In this chapter we shall assume that all forces may be regarded as concentrated forces. We will also consider systems consisting of a number of concentrated forces, possibly an infinite number.

A concentrated force is characterized not only by its *magnitude* and its *direction* but also by its location to a special point in space, the *point of application*. Furthermore, it has been known since the seventeenth century that the actions of two forces on a body are added according to the parallelogram law[1].

As an example we consider the body of Figure 1.1. We assume that a vertical force of magnitude R applied at P according to Figure 1.1(a) is required to keep the body at rest. From experiments, for instance by using dynamometers, it can be observed that the body will remain at rest if the force R is replaced by two suitably chosen forces F_1 and F_2 applied at the same point P, see Figure 1.1(b). By adding F_1 and F_2 according to the parallelogram law we obtain a vertical force of magnitude R, i.e. the diagonal of the parallelogram defined by F_1 and F_2 is vertical and its magnitude is equal to R. This is shown in Figure 1.1(c).

To sum up, a force is characterized by the properties of magnitude, direction and point of application. Furthermore, forces with a common point of application may be added according to the parallelogram law. With this background it seems natural to use vectors to describe forces mathematically. A vector is normally specified by its magnitude and its direction, and vector addition follows the parallelogram law. As the point of application is an essential property of a force, we have to modify the definition of a vector in this context by introducing what is called a *fixed vector*, i.e. a vector that is localized to a specific point in space. From now on we assume that a force can be described by a fixed vector.

1 The Dutch mathematician Simon Stevin (1584–1620) was aware of the parallelogram law and used it without explicitly stating it.

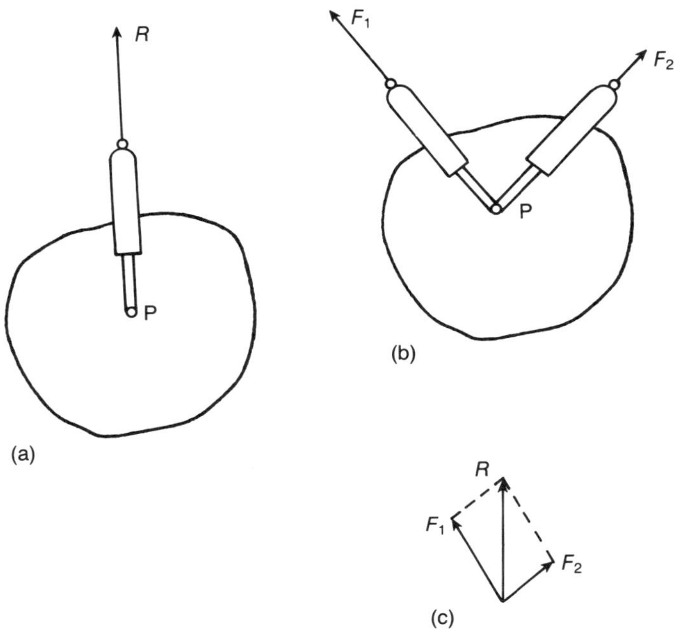

Figure 1.1

Consequently, we will use vector notation for forces, i.e. $\boldsymbol{F}$, while the corresponding scalar quantity F refers to the magnitude of the force:

$$F = |\boldsymbol{F}|$$

In diagrams, however, we will use the scalar symbol to denote a force. In this case the direction of the line segment corresponds to the direction of the force, cf. Figure 1.1. In three-dimensional situations precautions must be taken to avoid misunderstandings.

By using a vector as a mathematical model of a force we automatically state that all properties and laws that apply to vectors will also apply to forces. Addition of vectors according to the parallelogram law is one example. Inversely, a force may be resolved into vector components without affecting the body under consideration.

If more than two forces have a common point of application they can be replaced by one single force by repeated use of the parallelogram law. The forces $\boldsymbol{F}_1$, $\boldsymbol{F}_2$ and $\boldsymbol{F}_3$ of Figure 1.2 are added step by step. Hence,

$$\boldsymbol{R}_1 = \boldsymbol{F}_1 + \boldsymbol{F}_2$$
$$\boldsymbol{R} = \boldsymbol{R}_1 + \boldsymbol{F}_3 = \boldsymbol{F}_1 + \boldsymbol{F}_2 + \boldsymbol{F}_3$$

We see that adding more than two forces according to the parallelogram law is equivalent to vector addition. It is not necessary that the forces are located in the same plane. The force $\boldsymbol{R}$ is often called the *resultant* or the *force resultant*.

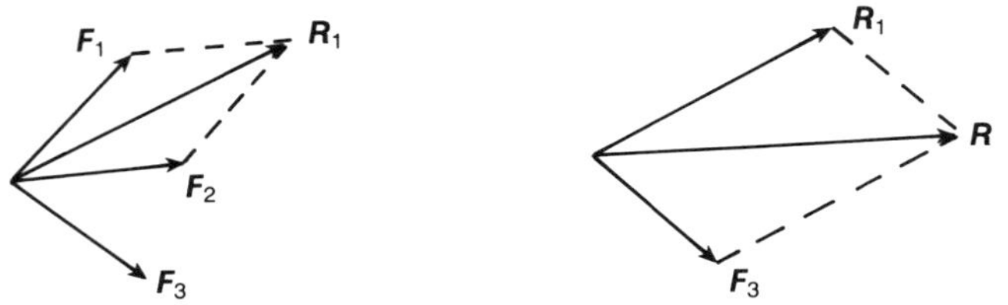

Figure 1.2

It should be noted that when using a mathematical model to describe the physical reality we transform the physical problem to a mathematical problem. The mathematical model is useful only as long as the results and predictions that follow from solving the mathematical problem can be confirmed by physical experiments. The fixed vector as a model of a force has proved to satisfy this requirement.

In many practical cases we do not need to know the point of application of a force exactly. Frequently, the net external effect on a body is determined by the magnitude, the direction and the *line of action* of the force. When we use the somewhat vague expression 'net external effect' we refer to the consequences for the body as a whole when a certain force is acting on it. The line of action is the line passing through the point of application parallel to the direction of the force. In the case shown in Figure 1.3 the line through points A and B is the line of action of the force $\boldsymbol{F}$.

The crate of Figure 1.4(a) is supported at C and is kept in its position by a rope DE while a force F is acting. It can be shown that the forces exerted by the support and by the rope on the crate are exactly the same regardless of the exact location of the point of application as long as it is somewhere on the line of action. Furthermore, the crate will still be at rest no matter where on the line of action the force F is applied. The cart of Figure 1.4(b) is pulled forward by a force F applied to a rope. The cart will move in the same way regardless of where we choose to apply the force.

If the net external effect of the force is determined by the line of action and not by the point of application, we may displace the force along its line of action, e.g. from point A to point B in Figure 1.3, without altering the net external effect on the body. Obviously, this is true in many cases but it does not apply generally. We shall give a few examples.

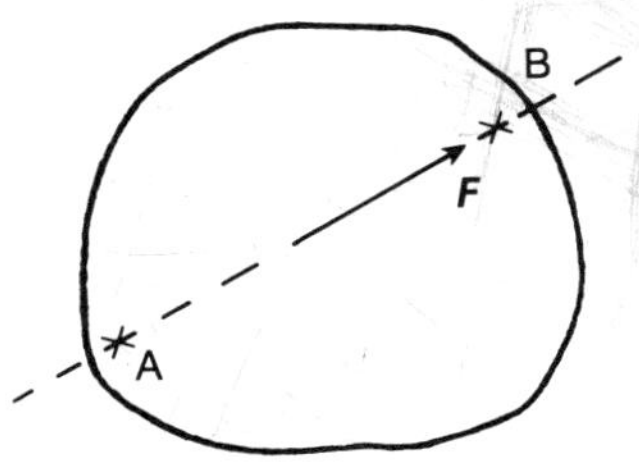

Figure 1.3

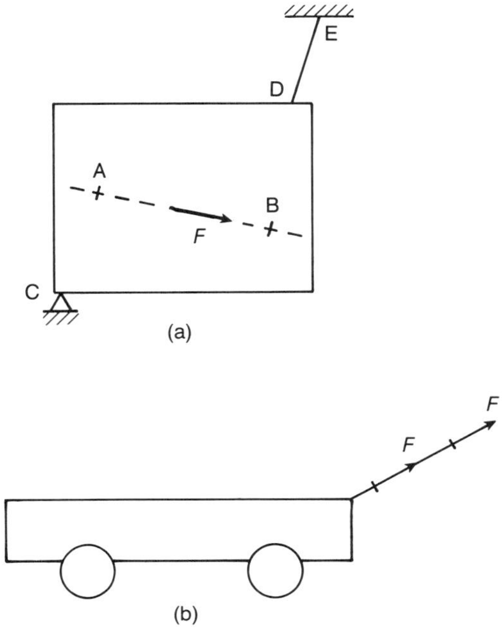

Figure 1.4

The elastic spring of Figure 1.5(a) is subjected to a force F directed along the spring. It is evident that the deformation of the spring is dependent on the exact location of the point of application. If the force is displaced along its line of action the effect on the spring will change radically.

The bar of Figure 1.5(b) has a varying cross-section. If a force F is applied at point A, there is a risk of fracture where the area of the cross-section is minimum. On the other hand, if the force is applied at B no such risk exists. The force cannot be displaced along its line of action without drastically changing the consequences. What are called the internal stresses in the bar depend on the location of the point of application. However, the risk that the bar will detach at its support to the left is still the same.

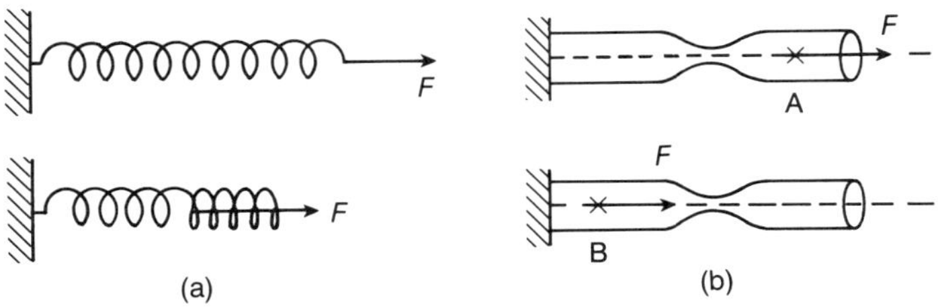

Figure 1.5

Obviously, if we are interested in deformations or some kind of internal stresses in a body we have to know the exact location of the point of application. If we are only interested in the net external effect on a *rigid* (undeformable) body, it seems that we only need to know the line of action. We shall investigate these circumstances later on. We shall see that the conditions that determine the state of rest or motion of a rigid body are not affected if a force is displaced along its line of action. In this chapter we assume that every force may be displaced along its line of action.

Hence, we have defined certain operations that may be applied to forces. We shall refer to these as *fundamental operations*:

1. Two forces with a common point of application may be added according to the parallelogram law. In the opposite way a force can be resolved into components along two arbitrary directions. With the notation according to Figure 1.6 we have

$$\boldsymbol{R} = \boldsymbol{F}_1 + \boldsymbol{F}_2$$

   where $\boldsymbol{R}$ is called the resultant of $\boldsymbol{F}_1$ and $\boldsymbol{F}_2$.
2. A force can be displaced along its line of action without affecting its net external effect on a rigid body, see Figure 1.7.

If two forces of the same magnitude are oppositely directed and share a common line of action, see Figure 1.8(a), they can be displaced until their points of application coincide, Figure 1.8(b). Obviously, the result of the operation is zero, Figure 1.8(c).

In a contrary way it is always possible to introduce two oppositely directed forces of the same magnitude with a common line of action. The net external effect on the body will not be altered.

In the remainder of this chapter we shall study systems of forces assuming that fundamental operations are permitted. First we shall consider two-dimensional force systems, i.e. systems where all forces are located in a plane, before we generalize to three dimensions.

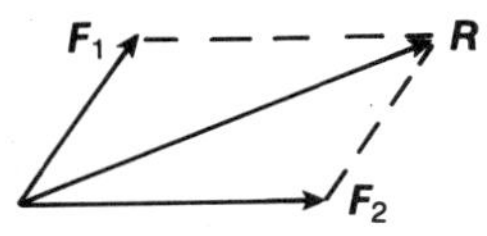

Figure 1.6

Figure 1.7

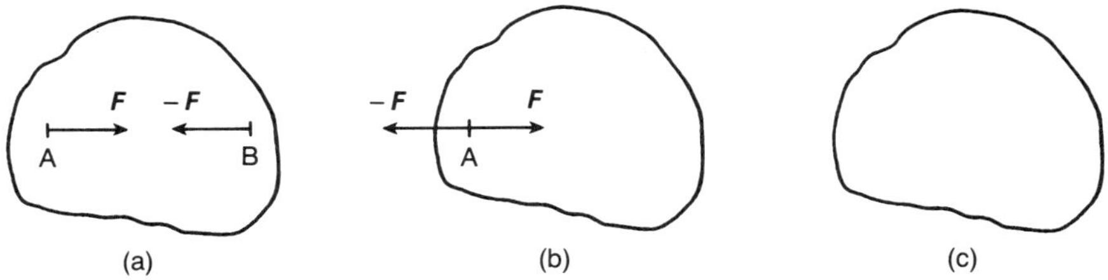

Figure 1.8

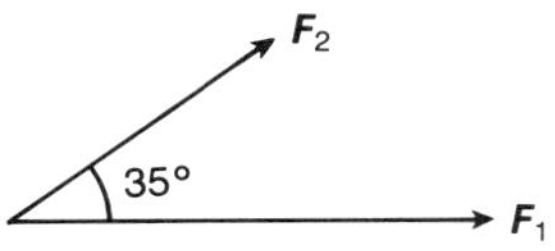

Figure 1.9

WORKED EXAMPLE 1.1

Two forces $\boldsymbol{F}_1$ and $\boldsymbol{F}_2$ of magnitudes $F_1 = 60$ N and $F_2 = 40$ N, respectively, share the same point of application, Figure 1.9.

Determine the magnitude of the resultant of the forces.

SOLUTION

The resultant $\boldsymbol{R} = \boldsymbol{F}_1 + \boldsymbol{F}_2$ is obtained from the parallelogram law. From the cosine law it follows, Figure 1.10, that

$$R^2 = F_1^2 + F_2^2 - 2F_1F_2\cos 145° \Rightarrow R \approx 96 \text{ N}$$

WORKED EXAMPLE 1.2

The magnitude of force $\boldsymbol{R}$, Figure 1.11, is 100 N. Resolve the force into components along directions OA and OB.

SOLUTION

The resultant $\boldsymbol{R} = \boldsymbol{F}_1 + \boldsymbol{F}_2$ is obtained by applying the parallelogram law as shown in Figure 1.12. From the sine law we find that

$$\frac{R}{\sin 130°} = \frac{F_1}{\sin 20°} = \frac{F_2}{\sin 30°} \Rightarrow \begin{cases} F_1 \approx 45 \text{ N} \\ F_2 \approx 65 \text{ N} \end{cases}$$

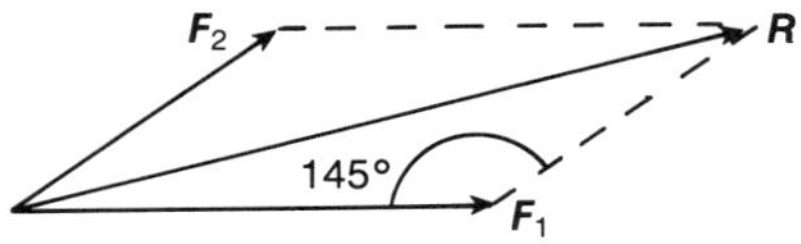

Figure 1.10

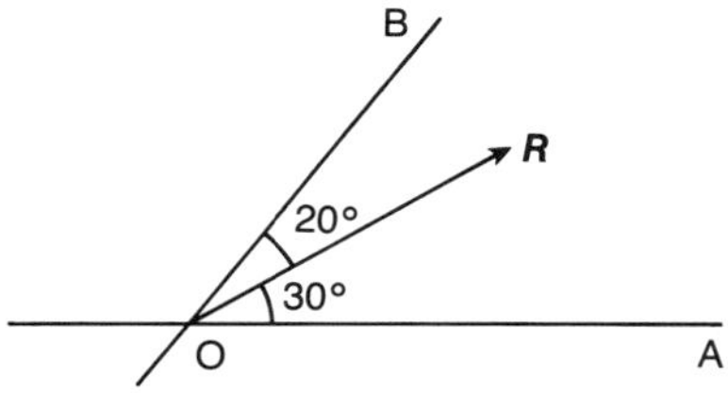

Figure 1.11

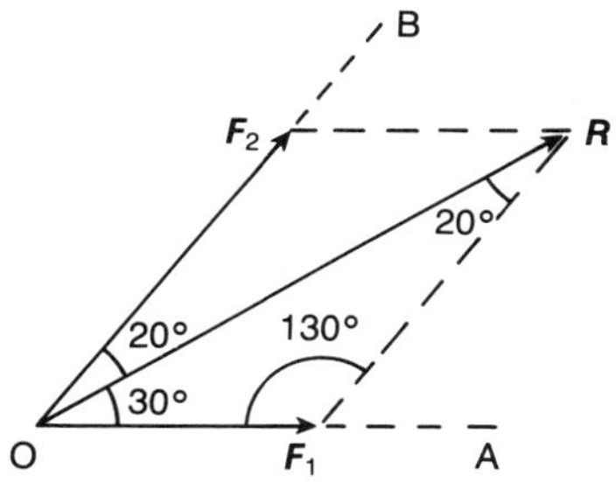

Figure 1.12

1.2 Forces in two dimensions

1.2.1 Coordinate systems, components

In the preceding section we saw that it is always possible to resolve a force into components along two given directions according to the parallelogram law. In a special case the directions are perpendicular. If we introduce a Cartesian coordinate system Oxy according to Figure 1.13 we can replace any force $\boldsymbol{F}$ by its components parallel to the coordinate axes. Hence,

$$\boldsymbol{F} = F_x\mathbf{e}_x + F_y\mathbf{e}_y = (F_x, F_y) \tag{1.1}$$

where $\mathbf{e}_x$ and $\mathbf{e}_y$ are unit vectors parallel to the coordinate axes. (F_x, F_y) is a simplified vector notation that will be used throughout this book except in cases where ambiguity may occur, for instance when more than one coordinate system is used in the same problem.

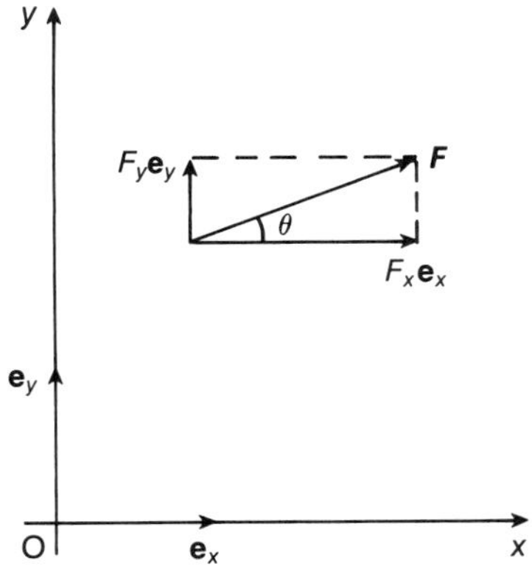

Figure 1.13

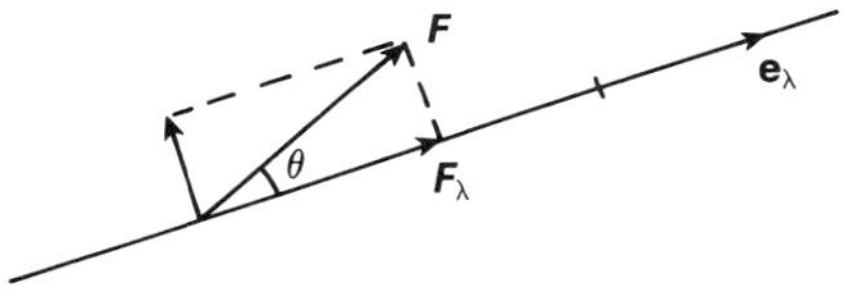

Figure 1.14

The scalar components F_x and F_y, the magnitude of the force $F = |\boldsymbol{F}|$ and the angle θ of Figure 1.13 must fulfil the following relations:

$$\begin{cases} F_x = F\cos\theta \\ F_y = F\sin\theta \end{cases} \qquad \begin{cases} F = \sqrt{F_x^2 + F_y^2} \\ \tan\theta = F_y/F_x \end{cases} \tag{1.2}$$

The force $\boldsymbol{F}$ of Figure 1.14 has been resolved into two perpendicular components, one of which is determined by the unit vector $\mathbf{e}_\lambda$. This vector component may be expressed as

$$\boldsymbol{F}_\lambda = F_\lambda \mathbf{e}_\lambda \tag{1.3}$$

where $F_\lambda = |\boldsymbol{F}|\cos\theta = \boldsymbol{F}\cdot\mathbf{e}_\lambda$. If the angle $\theta \leqslant 90^\circ$ the scalar component F_λ is equal to the magnitude of the vector component $\boldsymbol{F}_\lambda$.

WORKED EXAMPLE 1.3

The magnitude of force $\boldsymbol{F}$, Figure 1.15, is 53 N. Determine

(a) the x- and y-components,
(b) the component along line AB.

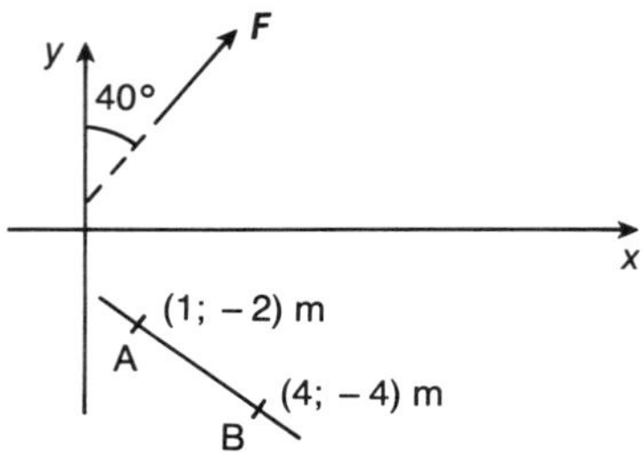

Figure 1.15

SOLUTION

(a) From Figure 1.16 it is easily seen that

$$\begin{cases} F_x = F \sin 40^\circ \approx 34 \text{ N} \\ F_y = F \cos 40^\circ \approx 41 \text{ N} \end{cases}$$

(b) First we determine a unit vector $\mathbf{e}_{AB}$ along AB. As the vector $\vec{AB} = |\vec{AB}|\,\mathbf{e}_{AB}$ we have

$$\mathbf{e}_{AB} = \frac{\vec{AB}}{|\vec{AB}|} = \frac{(3, -2)}{\sqrt{3^2 + (-2)^2}} = \frac{1}{\sqrt{13}}(3, -2)$$

According to Equation (1.3) the component of $\boldsymbol{F}$ along AB is

$$F_{AB} = \boldsymbol{F} \cdot \mathbf{e}_{AB} = F_x \frac{3}{\sqrt{13}} + F_y \frac{-2}{\sqrt{13}} \approx 5.8 \text{ N}$$

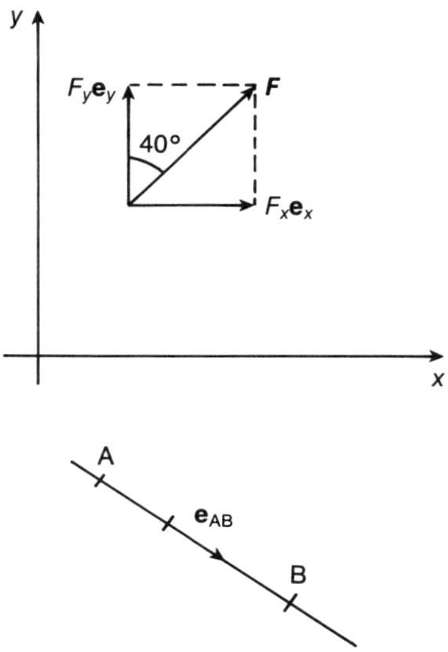

Figure 1.16

WORKED EXAMPLE 1.4

A force $\boldsymbol{F}$ of magnitude $F = 100$ N is resolved into components along directions OA and OB, Figure 1.17.

(a) Determine the angle α knowing that the component of $\boldsymbol{F}$ along OA is 70 N.
(b) Determine the component along OB.

SOLUTION

(a) The components F_1 and F_2 are shown in Figure 1.18(a). Using the sine law we obtain

$$\frac{F}{\sin 50^\circ} = \frac{F_1}{\sin \alpha}$$

where $F_1 = 70$ N. Hence,

$$\sin \alpha = \frac{F_1}{F} \sin 50^\circ = 0.7 \sin 50^\circ$$

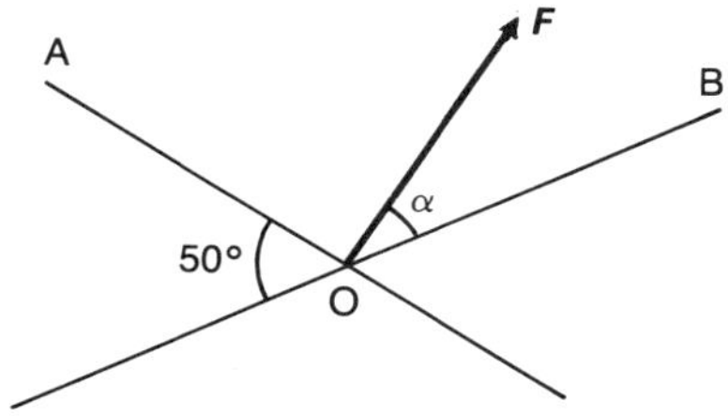

Figure 1.17

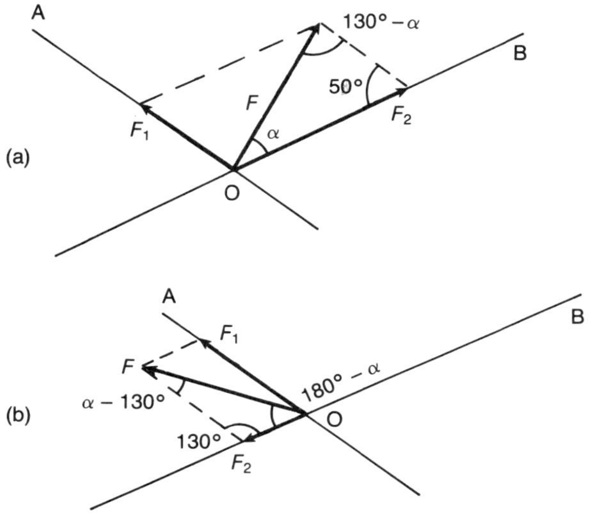

Figure 1.18

There are two solutions: $\alpha \approx 32^\circ$ and $\alpha \approx 148^\circ$. The significance of the second solution is displayed in Figure 1.18(b).

(b) In the case $\alpha \approx 32^\circ$ the sine law yields

$$\frac{F}{\sin 50^\circ} = \frac{F_2}{\sin(130^\circ - \alpha)} \Rightarrow F_2 \approx 129 \text{ N}$$

For $\alpha \approx 148^\circ$ we obtain, using Figure 1.18(b),

$$\frac{F}{\sin 130^\circ} = \frac{F_2}{\sin(\alpha - 130^\circ)} \Rightarrow F_2 \approx 39 \text{ N}$$

As F_2 is directed from B towards O the component along OB is -39 N.

ALTERNATIVE SOLUTION

We introduce x- and y-coordinates according to Figure 1.19.

The force $\boldsymbol{F}$ and its components along OA and OB may now be expressed as vectors:

$$\boldsymbol{F} = (F \cos \alpha, F \sin \alpha)$$
$$\boldsymbol{F}_1 = (-F_1 \cos 50^\circ, F_1 \sin 50^\circ), \text{ where } F_1 = 70 \text{ N}$$
$$\boldsymbol{F}_2 = (F_2, 0)$$

From $\boldsymbol{F} = \boldsymbol{F}_1 + \boldsymbol{F}_2$ it follows that

$$F \cos \alpha = -F_1 \cos 50^\circ + F_2 \tag{1.4}$$

$$F \sin \alpha = F_1 \sin 50^\circ \tag{1.5}$$

Equation (1.5) yields

$$\sin \alpha = \frac{F_1}{F} \sin 50^\circ \Rightarrow \alpha \approx 32^\circ \text{ or } \alpha \approx 148^\circ$$

Inserting into Equation (1.4) we have:

$$F_2 \approx 129 \text{ N or } F_2 \approx -39 \text{ N}$$

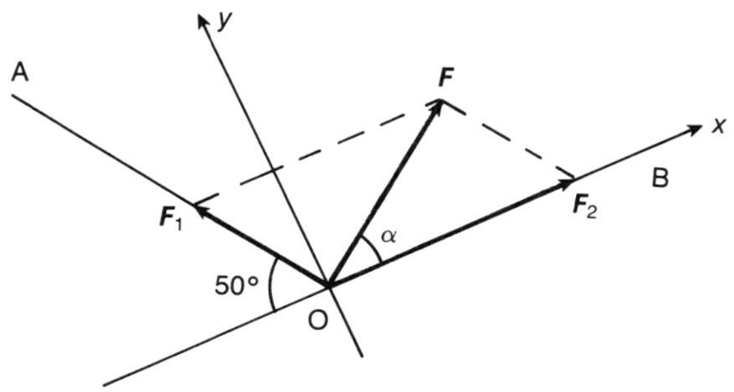

Figure 1.19

1.2.2 Moment

From everyday life we are familiar with a number of situations where the tendency of a force to rotate a certain body is of interest. As a typical example, consider the wrench of Figure 1.20. From experience we know that the value of the product Fd is important for the ability to loosen the nut. A force of magnitude $2F$ applied along a line at a distance $d/2$ from the centre of the nut will then have the same ability to rotate the nut.

In connection with the example above we define the concept *moment about an axis*, see Figure 1.21.

DEFINITION

The moment of the force F about the axis O, perpendicular to the plane of Figure 1.21, is

$$M_O = Fd \tag{1.6}$$

where d is called the *moment arm*, i.e. the distance from O to the line of action of the force. The term *torque* is frequently used for the moment of a force.

It is customary to define a positive reference direction in connection with moments. The force of Figure 1.21 has a tendency to rotate the body counterclockwise about the axis O. Let us choose counterclockwise as our positive direction of rotation. If another

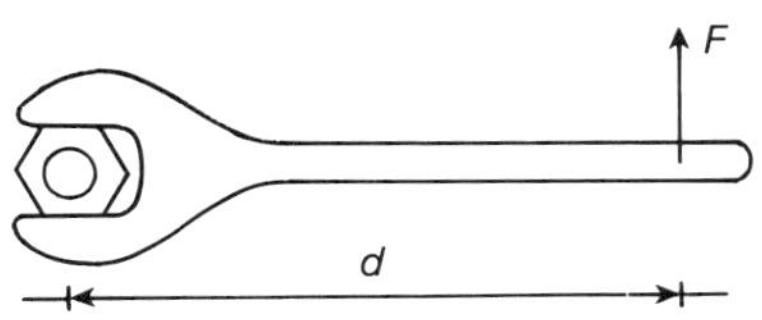

Figure 1.20

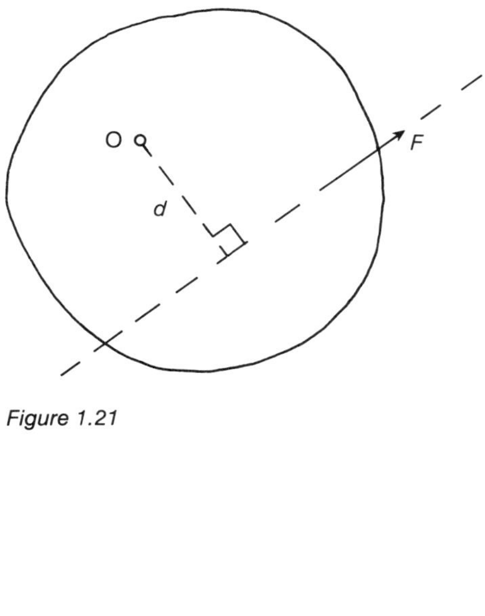

Figure 1.21

force is acting on the body with a tendency to rotate the body clockwise about O, then the moment of that force about O is negative.

We observe from Figure 1.21 that the moment M_O is unchanged if the force F is displaced along its line of action, since the moment arm d is not changed. Obviously, the moment is not affected by this fundamental operation.

In two-dimensional cases we often refer to the moment about a point O when we in fact mean the moment about an axis passing through the point O normal to the plane of the force. Normally, no misunderstandings should occur but it is important to know that the moment about a point is a more general concept that will be defined later in connection with forces in three dimensions.

Varignon's theorem[2]

We stated earlier that the net external effect on a body is not affected if a force is replaced by its components according to the parallelogram law. If this is correct, *the moment of a force about a certain axis is equal to the sum of the moments of the components.* This is the significance of Varignon's theorem, which can be proved in the following way.

Consider a force R with its point of application at P, see Figure 1.22. The force is resolved into two components F_1 and F_2 according to the parallelogram law. Hence,

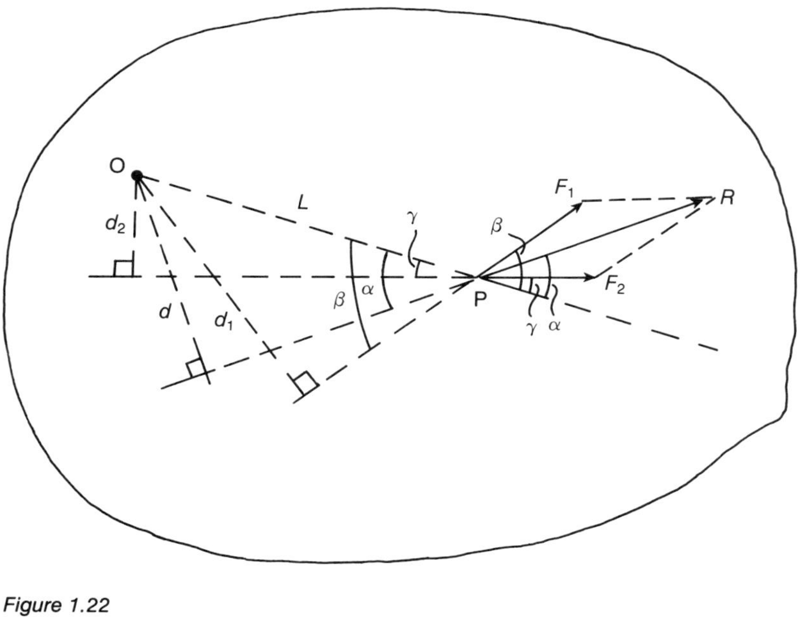

Figure 1.22

2 Varignon (1654–1722) is believed to have formulated the theorem in 1687, i.e. the same year as Newton's *Principia* was published. The work, however, was not published until after his death.

the component of R normal to line OP is equal to the sum of the corresponding components of F_1 and F_2, i.e.

$$R \sin \alpha = F_1 \sin \beta + F_2 \sin \gamma$$

If we multiply this equation by L and use some elementary geometrical relations, we find that

$$Rd = F_1 d_1 + F_2 d_2$$

In other words, the moment of a force R about O is equal to the sum of the moments of its components. Varignon's theorem is thereby proved. Consequently, we have shown that the moment of a force about an axis is not changed if the force is subjected to any of the fundamental operations defined in Section 1.1.

WORKED EXAMPLE 1.5

The magnitude of force $\mathbf{F}$, Figure 1.23, is 25 N. Determine the moment of $\mathbf{F}$ about an axis through the origin of the coordinate system.

SOLUTION

The force is resolved into components $F_x\mathbf{e}_x$ and $F_y\mathbf{e}_y$ parallel to the coordinate axes. Here, Figure 1.24,

$$F_x = F \cos 30^\circ \approx 21.7 \text{ N}$$
$$F_y = F \sin 30^\circ \approx 12.5 \text{ N}$$

If we choose clockwise as our positive reference direction we obtain, using Varignon's theorem,

$$M_O = F_x a - F_y b \approx 4.0 \text{ N m}$$

The significance of a and b is evident from the figure. Note that the contribution from F_y is negative since the corresponding moment is counterclockwise.

REMARK

M_O can also be determined as

$$M_O = Fd = F(a - b \tan 30^\circ) \cos 30^\circ \approx 4.0 \text{ N m}$$

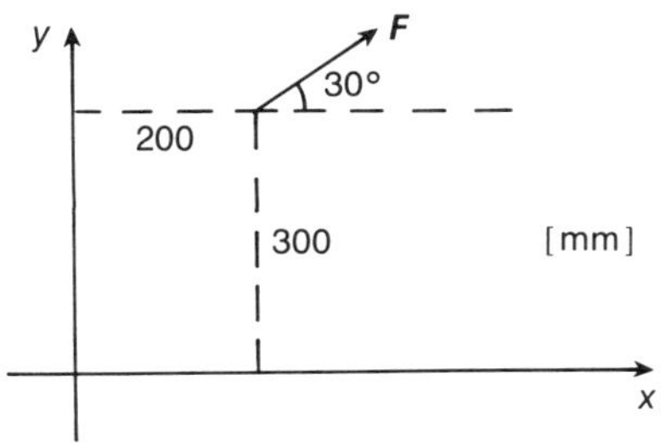

Figure 1.23

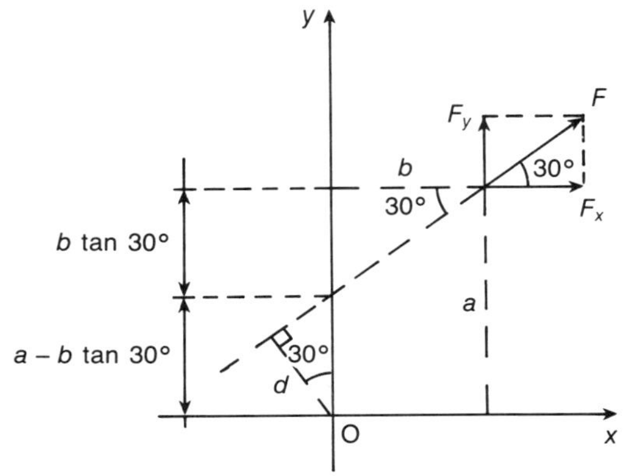

Figure 1.24

1.2.3 Sum of forces, sum of moments

Next, we consider a body which is acted upon by a system of forces. An example is shown in Figure 1.25, where the system consists of three forces. The number of forces can, of course, be arbitrary, say n. The forces are denoted by $\boldsymbol{F}_1, \boldsymbol{F}_2, \ldots, \boldsymbol{F}_n$.

We now define the *sum of forces* or the *force resultant* of the system and denote it symbolically by $\Sigma \boldsymbol{F}$:

$$\sum \boldsymbol{F} = \sum_{i=1}^{n} \boldsymbol{F}_i \tag{1.7}$$

Obviously, the sum of forces is obtained by adding the individual forces without taking into account that they may have different points of application. If a coordinate system is introduced according to Figure 1.25 we can express the individual forces as well as the sum of forces in terms of their components. Hence,

$$\sum F_x = \sum_{i=1}^{n} F_{ix} \text{ etc.}$$

Here ΣF_x denotes the x-component of the sum of forces.

In a similar way we define the *sum of moments* or the *moment resultant* of the system about the axis O.

$$\sum M_O = \sum_{i=1}^{n} M_{Oi} \tag{1.8}$$

Here M_{Oi} is the moment of the force $\boldsymbol{F}_i$ about the axis O. Furthermore, a positive reference direction must be defined, and the sign of each contribution of the type M_{Oi} must

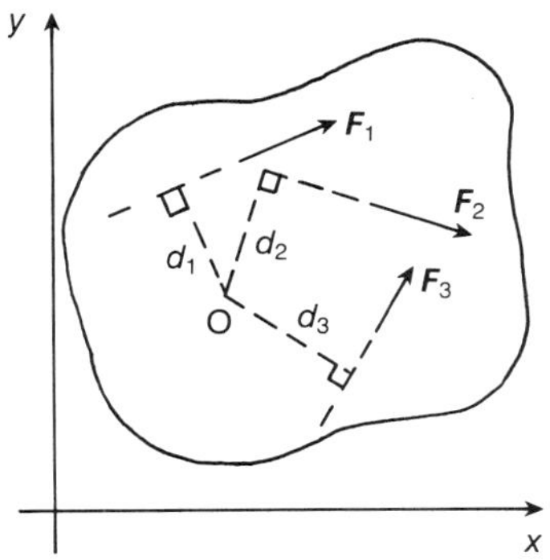

Figure 1.25

correspond to the choice of reference direction. For the force system of Figure 1.25 we have

$$\sum M_O = F_1 d_1 + F_2 d_2 - F_3 d_3$$

if we have chosen clockwise as the positive direction.

Finally, we note that *the sum of forces as well as the sum of moments are not affected when the force system is subjected to fundamental operations.*

WORKED EXAMPLE 1.6

A system of forces consists of three forces $\boldsymbol{F}_1$, $\boldsymbol{F}_2$ and $\boldsymbol{F}_3$ with magnitudes $F_1 = 54$ N, $F_2 = 28$ N and $F_3 = 16$ N, respectively. Directions and points of application are shown in Figure 1.26. Determine the sum of forces and the sum of moments about axes O and P.

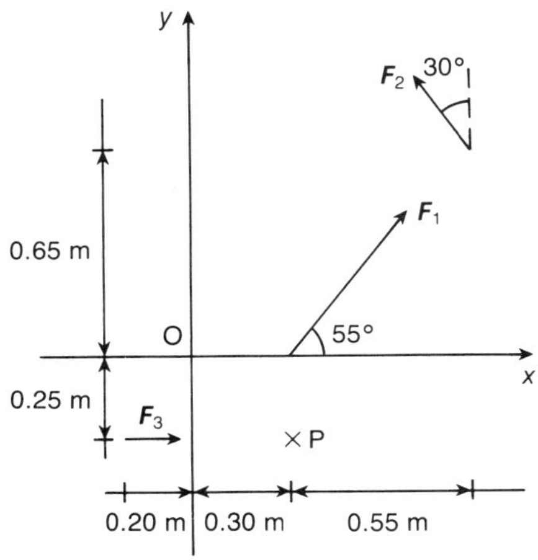

Figure 1.26

SOLUTION

To determine the sum of forces we use the *x*- and *y*-components of Equation (1.7). Resolving the forces into components and adding we have

$$\begin{cases} \sum F_x = F_1 \cos 55^\circ - F_2 \sin 30^\circ + F_3 \approx 33 \text{ N} \\ \sum F_y = F_1 \sin 55^\circ + F_2 \cos 30^\circ \approx 68 \text{ N} \end{cases}$$

which may be expressed as

$$\sum \mathbf{F} \approx (33,\ 68) \text{ N}$$

The sum of moments ΣM_O and ΣM_P are given by Equation (1.8):

$$\sum M_O = (54 \sin 55^\circ \cdot 0.30 + 28 \cos 30^\circ \cdot 0.85 + 28 \sin 30^\circ \cdot 0.65 + 16 \cdot 0.25) \text{ N m} \approx 47 \text{ N m counterclockwise}$$

$$\sum M_P = (-54 \cos 55^\circ \cdot 0.25 + 28 \cos 30^\circ \cdot 0.55 + 28 \sin 30^\circ \cdot 0.90) \text{ N m} \approx 18 \text{ N m counterclockwise}$$

1.2.4 Couple

Two parallel forces of the same magnitude but opposite directions are shown in Figure 1.27. This force system is called a *couple*. It is evident that the sum of forces ΣF of the system is zero. The sum of moments about an axis O, see Figure 1.27, can be expressed as

$$\sum M_O = F(a + d) - Fa = Fd$$

if we choose counterclockwise as the positive reference direction.

Obviously, the sum of moments depends on the magnitude F of the forces and the distance d between their lines of action. On the other hand the result is independent of the choice of the axis O. From Figure 1.27 we see that the sum of moments has a simple geometrical interpretation. It is easily seen that a parallelogram with two parallel sides determined by the forces has the 'area' Fd.

Consequently, all couples with the same value of the product Fd have the same sum of forces, namely zero, and the same sum of moments irrespective of the choice of axis. This means that fundamental quantities like the sum of forces and the sum of

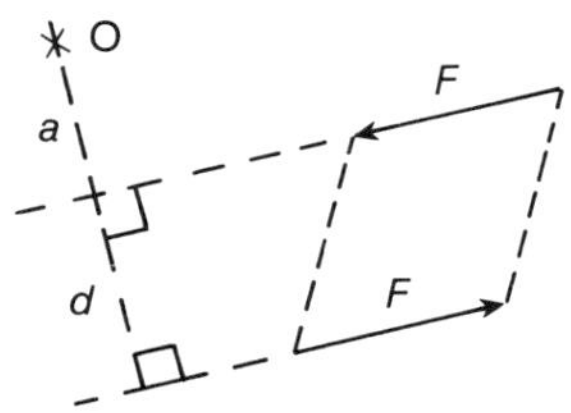

Figure 1.27

moments are not affected if a couple is replaced by another couple with the same value of the product Fd. A few examples are given in Figure 1.28, where a rough sketch of a box-spanner is shown. All couples have the same moment about any axis normal to the plane of the couple. This is especially true for the symmetry axis AB of the socket.

In a number of practical cases the moment of the couple is the only quantity of importance. The couple itself may be of no interest or even unknown. For this reason it has proved to be advantageous to introduce a new concept, namely the *couple-moment*. In the example of Figure 1.28 we state that the socket in each of the cases shown is subject to a couple-moment $M = Fd$ which describes the tendency to rotate the socket. In this case we need not bother about the separate forces that form the couple. What is important in this situation is the ability of the forces to loosen the nut, and this ability is determined by the couple-moment. All information available about the sum of forces and the sum of moments is included in M. However, it is important to realize that the various couples shown in Figure 1.28 may have different consequences in other senses, e.g. the internal stresses in the pin CD will vary from one case to another.

It is not even necessary that a single couple is responsible for the couple-moment. A system of couples is equivalent to a number of couple-moments that can be added. The result is one single couple-moment which is simply equal to the sum of moments of the original force system. This will be further explained in Section 1.2.5 below where we shall see that an arbitrary number of couples can be replaced by one single couple only by using fundamental operations.

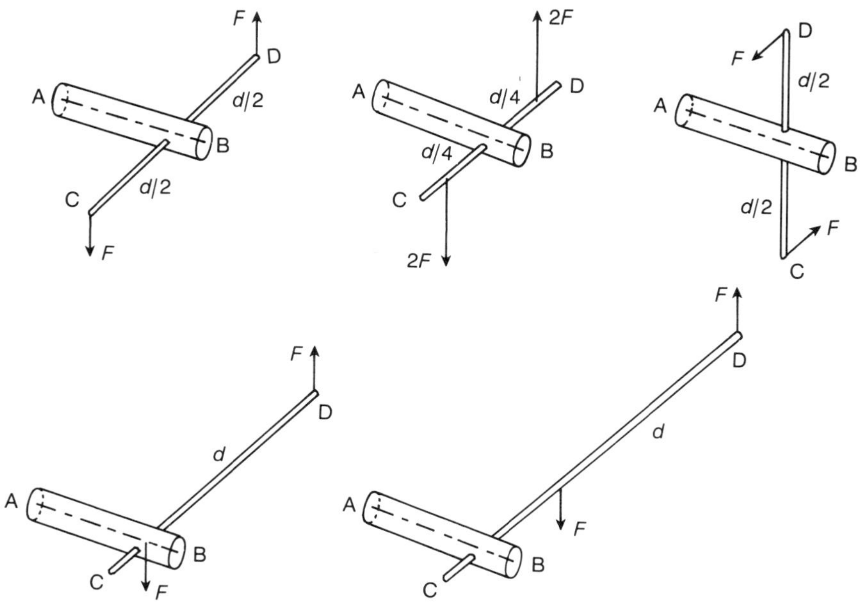

Figure 1.28

Let us consider a beam that is loaded by two forces according to Figure 1.29(a). Here S is an arbitrary cross-section of the beam. The two portions of the beam on each side of S will exert forces on each other. These forces will be distributed over the cross-sectional area in some way. Under certain conditions the force per unit area will vary according to Figure 1.29(b), where the two parts of the beam have been separated for clarity. The distributed forces may be considered as a very large number of couples, each of which may be replaced by a couple-moment. Adding these will result in a single couple-moment M, which is equal to the sum of moments and may be represented by the symbol of Figure 1.29(c).

In general, every force system with a vanishing sum of forces can be replaced by a couple-moment. In Section 1.2.5 we shall explore this property further in connection with the reduction of force systems.

From now on we shall permit force systems to include not only forces as in Figure 1.25, but also couple-moments. In many cases the underlying couple is not known. In a diagram couple-moments are represented by symbols according to Figure 1.30. The direction of the arrows indicates the tendency of the couple to rotate the body, clockwise or counterclockwise. Each couple-moment in Figure 1.30 may be considered as a symbol for a force system with the sum of forces equal to zero and the sum of moments about any axis equal to the couple-moment. It should be noted that a couple-moment does not have a special point of application in contrary to the forces that cause the moment.

It is obvious from the discussion above that there is no real need to distinguish between a couple and the corresponding couple-moment as long as we are interested only in quantities like the sum of forces and the sum of moments. All relevant information about the couple is included in the couple-moment. For this reason we shall allow ourselves to use the term 'couple' not only for a pair of forces but also for the corresponding couple-moment.

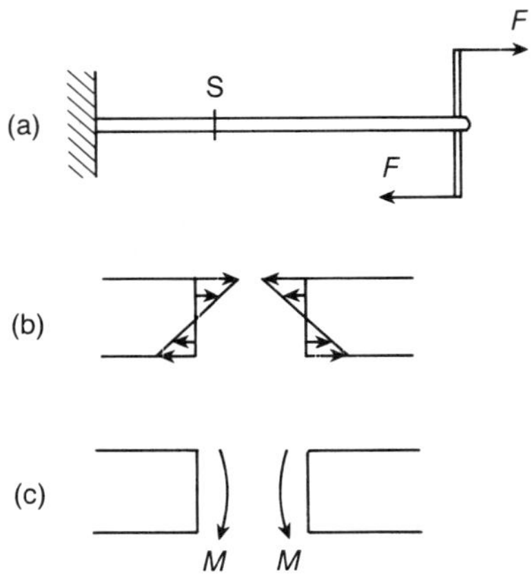

Figure 1.29

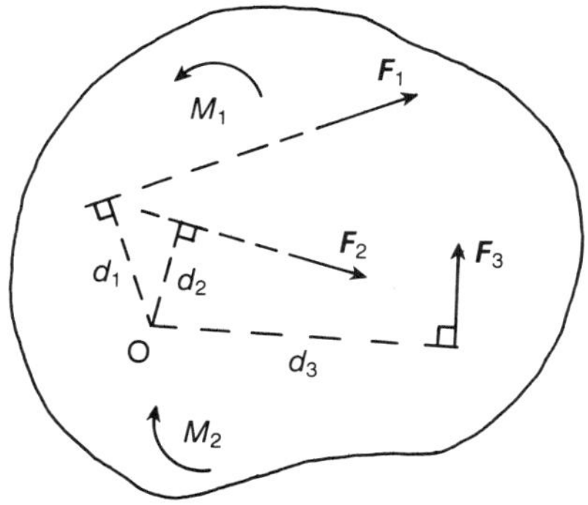

Figure 1.30

We now consider the generalized force system of Figure 1.30. The sum of forces and the sum of moments about an axis O are given by

$$\sum \boldsymbol{F} = \sum_{i=1}^{3} \boldsymbol{F}_i$$

$$\sum M_O = F_1 d_1 + F_2 d_2 - F_3 d_3 - M_1 + M_2$$

where the positive direction of moments is clockwise.

It should be noted that the couples do not contribute to the sum of forces, as a force system with a vanishing sum of forces is hidden behind each couple.

WORKED EXAMPLE 1.7

A body is subjected to two forces, $\boldsymbol{F}_1$ and $\boldsymbol{F}_2$, and two couples, $\boldsymbol{M}_1$ and $\boldsymbol{M}_2$, Figure 1.31. The magnitudes of the forces are $F_1 = 17$ N and $F_2 = 21$ N. The magnitudes of the couples are $M_1 = 8.0$ N m and $M_2 = 9.6$ N m.

Determine the sum of forces and the sum of moments about O.

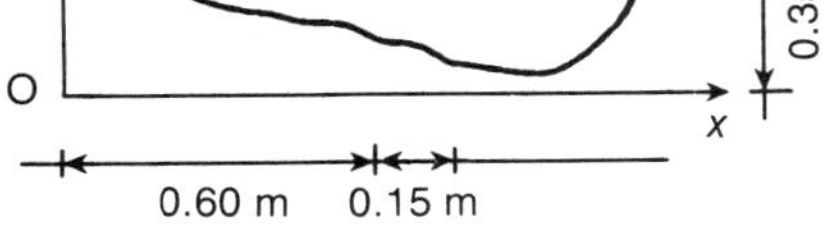

Figure 1.31

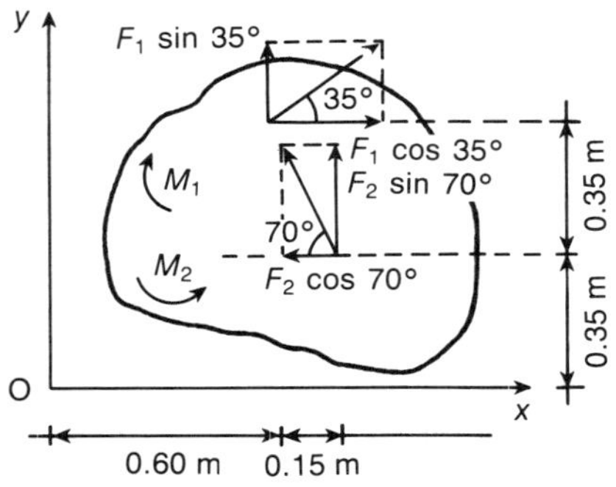

Figure 1.32

SOLUTION

The forces are resolved into x- and y-components as shown in Figure 1.32. We find immediately that

$$\sum F_x = F_1 \cos 35^\circ - F_2 \cos 70^\circ \approx 6.7 \text{ N}$$

$$\sum F_y = F_1 \sin 35^\circ + F_2 \sin 70^\circ \approx 29 \text{ N}$$

Hence, the sum of forces $\Sigma \boldsymbol{F} = (6.7, 29)$ N.

Using Varignon's theorem we obtain the sum of moments about O:

$$\sum M_O = (17 \sin 35^\circ \cdot 0.60 - 17 \cos 35^\circ \cdot 0.70 + 21 \sin 70^\circ \cdot 0.75$$
$$+ 21 \cos 70^\circ \cdot 0.35 - 8.0 + 9.6) \text{N m} \approx 15 \text{ N m counterclockwise}$$

1.2.5 Reduction of force systems

In this section we shall see how a general two-dimensional force system can be reduced, i.e. replaced by a simpler system which is equivalent to the original system in the sense that the sum of forces and the sum of moments are both equal. We recall that these quantities are unchanged when fundamental operations are performed on the forces.

The simplest type of a force system consists of a number of forces with a common point of application. These can be added immediately and the result of the operation will be one single force, the *resultant* of the system.

If the forces of a system have different points of application, the system can in most cases, not always however, be reduced by displacing forces along their lines of action. If two forces are displaced until their points of application coincide, they can then be added. If this procedure is repeated, the result may be one single force. This method is not very practical, however.

If a system consists of a number of parallel forces the above-mentioned method will not work. Consider two forces $\boldsymbol{F}_1$ and $\boldsymbol{F}_2$ according to Figure 1.33. It is of course

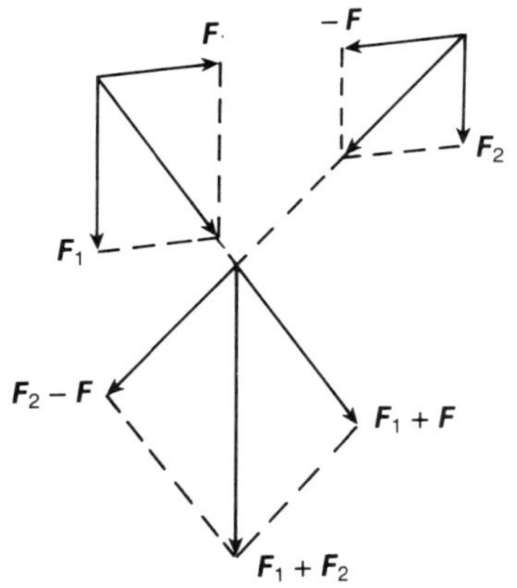

Figure 1.33

impossible to displace these forces along their lines of action until their points of application coincide. Introducing two forces $\boldsymbol{F}$ and $-\boldsymbol{F}$ with a common line of action and using the parallelogram law we can replace the original system by two non-parallel forces $\boldsymbol{F}_1+\boldsymbol{F}$ and $\boldsymbol{F}_2-\boldsymbol{F}$. Displacing these along their respective lines of action and using the parallelogram law we obtain one force $\boldsymbol{F}_1+\boldsymbol{F}_2$, whose line of action is determined according to Figure 1.33. This method will not work, however, if the sum of forces is zero. In this case we will end up with a couple, which cannot be further reduced.

This procedure can of course be repeated for a system consisting of more than two parallel forces. These *parallel force systems* can always be replaced by one single force, the *resultant* of the system, except in the case where the sum of forces is zero, in which case a couple will remain at the end. In fact, it is not even necessary that the force system is two dimensional. However, the above method is not very practical to use, especially when a large number of forces is involved. A more practical way to reduce a system of parallel forces is presented in Worked Example 1.8 below.

Using the above-mentioned methods it is in principle possible to reduce an arbitrary number of couples to one single couple. This is the justification for adding couples as we did in the preceding section. Remember that the sum of moments is not affected when fundamental operations are performed.

When reducing a force system it is sometimes useful to displace the point of application of a certain force. As we have mentioned earlier, displacing the force along its line of action is permitted, but a translation to a parallel position is not admitted in general. In Figure 1.34 it is shown how we can replace a force $\boldsymbol{F}$ applied at a point A by the same force $\boldsymbol{F}$ applied at point B and a couple M. In other words, by adding a suitably chosen couple we can translate any force $\boldsymbol{F}$ to an arbitrary parallel position.

Our next step will be to show how an arbitrary force system can be replaced by an equivalent system consisting of at most one force and one couple.

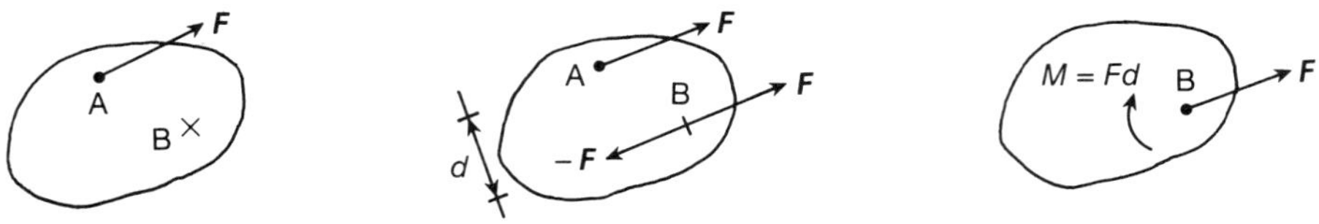

Figure 1.34

Consider an arbitrary force system consisting of a number of forces, say n, and a number of couples, say m. An example is shown in Figure 1.35(a), where $n = 3$ and $m = 2$. It is possible to translate each force to a parallel position with the point of application at an arbitrary location O provided that we add a couple for each force. The couple $F_1 d_1$ corresponds to the force $\boldsymbol{F}_1$ and so on. The result of the operation is shown in Figure 1.35(b).

Adding the forces we obtain one single force

$$\boldsymbol{R} = \sum_{i=1}^{n} \boldsymbol{F}_i \tag{1.9}$$

which is the force resultant of the system. Adding the couples, we obtain one single couple

$$M = \sum_{i=1}^{n} F_i d_i + \sum_{i=1}^{m} M_i \tag{1.10}$$

which is equal to the sum of moments or the moment resultant about O. The final result of the reduction (the force $\boldsymbol{R}$ and the couple M) is shown in Figure 1.35(c).

For simplicity we have assumed that all couples have the same orientation. Hence all terms of Equation (1.10) are positive. In a practical case we have to take into account that couples with different tendencies to rotate the body will correspond to different signs of the terms of Equation (1.10); see Worked Example 1.9 below.

The general force system of Figure 1.35(a) and the reduced system of Figure 1.35(c) have the same sum of forces and the same sum of moments about O. As the reduction procedure only is based on fundamental operations, the two systems must have the same sum of moments about any axis.

The axis O in Figure 1.35 was chosen arbitrarily. Consequently, this reduction is not unique. The force resultant $\boldsymbol{R}$ will always be the same according to Equation (1.9). The moment resultant M, however, will depend on which point we have chosen as the point of application of $\boldsymbol{R}$. This point, i.e. O in the example above, is called the *point of reduction*. Here we denote an axis as well as the point where the axis intersects the plane of the forces by the same letter O. Normally, there should be no risk of confusion.

If we choose another point of reduction, e.g. P of Figure 1.36(a), then the sum of moments of the force system is $\Sigma M_P = M - Rd$, where $R = |\boldsymbol{R}|$. If $d = M/R$, then $\Sigma M_P = 0$. In other words, by choosing the point of reduction P in a suitable way, we can replace the force system by one single force $\boldsymbol{R}$ according to Figure 1.36(b). This requires, however, that $\boldsymbol{R} \neq 0$. Therefore, a couple can never be reduced to one single force.

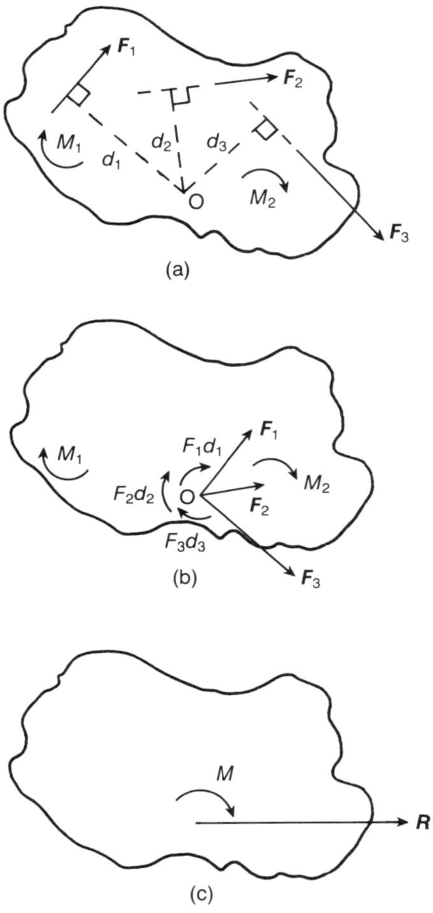

Figure 1.35

If $\boldsymbol{R} = 0$, the force system can always be reduced to a couple, i.e. every force system with a vanishing force resultant is equivalent to a couple.

According to Section I.4.3 we may consider a system of concentrated forces where the number of forces in a limiting case approaches infinity. In this case the forces are *distributed* over a certain region. The reduction procedure described above may equally well be applied to a system of forces that are distributed over a surface. Hence, distributed forces can always be replaced by a simpler system consisting of at most *one* force and *one* couple. This property will be utilized later on in a number of cases.

In statics we are particularly concerned with force systems which are equivalent to zero, i.e. the sum of forces and the sum of moments are both equal to zero.

The purpose of the discussion above has been to investigate the existing possibilities to reduce various force systems, not to demonstrate how the reduction is carried out in practice. Instead, this will be shown in the following worked examples.

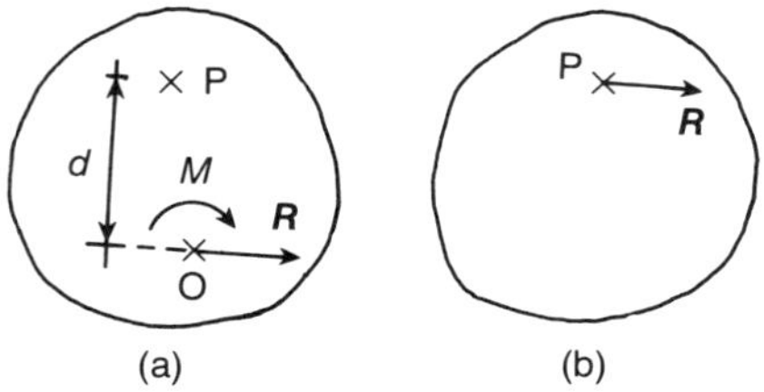

Figure 1.36

WORKED EXAMPLE 1.8

Replace the force system shown in Figure 1.37 by one single force, the resultant of the system.

SOLUTION

We introduce a coordinate system as shown in Figure 1.38. The force resultant $\boldsymbol{R}$ is equal to the sum of forces, i.e.

$$\boldsymbol{R} = (-2.0 - 1.5 + 2.3 - 1.8)\ \mathbf{e}_y\ \text{N} = -3.0\ \mathbf{e}_y\ \text{N}$$

The moment of the resultant about an arbitrary axis must be equivalent to the sum of moments of the original system. Let us choose the axis O. If we take clockwise to be the positive reference direction, then

$$\sum M_O = Rx = (1.5 \cdot 1.4 - 2.3 \cdot 2.4 + 1.8 \cdot 3.3)\ \text{N m}$$

$$\Rightarrow x = \frac{2.52}{3.0}\ \text{m} = 0.84\ \text{m}$$

which determines the line of action.

WORKED EXAMPLE 1.9

The magnitudes of the forces and couples shown in Figure 1.39 are:

$$F_1 = 32\ \text{N} \qquad F_2 = 18\ \text{N} \qquad F_3 = 41\ \text{N} \qquad M_1 = 11\ \text{N m} \qquad \text{M}_2 = 2.3\ \text{N m}$$

(a) Replace the force system by a force $\boldsymbol{R}$ applied at P and a couple.
(b) Replace the force system by a single force and determine its line of action.

SOLUTION

(a) The force $\boldsymbol{R}$, Figure 1.40 (see p. 38), is equal to the sum of forces, i.e.

$$R_x = \sum F_x = -F_1 \cos 65^\circ + F_2 \cos 80^\circ - F_3 \sin 75^\circ \approx -50\ \text{N}$$

$$R_y = \sum F_y = F_1 \sin 65^\circ + F_2 \sin 80^\circ + F_3 \cos 75^\circ \approx 57\ \text{N}$$

or $\boldsymbol{R} \approx (-50, 57)$ N

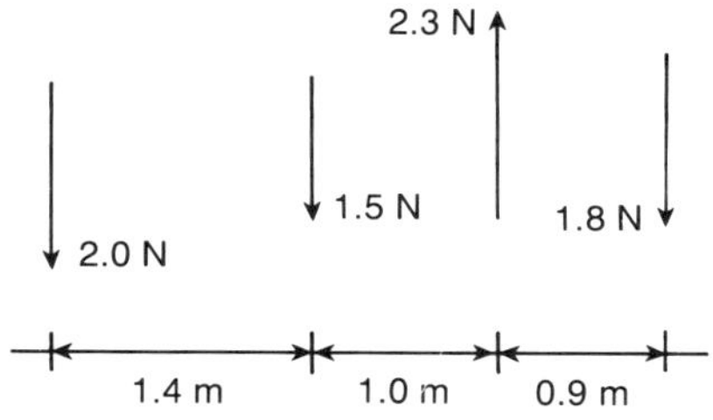

Figure 1.37

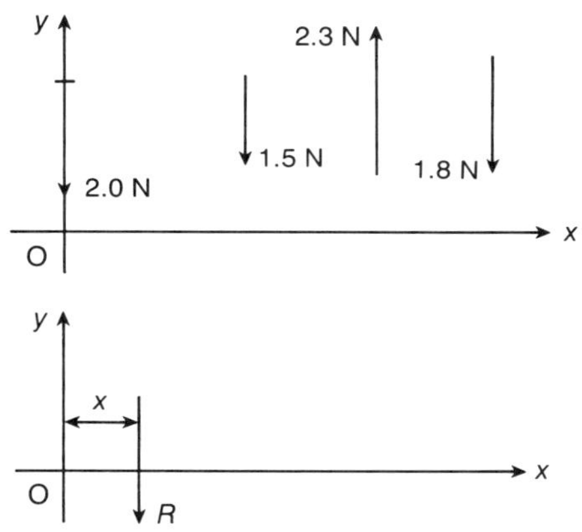

Figure 1.38

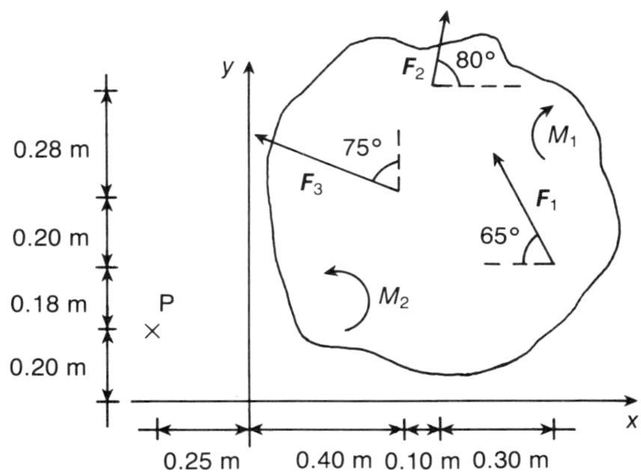

Figure 1.39

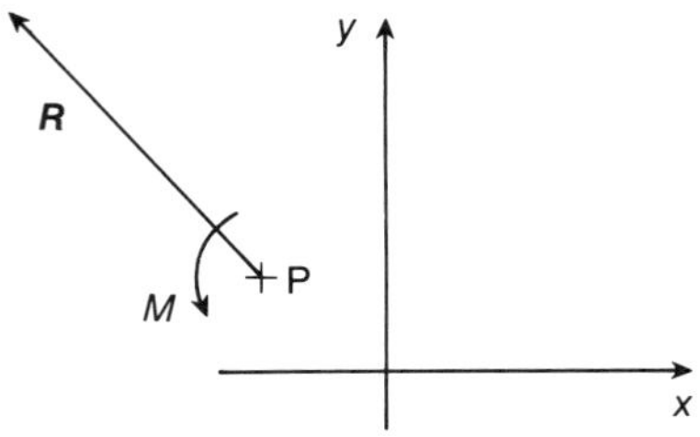

Figure 1.40

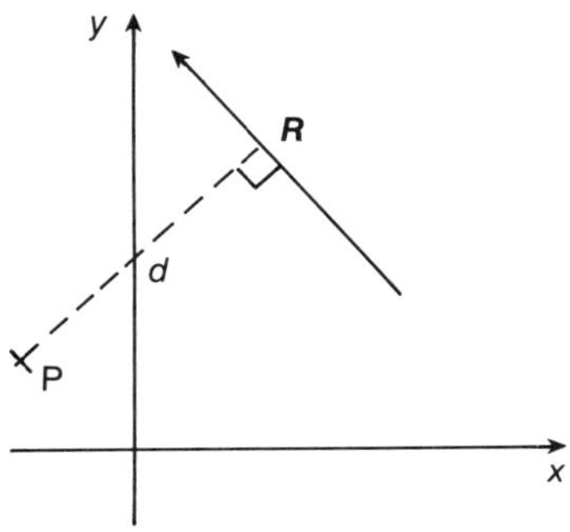

Figure 1.41

The couple is equal to the sum of moments of the original system about P. Hence,

$$M = (32 \sin 65^\circ \cdot 1.05 + 32 \cos 65^\circ \cdot 0.18 + 18 \sin 80^\circ \cdot 0.75 - 18 \cos 80^\circ \cdot 0.66 + 41 \cos 75^\circ \cdot 0.65 + 41 \sin 75^\circ \cdot 0.38 - 11 + 2.3)\ \text{N m} \approx 57\ \text{N m counterclockwise}$$

(b) If the force system is replaced by one single force, this must be equal to the sum of forces, i.e. equal to the force **R** determined above. Hence,

$$\boldsymbol{R} \approx (-50, 57)\ \text{N}$$

To determine the line of action we utilize the fact that the sum of moments about P must be the same after the reduction. Thus the distance d shown in Figure 1.41 must satisfy the equation

$$M = |\boldsymbol{R}|\, d \Rightarrow d = \frac{M}{|\boldsymbol{R}|} \approx 0.75\ \text{m}$$

Thereby the line of action is determined.

PROBLEMS

○ **1.1** Determine the magnitude of the resultant of the two forces shown.

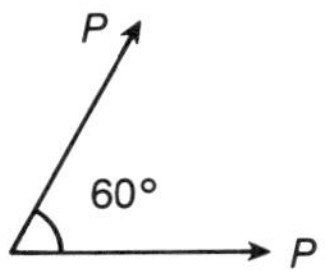

○ **1.2** A crate of mass 2.0 kg is resting on a 25° incline. Resolve the gravitational force into components parallel to the incline and perpendicular to it.

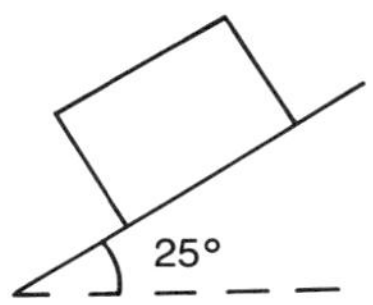

○ **1.3** Determine the magnitude of the resultant of the two forces shown. Also determine the angle formed by the resultant and the 3 N force.

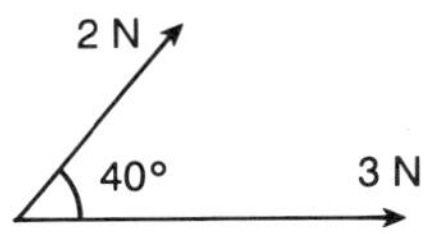

○ **1.4** A load *D* of mass 80 kg is supported by two bars AB and BC. Resolve the gravitational force on the load into components parallel to the two bars.

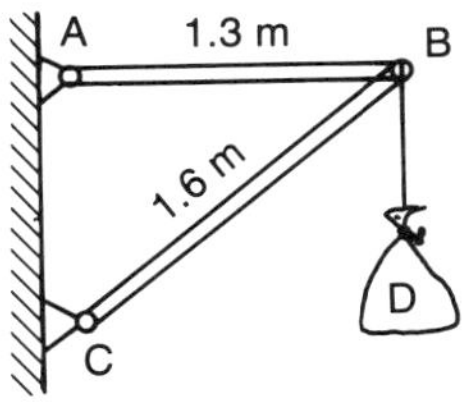

1.5 The force shown in the next figure is resolved into two components, one along OA and one of magnitude 18 N.
(a) Determine the component along OA.
(b) Determine the angle between the 18 N component and the force *F*.

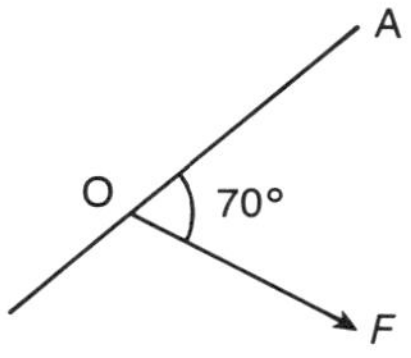

1.6 A car is towed by two ropes attached at A. The tensions in the ropes (denoted by F_1 and F_2) have a resultant of 1.5 kN along the forward direction of the car.
(a) Determine F_1 and F_2 for the case $\alpha = 30°$.
(b) Consider the case where F_2 has its minimum value. Determine F_2 and the corresponding values of α and F_1.

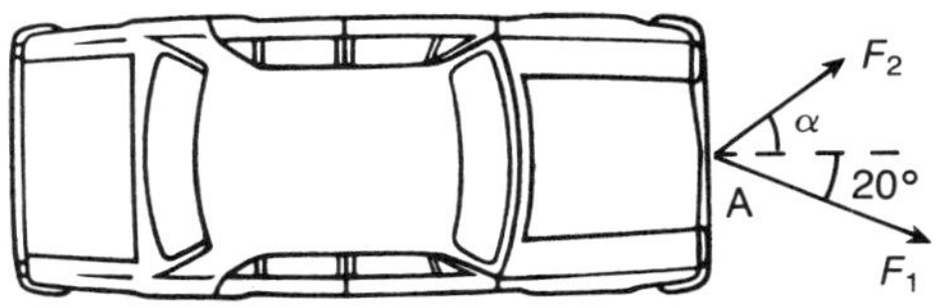

○ **1.7** The spring constant of the spring shown is 1.3 kN/m. The unstretched length is 1.5 m. Write the vector expression, using *x*- and *y*-components, for the force exerted by the spring on
(a) support A,
(b) support B.

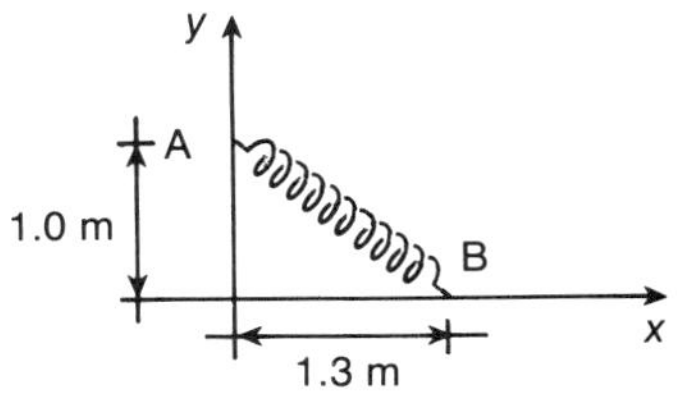

1.8 A force of magnitude $F = 50$ N is applied to a bar ABC as shown overleaf.
(a) Determine the *x*- and *y*-components of the force.
(b) Resolve the force into two components, one parallel to direction AB and the other one normal to it. Calculate the magnitudes of these components.

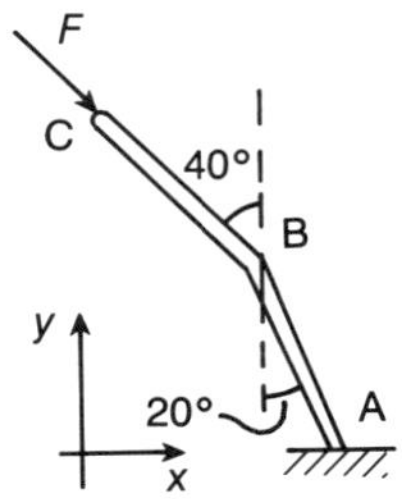

○ **1.9** Determine the component along line AB of the force shown.

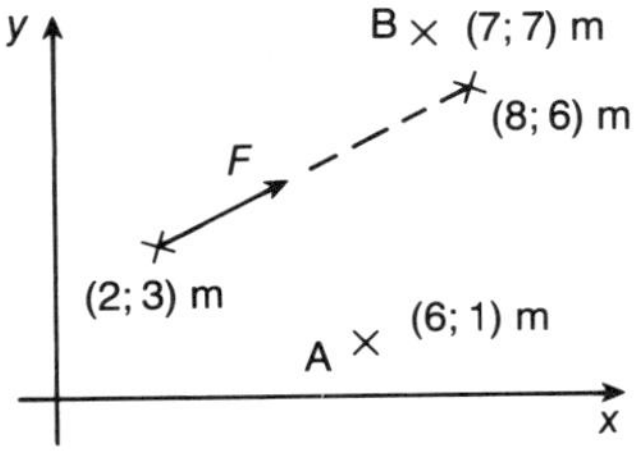

○ **1.10** Determine the moment about O of the force shown.

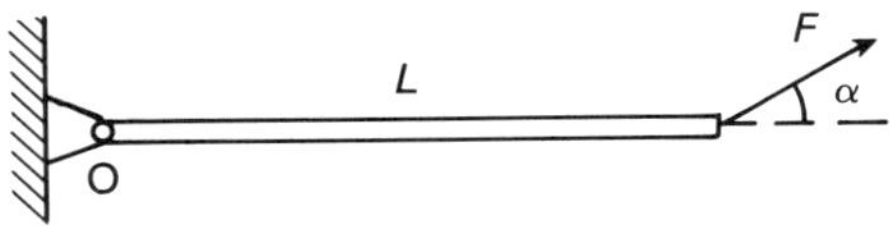

○ **1.11** Determine the moment of the force shown about A, B and C, respectively.

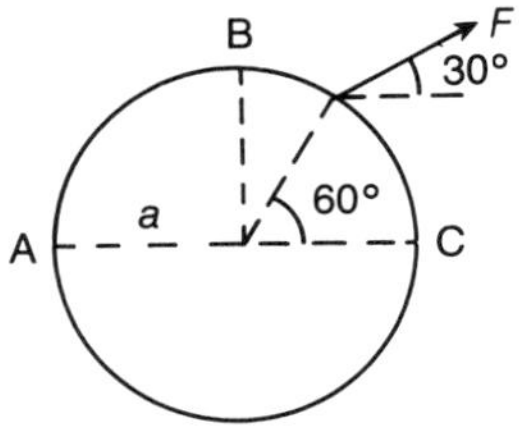

○ **1.12** In order to move a crate a force is applied as shown. Determine the moment of the force about A.

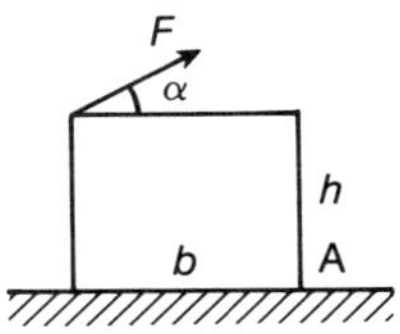

○ **1.13** The cross-section of what is called an I-beam is shown. A force of magnitude $F = 80$ N is applied at the flange of the beam. Determine the moment about axis C (the centroidal line of the beam).

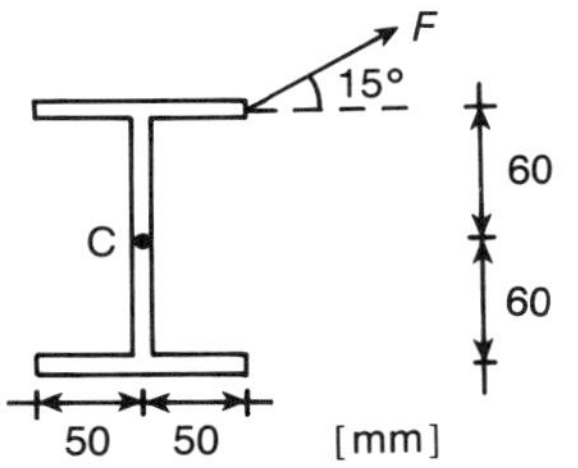

○ **1.14** Determine the moment of the 8 N force about
(a) O (b) B (c) C (d) D (e) E

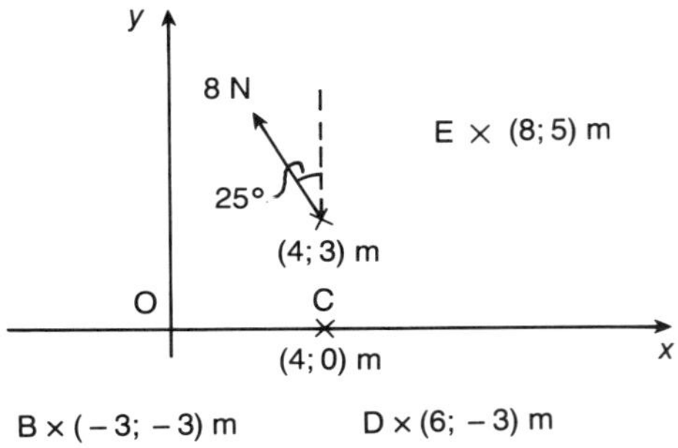

1.15 In order to loosen a bolt a force F is applied to the handle of the spanner as shown.
(a) Determine the moment M_O about the axis of the bolt when $\alpha = 75°$.
(b) Determine the maximum and minimum values of $|M_O|$ and the corresponding values of α.

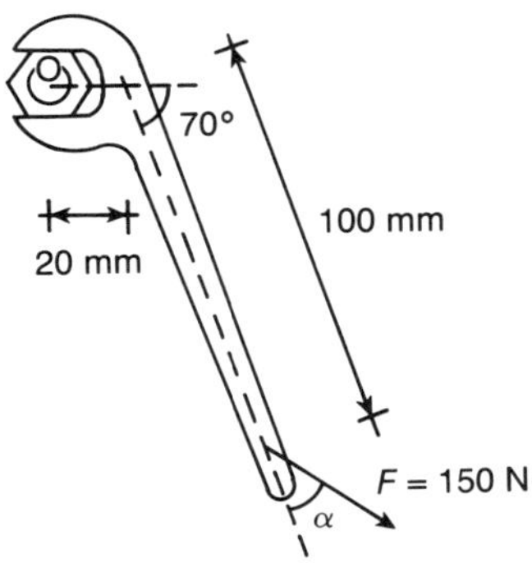

1.16 In order to loosen a bolt a force $F = 300$ N is applied to a tyre wrench as shown. Determine the moment of the force about
(a) axis O,
(b) axis P.

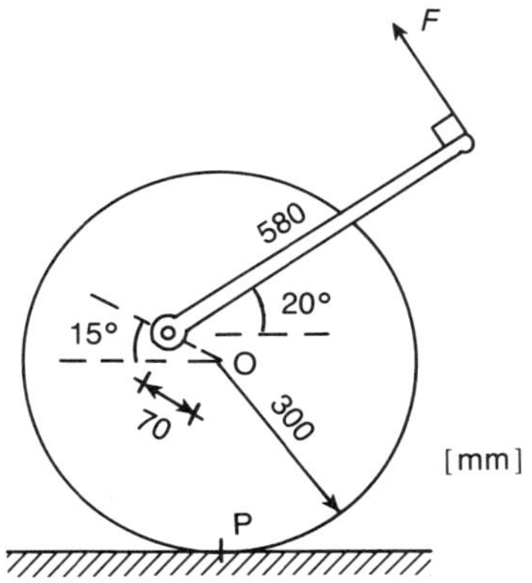

○ **1.17** Four forces are applied at point A as shown. Determine the x- and y-components of the resultant and its magnitude.

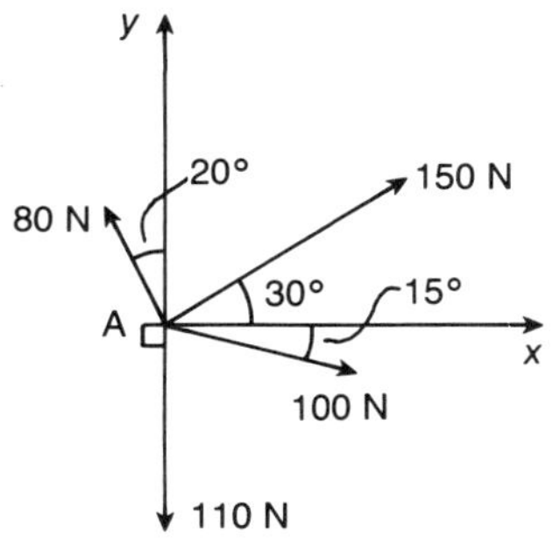

○ **1.18** For the force system shown, determine
(a) the sum of forces,
(b) the sum of moments about O,
(c) the sum of moments about A,
(d) the sum of moments about B,
(e) the sum of moments about C.

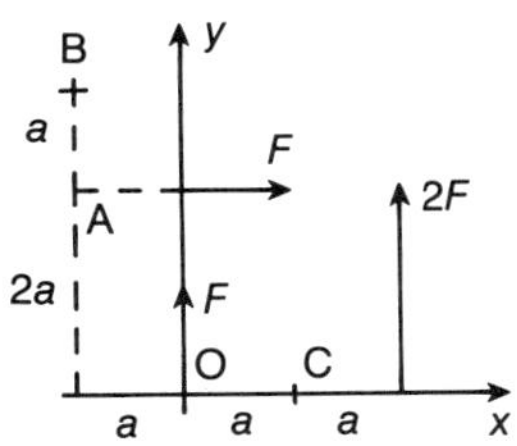

○ **1.19** For the force system shown, determine
(a) the sum of forces,
(b) the sum of moments about O,
(c) the sum of moments about A.

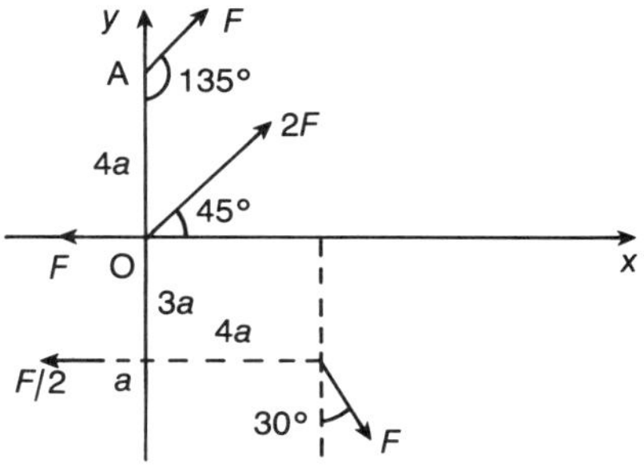

○ **1.20** A nut can be loosened in various ways, some of which are shown here. Determine for each case the sum of moments about O.

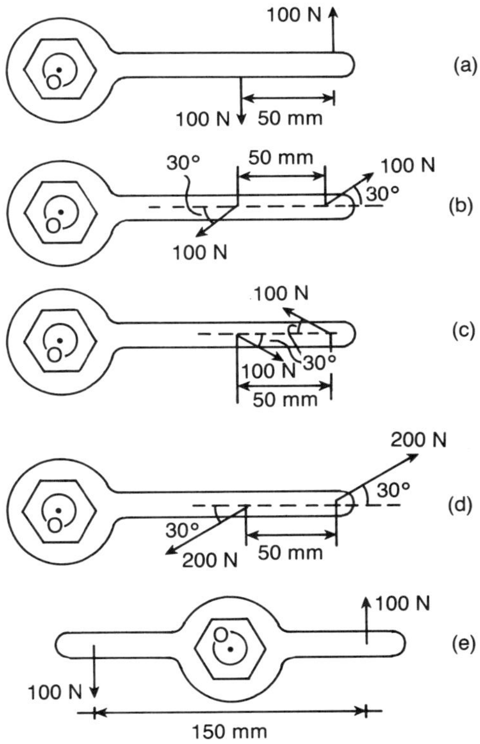

continued overleaf

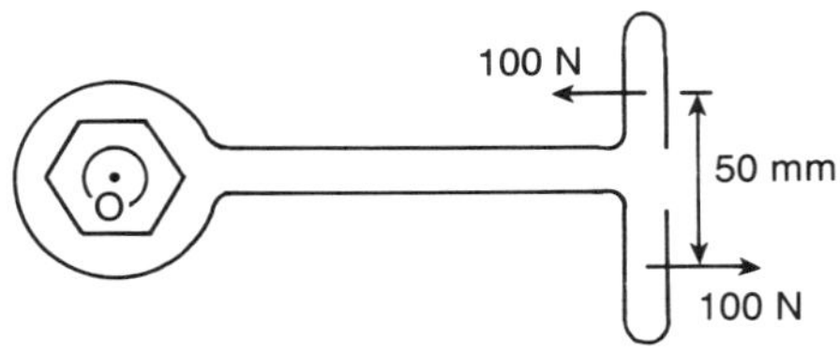

(f)

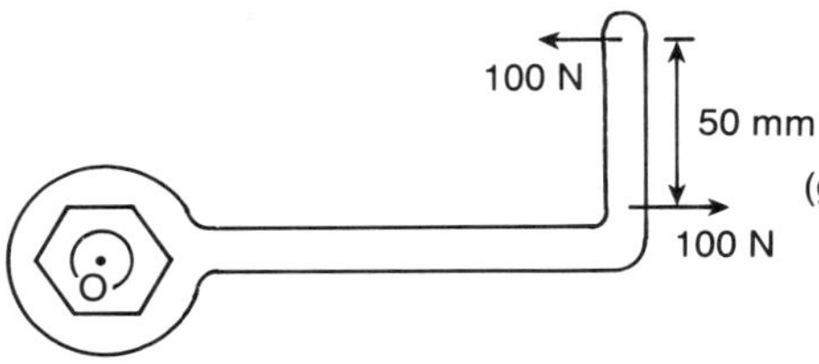

(g)

○ **1.21** A tape is running over two spools which are mounted on a plate. The plate is supported at A and B, where two oppositely directed forces of equal magnitude *F* are applied as shown. Two tensile forces, each of magnitude 8 N, are applied to the tape. Determine the value of *F*, if the sum of moments of the force system is zero.

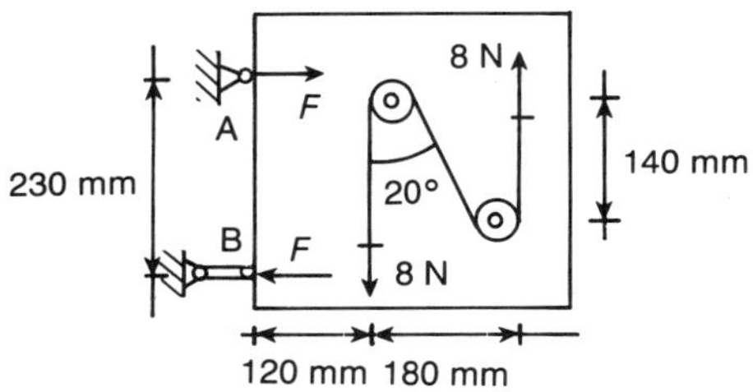

1.22 For the force system shown, determine
(a) the sum of forces,
(b) the sum of moments about the *z*-axis.

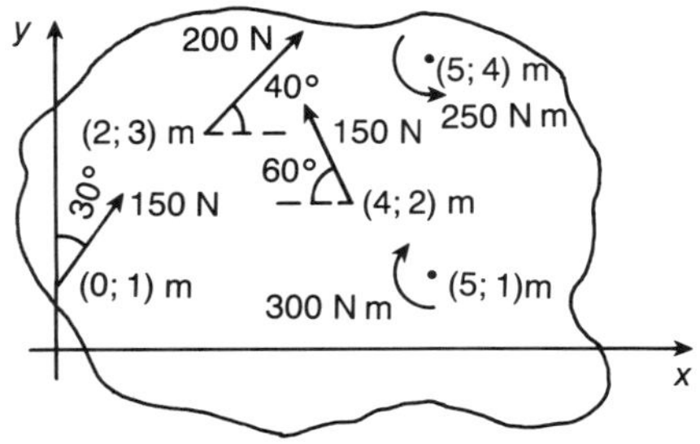

○ **1.23** Four different force systems acting on a cantilever beam are shown here. Replace each system by one single force, the resultant. Specify the magnitude of the resultant and its line of action.

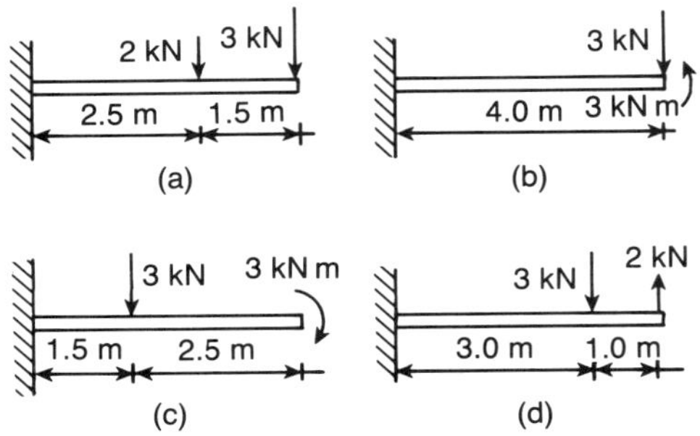

○ **1.24** A force system is applied to a beam as shown. Determine
(a) the sum of forces,
(b) the sum of moments about A.

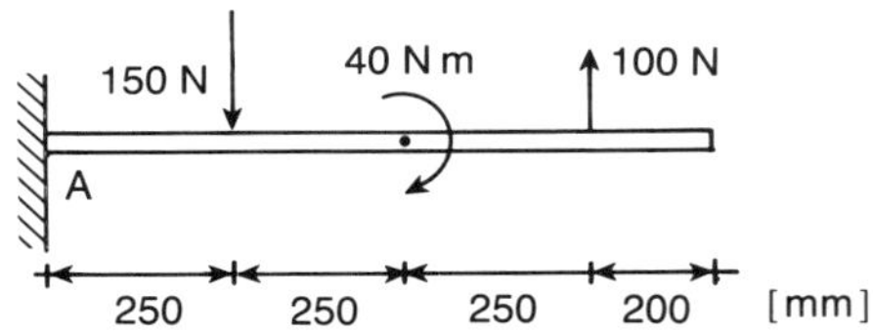

○ **1.25** A belt is exerting forces on a pulley as shown. Replace the forces by a force at the axis O and a couple.

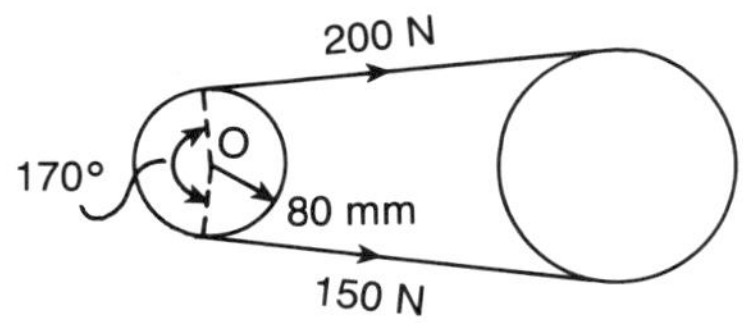

○ **1.26** Replace the two forces acting on the spanner by a force at O and a couple.

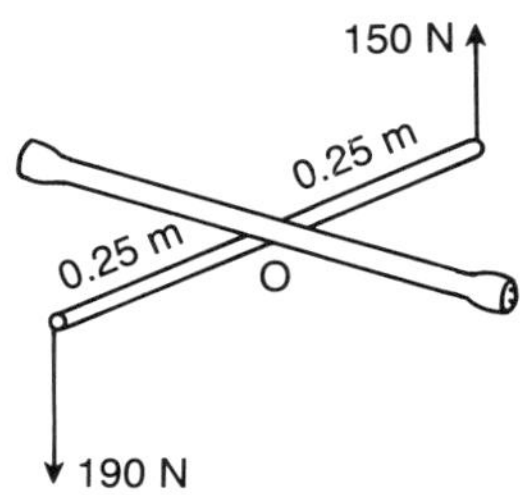

1.27 A force system is acting on a body as shown.
(a) Replace the system by a force at O and a couple.
(b) Replace the system by its resultant and determine the line of action.

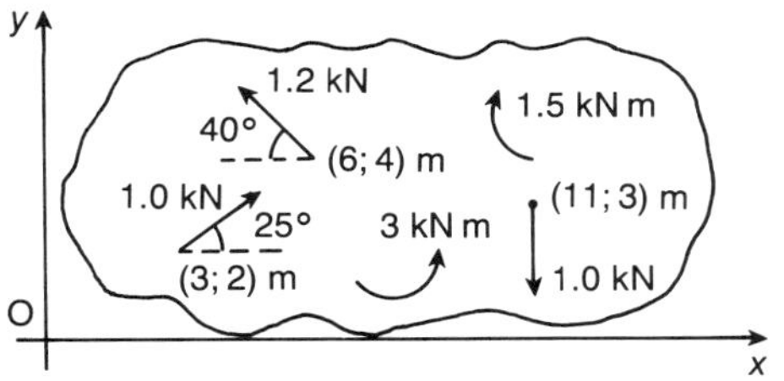

1.28 A force $F = 57$ N and a couple $M = 46$ N m act on a body as shown. Replace the system by its resultant and determine the line of action.

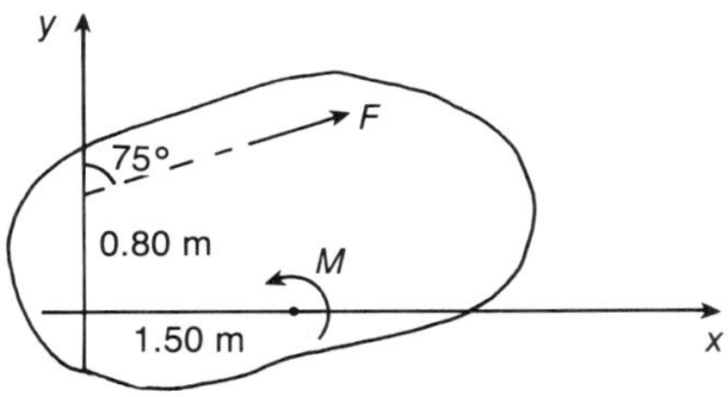

○ **1.29** A bar is kept in the position shown by a cord. The tension in the cord is 200 N. Replace the tension by a force at O and a couple.

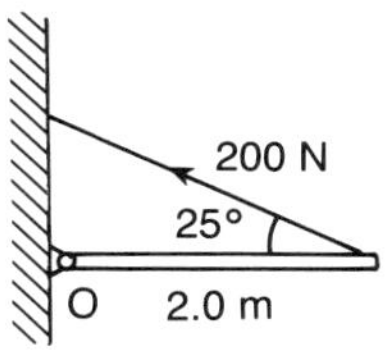

○ **1.30** A bar is kept in the position shown by a cord. The tension in the cord is 200 N. Replace the tension by a force at O and a couple.

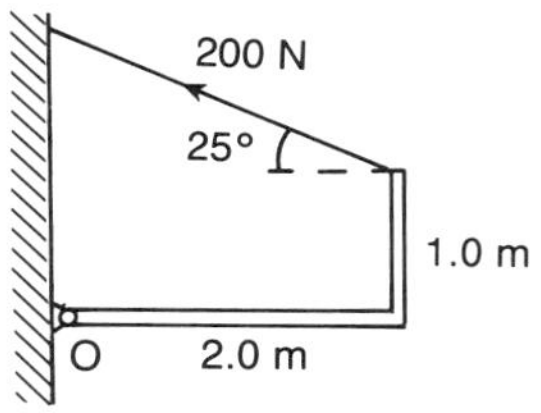

◎ **1.31** A swing with two children is subjected to gravitational forces as shown. Determine the angle α, if the line of action of the resultant of the gravitational forces passes through axis O.

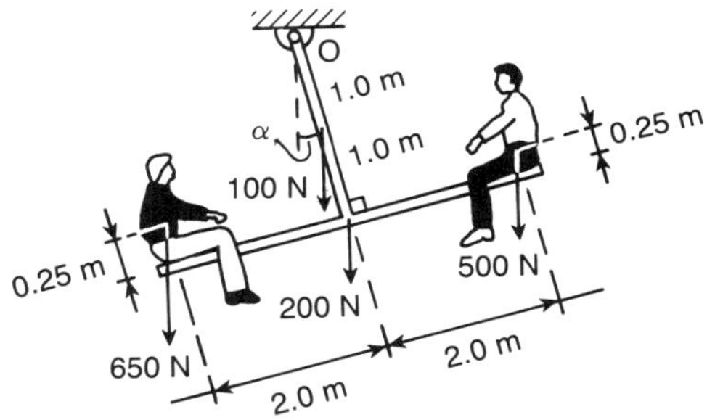

1.32 Three forces act on the cross-section of a beam as shown. Replace the force system by its resultant and determine the line of action.

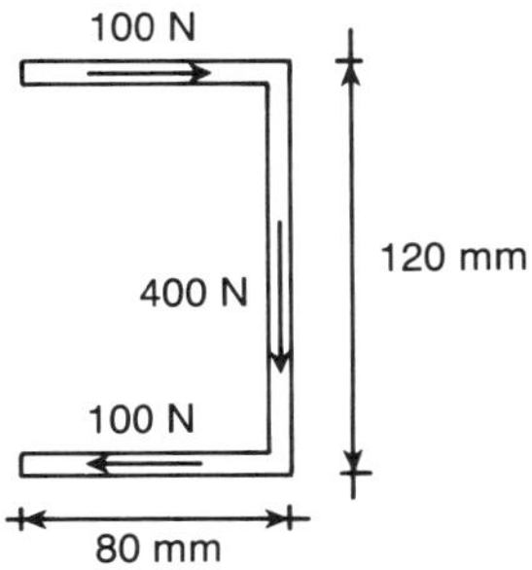

1.3 Forces in three dimensions

1.3.1 Vectors in three dimensions

Describing forces by vectors in a three-dimensional case requires a set of three unit vectors. The corresponding directions are normally denoted by *x*, *y* and *z*, but other

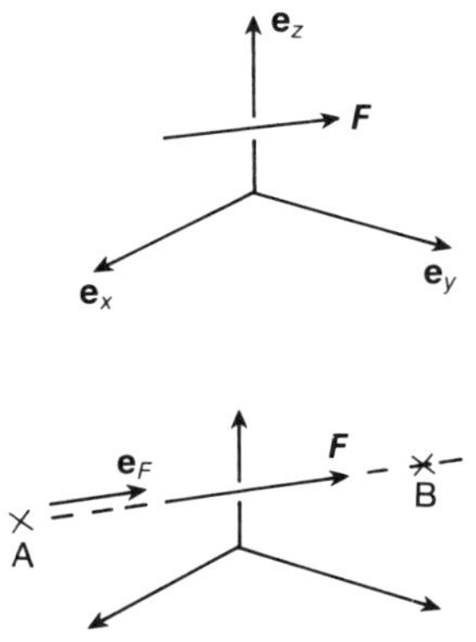

Figure 1.42

combinations may occur and are sometimes preferable, especially in dynamics. In most cases, however, we use the Cartesian *xyz*-system with corresponding unit vectors $\mathbf{e}_x$, $\mathbf{e}_y$ and $\mathbf{e}_z$.

A force $\boldsymbol{F}$, see Figure 1.42, may be represented by its scalar components F_x, F_y and F_z meaning that

$$\boldsymbol{F} = F_x\,\mathbf{e}_x + F_y\,\mathbf{e}_y + F_z\,\mathbf{e}_z \tag{1.11}$$

If the components of the force are unknown, but the magnitude F and the line of action are known, we can start by expressing the vector as $\boldsymbol{F} = F\mathbf{e}_F$. Here $\mathbf{e}_F$ is a unit vector which is parallel to the force. If A and B are two points on the line of action such that the vector $\overrightarrow{\mathrm{AB}}$ is parallel to $\boldsymbol{F}$, see Figure 1.42, then $\mathbf{e}_F$ is obtained from

$$\mathbf{e}_F = \frac{\overrightarrow{\mathrm{AB}}}{|\overrightarrow{\mathrm{AB}}|} \tag{1.12}$$

If the coordinates of A and B are known the vector $\overrightarrow{\mathrm{AB}}$ can easily be expressed in terms of its scalar components in the *xyz*-system. Hence,

$$\boldsymbol{F} = F\mathbf{e}_F = F\,\frac{\overrightarrow{\mathrm{AB}}}{|\overrightarrow{\mathrm{AB}}|} \tag{1.13}$$

is expressed in terms of scalar components. This result is useful in problem solving as we shall see later on.

WORKED EXAMPLE 1.10

A load of mass 80 kg is attached to a rope which runs through a smooth loop B and is fastened at A, Figure 1.43. Determine the vector expression for the force exerted by the rope on the support at A.

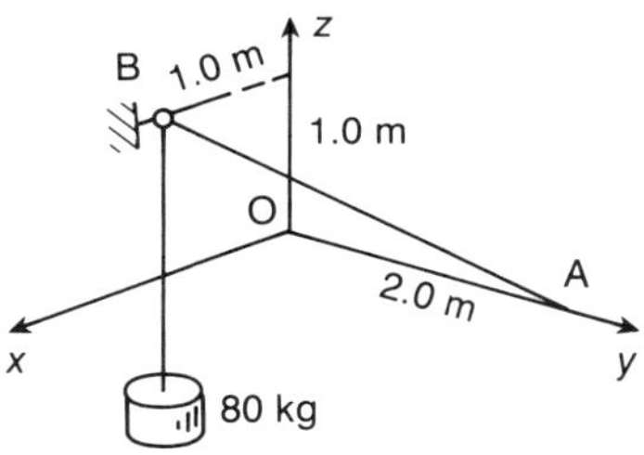

Figure 1.43

SOLUTION

The magnitude of the force is $F = 80 \cdot 9.81$ N. The vector $\vec{AB}$ along the line of action is given by (1, −2, 1) m. The unit vector $\mathbf{e}_F$ is obtained from Equation (1.12) as

$$\mathbf{e}_F = \frac{\vec{AB}}{|\vec{AB}|} = \frac{(1, -2, 1)}{\sqrt{6}}$$

Hence, the force exerted by the rope on A is

$$\boldsymbol{F} = F\mathbf{e}_F = \frac{80 \cdot 9.81}{\sqrt{6}}(1, -2, 1)\ \text{N} = (0.32, -0.64, 0.32)\ \text{kN}$$

Another common problem, where vector algebra may be useful, appears when we want to determine the scalar component of a force, represented by a vector $\boldsymbol{F}$, along a certain direction $\boldsymbol{e}_\lambda$. In other words, we are interested in the projection of the force on a line λ.

If the vector component is expressed as $\boldsymbol{F}_\lambda = F_\lambda \mathbf{e}_\lambda$, then the scalar component F_λ is given by $F_\lambda = \boldsymbol{F} \cdot \mathbf{e}_\lambda$ (the scalar product $\boldsymbol{F} \cdot \mathbf{e}_\lambda$ is defined in Appendix A1.1.8). In many cases it is practical to find $\mathbf{e}_\lambda$ from the relation

$$\mathbf{e}_\lambda = \frac{\vec{DC}}{|\vec{DC}|}$$

where D and C are two arbitrary points on the line λ. If $\boldsymbol{F}$ and λ are located in the same plane, λ and the line of action of $\boldsymbol{F}$ intersect, see Figure 1.44(a).

However, the projection of a vector on a certain line is defined even if the vector $\boldsymbol{F}$ and the line λ are not located in the same plane, i.e. λ and the line of action of $\boldsymbol{F}$ do not intersect. The expression $\boldsymbol{F} \cdot \mathrm{e}_\lambda$ is still valid for the projection F_λ. This case is illustrated in Figure 1.44(b).

WORKED EXAMPLE 1.11

A force of magnitude P is applied at the corner A of the rectangular parallelepiped shown in Figure 1.45. Determine the scalar component along line BC.

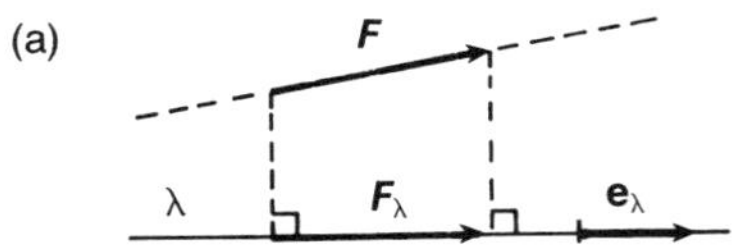

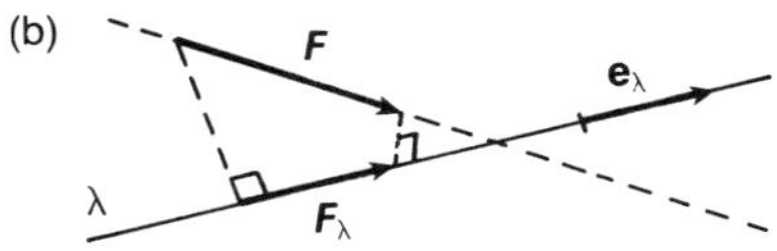

Figure 1.44

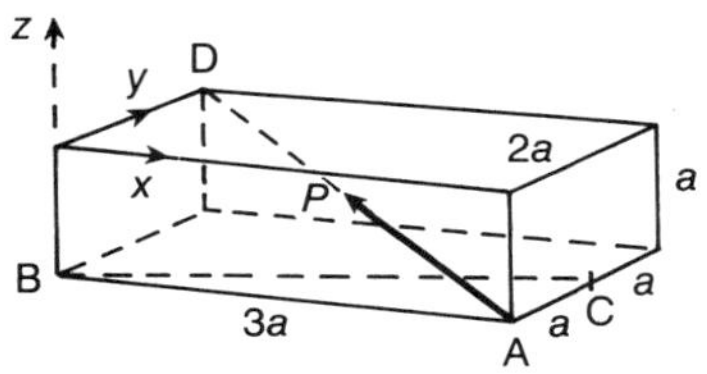

Figure 1.45

SOLUTION

Introducing a unit vector $\mathbf{e}_F$ along the diagonal AD we can express the force as $\boldsymbol{F} = P\mathbf{e}_F$, where

$$\mathbf{e}_F = \frac{\overrightarrow{AD}}{|\overrightarrow{AD}|} = \frac{(-3a, 2a, a)}{\sqrt{14a^2}}$$

The force is projected on the line λ passing through B and C. The corresponding unit vector $\mathbf{e}_\lambda$ is obtained from

$$\mathbf{e}_\lambda = \frac{(3a, a, 0)}{\sqrt{10a^2}}$$

From the relation $F_\lambda = \boldsymbol{F} \cdot \mathbf{e}_\lambda$ we obtain

$$F_\lambda = \frac{P}{\sqrt{14}\sqrt{10}}(-3, 2, 1) \cdot (3, 1, 0) = -P\sqrt{\frac{7}{20}} \approx -0.59P$$

The negative sign means that the vector component $\boldsymbol{F}_\lambda$ is directed from C towards B, i.e. in the direction corresponding to the unit vector $-\mathbf{e}_\lambda$. The magnitude of the component is $0.59P$.

1.3.2 Moment

As a starting point we shall attempt to express our everyday life experience of moments in terms of vector components. In Section 1.2.2 we introduced the moment of a force about an axis starting from our conception of the tendency of a force to rotate a certain body. This conception, which is based on practical experience, was used as a basis for a mathematical description of the moment. However, our treatment was restricted to two-dimensional force systems, where the axis of interest was perpendicular to the plane of the forces.

Our next step will be to consider the case where this restriction does not apply; for instance, we may be interested in the moment of a force about an axis which is not perpendicular to the force. Our starting point will once again be our everyday life experiences of the tendency of a force to rotate a body about an axis:

- The moment of a force about an axis which is parallel to the force is zero, Figure 1.46(a).
- The moment of a force about an axis is zero if the line of action of the force and the axis intersect, Figure 1.46(b).
- If a force is perpendicular to an axis, then according to Section 1.2.2 the moment about the axis is Fd. Here d is the distance from the line of action to the axis, Figure 1.46(c). The moments of two forces may be cooperating or counteracting. If we introduce a positive direction of reference, the moments may be added provided that the signs of the moments are taken into account.

Starting from the items above we now proceed to find an expression for the moment M_x about an axis of a force $\boldsymbol{F} = (F_x, F_y, F_z)$ which is applied at a point with a position vector $\boldsymbol{r} = (x, y, z)$. A right-handed coordinate system $Oxyz$ is introduced so that the axis coincides with the x-axis. The force $\boldsymbol{F}$ and its point of application may be arbitrary.

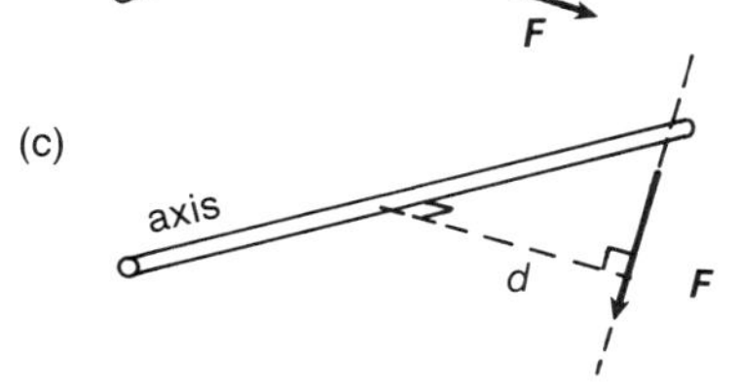

Figure 1.46

The situation is presented in Figure 1.47 where it has been assumed that the components of $\boldsymbol{F}$ and $\boldsymbol{r}$ are all positive.

The force is resolved into components of magnitudes F_x, F_y and F_z as shown. The two components F_y and F_z are perpendicular to the x-axis and their moment arms with respect to the axis are z and y, respectively. Obviously, the moments are counteracting. We now choose our positive direction of reference as the direction of rotation which, according to the right-hand rule, corresponds to $\mathbf{e}_x$. Hence, $M_x = F_z y - F_y z$. It is left to the reader to show that the expression for M_x is valid even if one or more of the scalar components of $\boldsymbol{F}$ and $\boldsymbol{r}$ are negative. Using cyclic permutations we easily obtain the expressions for the moments about the y- and z-axes.

To sum up, the moments about the coordinate axes of a force $\boldsymbol{F}$ applied at a point $(x; y; z)$ are

$$\begin{aligned} M_x &= F_z y - F_y z \\ M_y &= F_x z - F_z x \\ M_z &= F_y x - F_x y \end{aligned} \qquad (1.14)$$

Moment about a point

Consider Equations (1.14). The expressions for M_x, M_y and M_z have one remarkable property in common: they are all components of the vector $\boldsymbol{r} \times \boldsymbol{F}$. We shall denote this vector by $\boldsymbol{M}_\mathrm{O}$. The scalar components of $\boldsymbol{M}_\mathrm{O}$ are denoted by $(\boldsymbol{M}_\mathrm{O})_x$, $(\boldsymbol{M}_\mathrm{O})_y$ and $(\boldsymbol{M}_\mathrm{O})_z$, i.e. $(\boldsymbol{M}_\mathrm{O})_x = \boldsymbol{M}_\mathrm{O} \cdot \mathbf{e}_x$, etc. Hence,

$$\begin{aligned} M_x &= (\boldsymbol{M}_\mathrm{O})_x = \boldsymbol{M}_\mathrm{O} \cdot \mathbf{e}_x \\ M_y &= (\boldsymbol{M}_\mathrm{O})_y = \boldsymbol{M}_\mathrm{O} \cdot \mathbf{e}_y \\ M_z &= (\boldsymbol{M}_\mathrm{O})_z = \boldsymbol{M}_\mathrm{O} \cdot \mathbf{e}_z \end{aligned} \qquad (1.15)$$

The index O used in the symbol $\boldsymbol{M}_\mathrm{O}$ refers to the origin of the coordinate system, i.e. the point where the three axes intersect. Obviously, all relevant information about *three* physical quantities (the moments M_x, M_y and M_z) are included in *one* mathematical quantity (the vector $\boldsymbol{M}_\mathrm{O}$).

We have here an example where a mathematical model can be used to represent our experiences from physical experiments. In order to find a more efficient mathematical

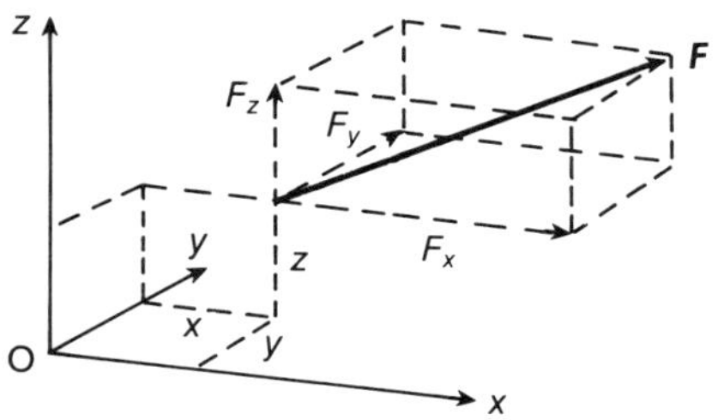

Figure 1.47

description of the physical reality we take one additional step. The point O above is the point where three particular axes, denoted by x, y and z, intersect. The denotations themselves are not important, however, and any point in space could have been considered instead of O. If a certain axis is of special physical interest, it is always possible to introduce a coordinate system with one of its axes along the axis in question.

With this background we now define a quantity known as the *moment about a point*. We consider an arbitrary point A and a force $\boldsymbol{F}$ applied at a point P according to Figure 1.48.

The *moment about the point* A is denoted by $\boldsymbol{M}_A$ and is defined as

$$\boldsymbol{M}_A = \boldsymbol{r} \times \boldsymbol{F} \tag{1.16}$$

where $\boldsymbol{r} = \overrightarrow{AP}$. Point A is sometimes referred to as the *moment point*. The vector product $\boldsymbol{r} \times \boldsymbol{F}$ is defined in Appendix A1.1.8.

From the definition and Figure 1.48 we conclude that the vector $\boldsymbol{M}_A$ has the following properties:

- The magnitude of the vector is $|\boldsymbol{M}_A| = Fr \sin \alpha = Fd$, where d is the moment arm with respect to an axis normal to the plane A–$\boldsymbol{F}$, i.e. the plane containing $\boldsymbol{F}$ and A.
- The vector $\boldsymbol{M}_A$ is perpendicular to the plane A–$\boldsymbol{F}$.
- The direction of $\boldsymbol{M}_A$ is obtained from the right-hand rule.
- $\boldsymbol{M}_A$ is not affected if the force is displaced along its line of action. This follows from the points above.

REMARK

In a diagram the line segment representing $\boldsymbol{M}_A$ is often drawn from point A. This does not mean that the moment has some kind of 'point of application' at A. It is merely a convention. In order to distinguish between forces and moments we shall represent moments by double arrowheads as shown in Figure 1.48.

It is important that the vector character of $\boldsymbol{M}_A$ is absolutely clear to the reader. The moment about the point A, i.e. the vector $\boldsymbol{M}_A$, should not be confused with the

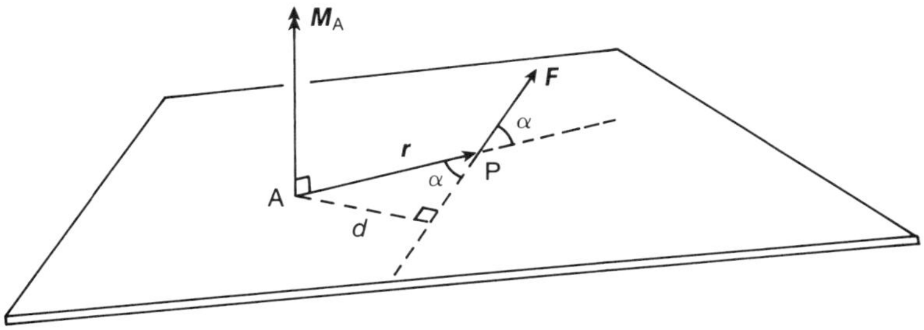

Figure 1.48

Figure 1.49

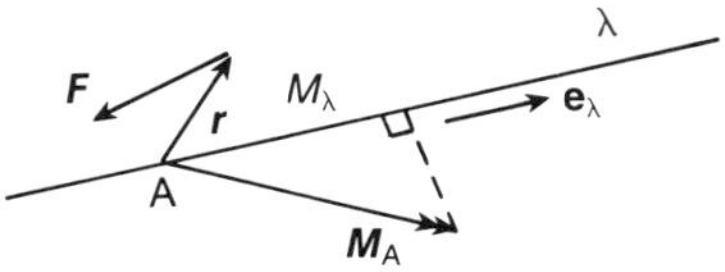

Figure 1.50

moment about an axis, e.g. M_x, which is a scalar quantity. The distinction as well as the relationship between the two concepts must be clear.

We now consider a common application of the theory above. Let us assume that we want to determine the moment M_λ of a force F about an axis λ as shown in Figure 1.49.

Starting from Equations (1.15) we can express any of the three relations, e.g. $M_x = (\boldsymbol{M}_O)_x = \boldsymbol{M}_O \cdot \mathbf{e}_x$, in words as follows:

The moment (M_x) of a force about a certain axis (the x-axis) is equal to the projection ($\boldsymbol{M}_O \cdot \mathbf{e}_x$) on the axis of the moment ($\boldsymbol{M}_O$) about a point (O) on the axis. In the more general problem stated above we proceed as follows:

1. Choose a point A on the axis λ, see Figure 1.50.
2. Determine the moment $\boldsymbol{M}_A$ of F about the point A according to Equation (1.16).
3. The moment M_λ is obtained from $M_\lambda = \boldsymbol{M}_A \cdot \mathbf{e}_\lambda$, where $\mathbf{e}_\lambda$ is a unit vector along the axis λ. When defining $\mathbf{e}_\lambda$ we have to choose a positive direction of the axis.

WORKED EXAMPLE 1.12

The axis λ passes through points A and B in Figure 1.51 with coordinates (2; −1; 1) m and (−1; 2; 3) m, respectively. A force $\boldsymbol{F} = (1, 2, -2)$ N is applied at a point P with coordinates (1; 1; 1) m. Determine the moment of the force about the axis λ.

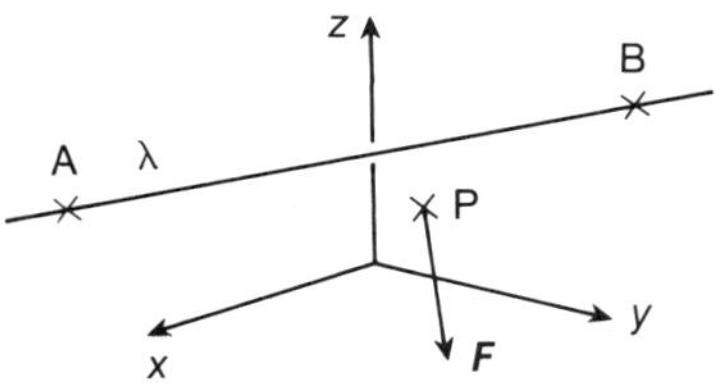

Figure 1.51

SOLUTION

We choose a point on the axis, say A. The moment about A is obtained from Equation (1.16):

$$\boldsymbol{M}_A = \overrightarrow{AP} \times \boldsymbol{F}$$

where $\overrightarrow{AP} = (-1, 2, 0)$ m. We then have

$$\boldsymbol{M}_A = \begin{vmatrix} \mathbf{e}_x & \mathbf{e}_y & \mathbf{e}_z \\ -1 & 2 & 0 \\ 1 & 2 & -2 \end{vmatrix} \text{ N m} = (-4, -2, -4) \text{ N m}$$

where we have utilized that a vector product can be represented by a determinant. The reader is referred to Appendix A1.1.8 for details. The unit vector $\mathbf{e}_\lambda$ is given by

$$\mathbf{e}_\lambda = \frac{\overrightarrow{AB}}{|\overrightarrow{AB}|} = \frac{(-3, 3, 2)}{\sqrt{22}}$$

Finally, the moment about the axis λ is obtained from

$$M_\lambda = \boldsymbol{M}_A \cdot \mathbf{e}_\lambda = (-4, -2, -4) \cdot \frac{(-3, 3, 2)}{\sqrt{22}} \text{ N m} \approx -0.426 \text{ N m}$$

The minus sign indicates that the tendency of rotation corresponds to the direction of $-\mathbf{e}_\lambda$, i.e. in the direction B → A according to the right-hand rule.

REMARK

The choice of a point, in our case A, on the axis is quite arbitrary. This can easily be seen by calculating M_λ in two ways, by using A and B as points on the axis. We then have

$$M_{A\lambda} = \boldsymbol{M}_A \cdot \mathbf{e}_\lambda = (\overrightarrow{AP} \times \boldsymbol{F}) \cdot \mathbf{e}_\lambda$$
$$M_{B\lambda} = \boldsymbol{M}_B \cdot \mathbf{e}_\lambda = (\overrightarrow{BP} \times \boldsymbol{F}) \cdot \mathbf{e}_\lambda$$

The difference $M_{A\lambda} - M_{B\lambda}$ is given by

$$M_{A\lambda} - M_{B\lambda} = [(\overrightarrow{AP} - \overrightarrow{BP}) \times \boldsymbol{F}] \cdot \mathbf{e}_\lambda = (\overrightarrow{AB} \times \boldsymbol{F}) \cdot \mathbf{e}_\lambda = (\mathbf{e}_\lambda \times \overrightarrow{AB}) \cdot \boldsymbol{F} = 0$$

where we have utilized the fact that $\overrightarrow{AB}$ and $\mathbf{e}_\lambda$ are collinear vectors. Hence, $M_{B\lambda} = M_{A\lambda}$. We have thus found that the moment M_λ about the axis does not depend on which point on the axis we have chosen in the calculation. However, the moment vectors $\boldsymbol{M}_A$ and $\boldsymbol{M}_B$ are not equal.

Varignon's theorem

In Section 1.2.2 we showed that the moment of a force about an axis does not change if we first resolve the force into components according to the parallelogram law and then add the moments of the components. The corresponding theorem, which is known as *Varignon's theorem*, can easily be generalized to three dimensions as follows.

Two forces $\boldsymbol{F}_1$ and $\boldsymbol{F}_2$ are applied at the same point P according to Figure 1.52. Their sum is $\boldsymbol{R} = \boldsymbol{F}_1 + \boldsymbol{F}_2$. A point O is chosen arbitrarily. Then, the moment of $\boldsymbol{R}$ about O is equal to the sum of the moments of $\boldsymbol{F}_1$ and $\boldsymbol{F}_2$ about O.

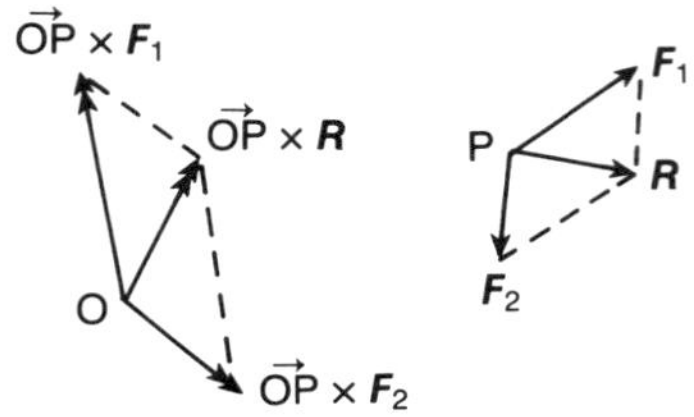

Figure 1.52

The proof is almost trivial since vector multiplication obeys the distributive law, i.e.

$$M_O = \vec{OP} \times R = \vec{OP} \times (F_1 + F_2) = \vec{OP} \times F_1 + \vec{OP} \times F_2$$

Obviously, Varignon's theorem applies to the moment M_O about a point O. It then follows that it also applies to the moment about an axis as this can always be regarded as a scalar component of a moment vector about a point.

1.3.3 Sum of forces, sum of moments

In the preceding section we discussed the sum of the moments of two forces acting at the same point. In analogy to the case of two-dimensional force systems, Section 1.2.3, we shall now introduce the sum of forces and the sum of moments of a general force system, i.e. a system of forces with different points of application.

Here we shall only define these quantities. Later on, in statics as well as in dynamics, we shall see that they are involved in several relations of great importance.

We consider a system of n forces as shown in Figure 1.53: $F_1, F_2, \ldots, F_n$.

The *sum of forces* denoted by ΣF is defined as

$$\sum F = \sum_{i=1}^{n} F_i \tag{1.17}$$

Note that the forces are added without regard to their points of application.

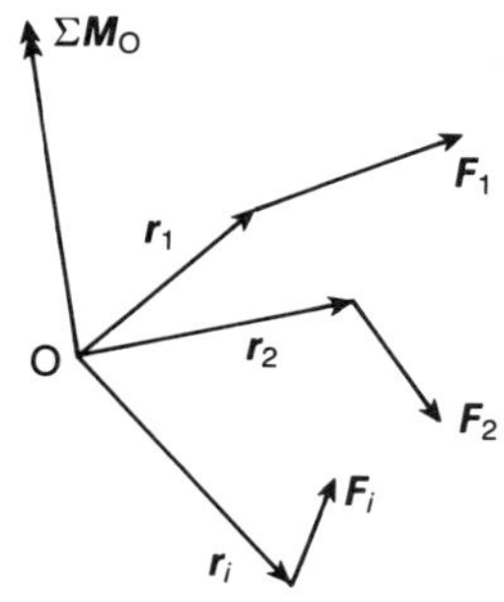

Figure 1.53

The *sum of moments* about a point O is denoted by ΣM_O. It is defined by

$$\sum M_O = \sum_{i=1}^{n} r_i \times F_i \tag{1.18}$$

where r_i is the position vector of the point of application of the force F_i with respect to O. In a diagram the vector ΣM_O normally is represented according to Figure 1.53.

1.3.4 Couple

In Section 1.2.4 the *couple* was introduced in connection with two-dimensional force systems. However, two parallel forces must always be located in the same plane. Consequently, it is not possible to generalize the treatment of one single couple any more even if we are concerned with three-dimensional force systems in this section. On the other hand, different couples may be located in different planes. Therefore a general vector description is necessary here.

We consider two forces F and $-F$ applied at points A and B, respectively. O is an arbitrary point, which in the general case is not located in the plane of the two forces. Using the notation of Figure 1.54(a) we find the sum of moments about O:

$$\sum M_O = r_A \times F + r_B \times (-F) = (r_A - r_B) \times F = r \times F$$

where r is defined according to Figure 1.54. The sum of moments is thus independent of the location of the moment point O. We shall denote this quantity by M. In a diagram the corresponding vector symbol may be placed anywhere. As we pointed out

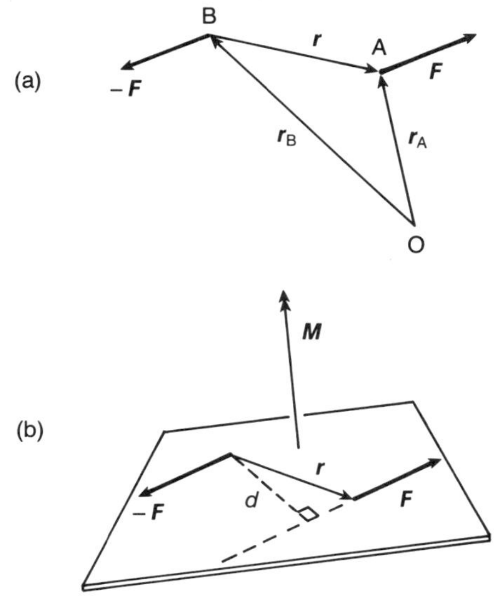

Figure 1.54

in Section 1.2.4, we shall use the term 'couple' not only for the pair of forces (F and $-F$) but also for the corresponding sum of moments ($\boldsymbol{M}$).

The vector $\boldsymbol{M}$ will always be perpendicular to the plane of the forces, see Figure 1.54(b). The magnitude of $\boldsymbol{M}$ is given by $|\boldsymbol{M}| = Fd$ in the same way as in Section 1.2.4.

WORKED EXAMPLE 1.13

Determine the resulting couple of the system of three pairs of equal and opposite forces shown in Figure 1.55.

SOLUTION

Using the coordinate system shown we may express the three couples in vector notation:

The 1 N forces $\quad \boldsymbol{M}_1 = 1\text{ N}\cdot 2\text{ m}(-\mathbf{e}_z)$

The 2 N forces $\quad \boldsymbol{M}_2 = 2\text{ N}\cdot 1\text{ m}(-\mathbf{e}_y)$

The 3 N forces $\quad \boldsymbol{M}_3 = 3\text{ N}\cdot 1\text{ m}(+\mathbf{e}_3)$

Here, $\mathbf{e}_3 = \cos\alpha\cdot\mathbf{e}_x + \sin\alpha\cdot\mathbf{e}_y$, where $\tan\alpha = 2/3$.

Hence, $\mathbf{e}_3 = (3/\sqrt{13})\,\mathbf{e}_x + (2/\sqrt{13})\,\mathbf{e}_y$ which gives

$$\boldsymbol{M} = \boldsymbol{M}_1 + \boldsymbol{M}_2 + \boldsymbol{M}_3 = (3\cdot 3/\sqrt{13},\ -2 + 3\cdot 2/\sqrt{13},\ -2)\text{ N m}$$

ANSWER

The resulting couple is $\boldsymbol{M} = (2.50, -0.34, -2.00)$ N m. The direction of the resulting couple $\boldsymbol{M}$ is shown in Figure 1.56. Note that there is no specified point of application; the couple is a 'free vector'.

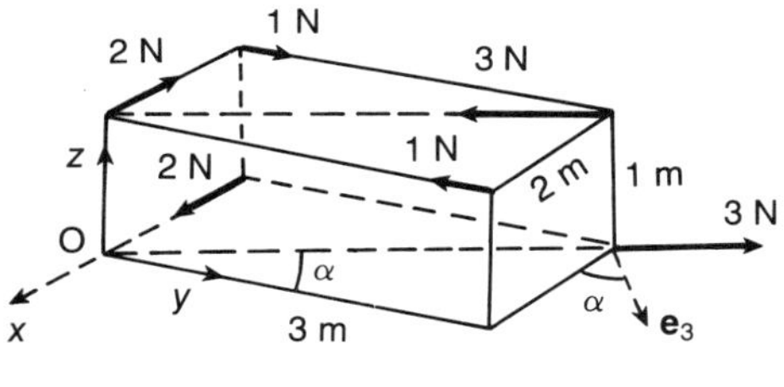

Figure 1.55

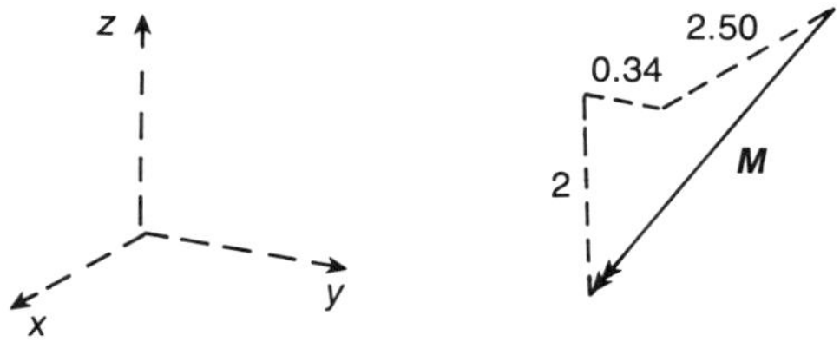

Figure 1.56

WORKED EXAMPLE 1,14

The input shaft of a gear box is subjected to a 20 N m couple. The system driven by the gear box exerts a couple of 90 N m on the output shaft as shown in Figure 1.57. What is the resulting couple acting on the gear box?

SOLUTION

In vector notation the two couples may be expressed as

$$\boldsymbol{M}_1 = 20\ \text{N m}(-\mathbf{e}_x) \qquad \boldsymbol{M}_2 = 90\ \text{N m}(-\mathbf{e}_y)$$

The locations of the couples in Figure 1.58 are of no importance concerning the total effect on the gear box. The quantity of interest here is the vector

$$\boldsymbol{M} = \boldsymbol{M}_1 + \boldsymbol{M}_2 = (-20,\ -90,\ 0)\ \text{N m}$$

Translation of a force

The net external effect on a rigid body is unaffected if a force is displaced along its line of action. Such a displacement is one of the fundamental operations defined in Section 1.1. This operation does not change the contribution of the force to the sum of forces ΣF or the sum of moments ΣM.

A translation of a force to a parallel position, however, will affect the moment of the force. To make up for this we have to make a correction – the easiest way is to add

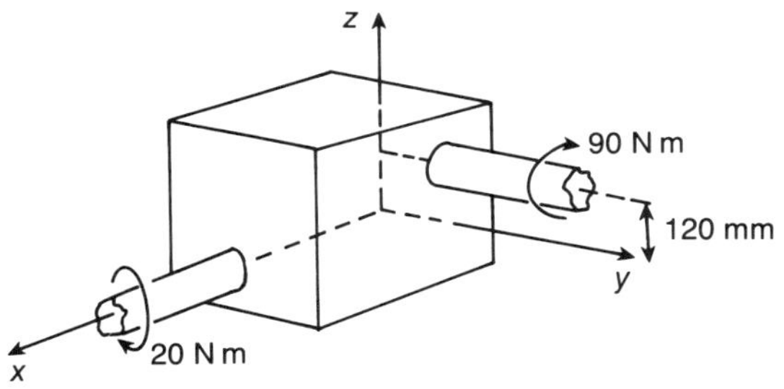

Figure 1.57

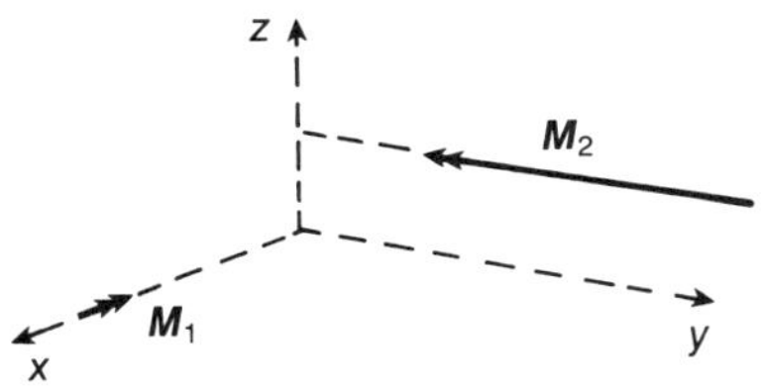

Figure 1.58

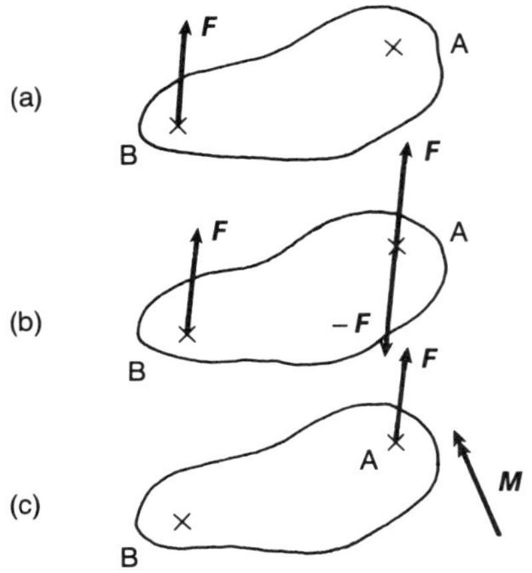

Figure 1.59

a couple. To see this we consider the force $\boldsymbol{F}$ of Figure 1.59. The force is applied at B, and our purpose is to transfer it to the parallel position at A.

Our first step is to perform a fundamental operation by introducing the two forces $\boldsymbol{F}$ and $-\boldsymbol{F}$ at A. The resulting system is a force $\boldsymbol{F}$ applied at A and a couple equal to $\overrightarrow{AB} \times \boldsymbol{F}$.

WORKED EXAMPLE 1.15

A 60 N force is applied to the adjustable spanner as shown in Figure 1.60. Replace the force by a force applied at A and a couple.

SOLUTION

The point of application is denoted by B. A coordinate system A*xyz* is introduced as shown in Figure 1.61.

The force $\boldsymbol{F} = (60\ \mathrm{N})\mathbf{e}_x$ is translated to A.

The couple $\boldsymbol{M}$ that needs to be added is obtained from $\overrightarrow{AB} \times \boldsymbol{F}$.

Using the fact that $\overrightarrow{AB} = (0, 180, 75)$ mm, we obtain

$\boldsymbol{M} = \overrightarrow{AB} \times \boldsymbol{F} = (0, 4.5, -10.8)$ N m

REMARK

The components of $\boldsymbol{M}$ are shown in Figure 1.62. The following deductions can be made:

- The component M_z of magnitude 10.8 N m represents the turning action of the force $\boldsymbol{F}$ (the torsional moment).
- The component M_y of magnitude 4.5 N m represents a bending action of the force $\boldsymbol{F}$ (the bending moment).
- The force $\boldsymbol{F}$ represents what is called a shearing action (shear force).

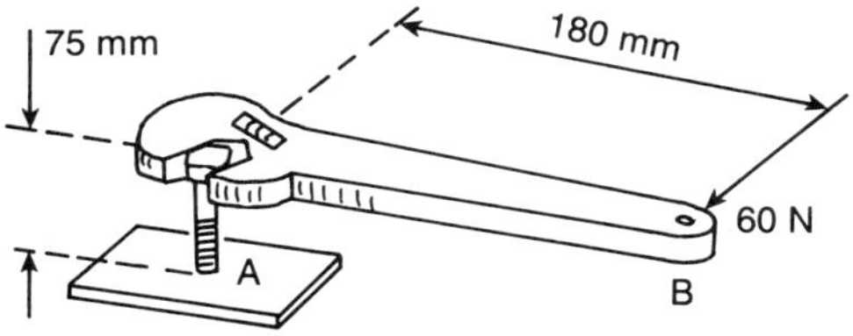

Figure 1.60

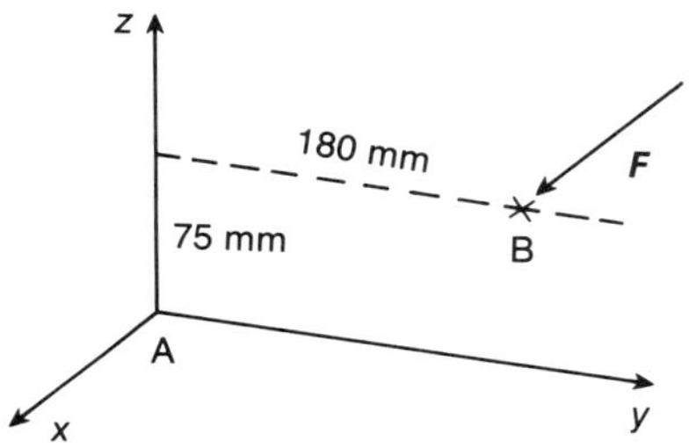

Figure 1.61

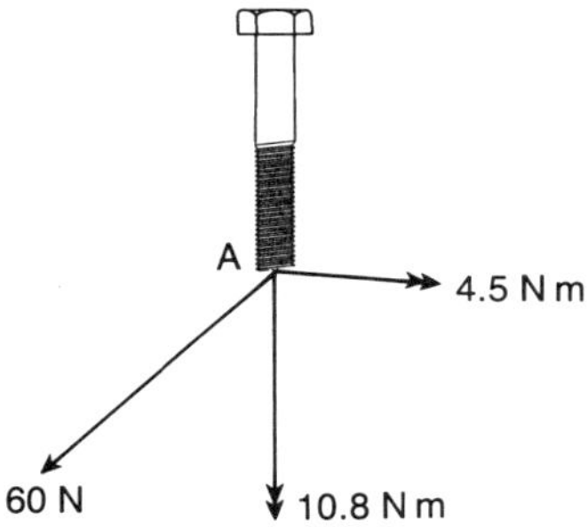

Figure 1.62

1.3.5 Reduction of force systems

In the preceding section we saw that any force $\boldsymbol{F}$ may be replaced by a force $\boldsymbol{F}$ with an arbitrary point of application and a couple. At a first glance this operation does not seem to simplify the situation. When we consider a system of forces, however, it may lead to a considerable simplification as the procedure can be utilized to reduce a system of forces.

We consider a force system consisting of an arbitrary number of forces. For simplicity only three forces are shown in Figure 1.63(a). Our purpose is to translate all forces to point A. In doing this we obtain three couples $\boldsymbol{M}_1$, $\boldsymbol{M}_2$ and $\boldsymbol{M}_3$. The

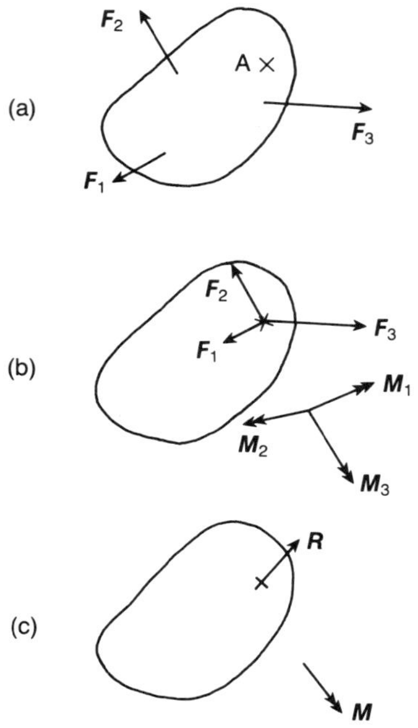

Figure 1.63

corresponding symbols may be placed anywhere as shown in Figure 1.63(b). Adding the forces we obtain the resultant force:

$$\boldsymbol{R} = \boldsymbol{F}_1 + \boldsymbol{F}_2 + \boldsymbol{F}_3 + \ldots = \sum \boldsymbol{F}$$

Adding the couples gives the resulting couple:

$$\boldsymbol{M} = \boldsymbol{M}_1 + \boldsymbol{M}_2 + \boldsymbol{M}_3 + \ldots = \sum \boldsymbol{M}_{\mathrm{A}}$$

The result of the reduction is shown in Figure 1.63(c).

In many cases the vector symbol of the moment $\boldsymbol{M}$ is located at point A in a diagram. However, it should be noted that this is not necessary since a couple does not have any specific point of application.

Except for forces ($\boldsymbol{F}_1$, $\boldsymbol{F}_2$, etc.) the original system may also include couples. In the above procedure these can be added directly to the other couples. In the resulting system ($\boldsymbol{R}, \boldsymbol{M}$) they will only be included in $\boldsymbol{M}$.

A reduction of a force system, performed according to the procedure described here, can be made with an arbitrary choice of the reduction point A. The resultant $\boldsymbol{R}$ will be independent of the location of this point as the relation $\boldsymbol{R} = \Sigma \boldsymbol{F}$ still holds. The couple $\boldsymbol{M}$, however, will vary depending on the choice of A. This can be illustrated as follows.

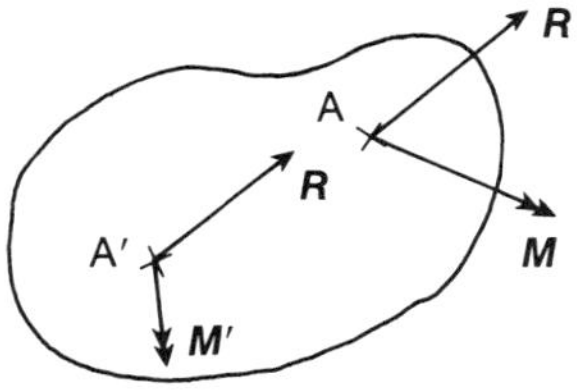

Figure 1.64

Let us assume that the reduction point A′ corresponds to $\boldsymbol{R}$ and $\boldsymbol{M}'$ as shown in Figure 1.64. If at the same time the reduction point A corresponds to $\boldsymbol{R}$ and $\boldsymbol{M}$, then the following relation must hold:

$$\boldsymbol{M}' = \boldsymbol{M} + \overrightarrow{\mathrm{A'A}} \times \boldsymbol{R} \tag{1.19}$$

since the two force systems are equivalent. Equation (1.19) expresses the fact that their sums of moments about A′ are equal.

WORKED EXAMPLE 1.16

Three forces, each of magnitude F, are acting as shown in Figure 1.65. Replace the system by an equivalent system consisting of a force applied at point A and a couple.

SOLUTION

The individual forces and the corresponding points of application are

$\boldsymbol{F}_1 = F(0, -1, 0)$ applied at $(0; a; 0)$

$\boldsymbol{F}_2 = \dfrac{F}{\sqrt{5}}(0, 1, 2)$ applied at $(-2a; 0; 0)$

$\boldsymbol{F}_3 = \dfrac{F}{\sqrt{5}}(-2, -1, 0)$ applied at $(0; a; 2a)$

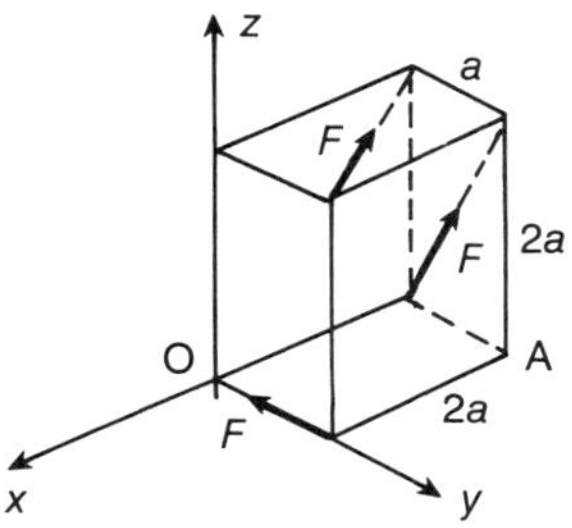

Figure 1.65

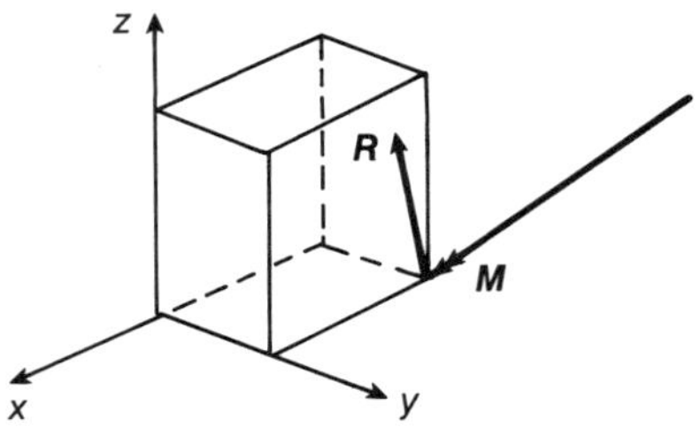

Figure 1.66

The resultant force is obtained from

$$\boldsymbol{R} = \sum \boldsymbol{F} = \frac{F}{\sqrt{5}}(-2, -\sqrt{5}, 2) \approx F(-0.89, -1, 0.89)$$

The couple must be equal to $\Sigma \boldsymbol{M}_A$ for the original system:

$$\sum \boldsymbol{M}_A = \boldsymbol{r}_1 \times \boldsymbol{F}_1 + \boldsymbol{r}_2 \times \boldsymbol{F}_2 + \boldsymbol{r}_3 \times \boldsymbol{F}_3$$

where $\boldsymbol{r}_1$, $\boldsymbol{r}_2$ and $\boldsymbol{r}_3$ are vectors from A to the points of application of $\boldsymbol{F}_1$, $\boldsymbol{F}_2$ and $\boldsymbol{F}_3$ respectively. It is easily seen that $\boldsymbol{r}_1 = (2, 0, 0)a$, $\boldsymbol{r}_2 = (0, -1, 0)a$ and $\boldsymbol{r}_3 = (2, 0, 2)a$. After some algebra we have

$$\sum \boldsymbol{M}_A = aF\left(0, -\frac{4}{\sqrt{5}}, -2 - \frac{2}{\sqrt{5}}\right) \approx aF(0, -1.79, -2.89)$$

Hence, the resulting force system with the point of reduction at A, Figure 1.66, is

$$\boldsymbol{R} = F(-0.89, -1, 0.89)$$
$$\boldsymbol{M} = aF(0, -1.79, -2.89)$$

1.3.6 Special force systems

In some cases a force system upon reduction will transform into a system that is even simpler than a resultant force ($\boldsymbol{R}$) and a couple ($\boldsymbol{M}$). A brief survey of special cases is given here:

1. $\boldsymbol{R} = 0$, a couple remains, see Figure 1.67(a).
   This means that $\Sigma F = 0$, $\Sigma M_A = \Sigma M_B = \ldots = M$.
2. There is a point A such that $\Sigma M_A = 0$. A force with its line of action passing through A remains, see Figure 1.67(b).
   This means that $\Sigma M_A = \Sigma M_B = \ldots = 0$, where points A, B, ... are all located on the line of action of $\boldsymbol{R}$.
3. $\boldsymbol{R} = 0$ and $\boldsymbol{M} = 0$. This system is equivalent to zero and is of fundamental importance in statics.
   This means that $\Sigma F = 0$ and $\Sigma M_A = 0$ for any point A.

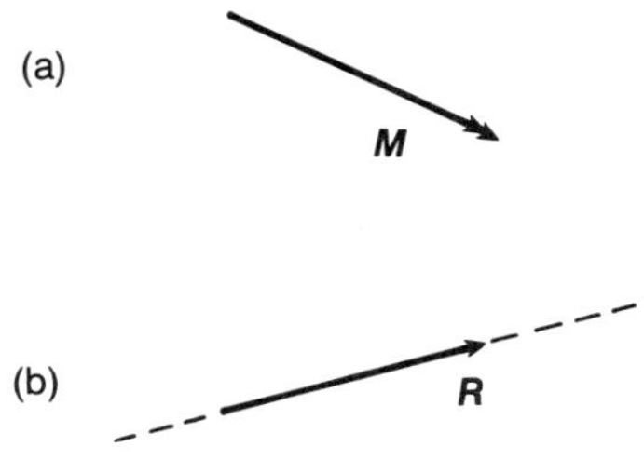

Figure 1.67

Force systems equivalent to zero will be treated in a great number of cases during the remainder of the study of statics, when the real theory of equilibrium will be considered.

PROBLEMS

○ **1.33** The z-component of $\boldsymbol{F}$ is 60 N.

(a) Determine the magnitude of $\boldsymbol{F}$ and the components F_x and F_y.

(b) Determine the angles which the force makes with the x- and y-axes.

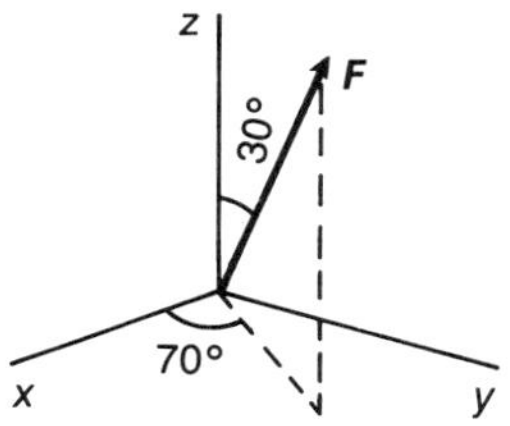

○ **1.34** A force of magnitude P is acting along the diagonal AG of a cube. Determine the component of the force along AB, DC, EG, HF, BD, EB and BE, respectively.

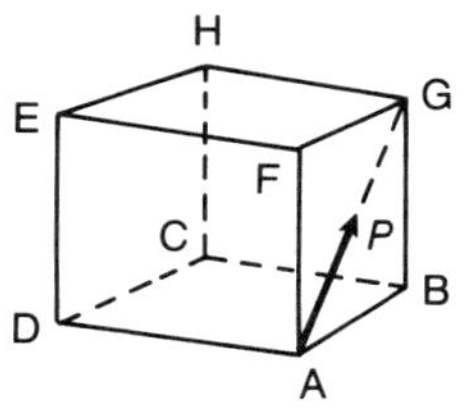

○ **1.35** A force $\boldsymbol{P}$ of magnitude P is acting as shown. Express $\boldsymbol{P}$ in terms of its components in

(a) the xyz-system

(b) the $x'y'z'$-system

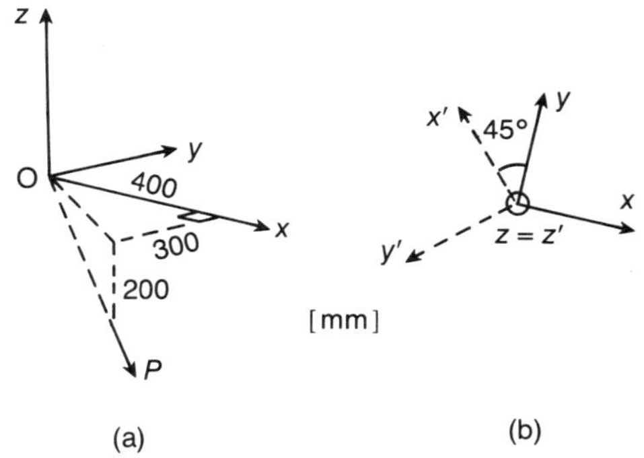

○ **1.36** Consider force $\boldsymbol{P}$ of problem 1.35. A point C has coordinates (a; a; a) in the xyz system. What is the component of $\boldsymbol{P}$ along OC?

○ **1.37** Consider Worked Example 1.10 and the force $\boldsymbol{F}$ determined there. Determine the following components of $\boldsymbol{F}$:

(a) the vertical component

(b) the horizontal component

(c) the component along BC, where the coordinates of C are (0.5; 1.0; 1.5) m

(d) the component along OC.

1.38 The mass of the bar shown overleaf is 11.8 kg. Resolve the gravitational force into two vector components, one parallel to the bar and one normal to it. What are the magnitudes of these components?

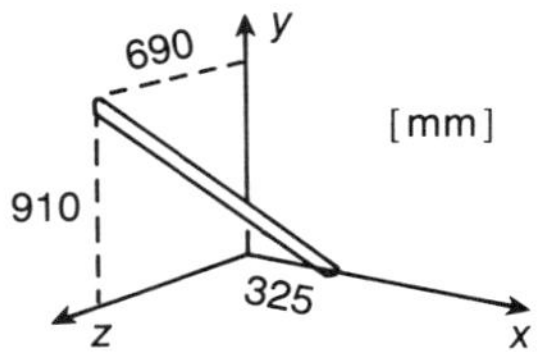

1.39 A puck is moving on a 30° incline. A normal force of 920 mN and a friction force of 650 mN are exerted by the incline on the puck as shown. Determine the vector $\boldsymbol{F}$ representing the total force exerted on the puck by the incline. Express $\boldsymbol{F}$ in the form $F_x\mathbf{e}_x + F_y\mathbf{e}_y + F_z\mathbf{e}_z$.

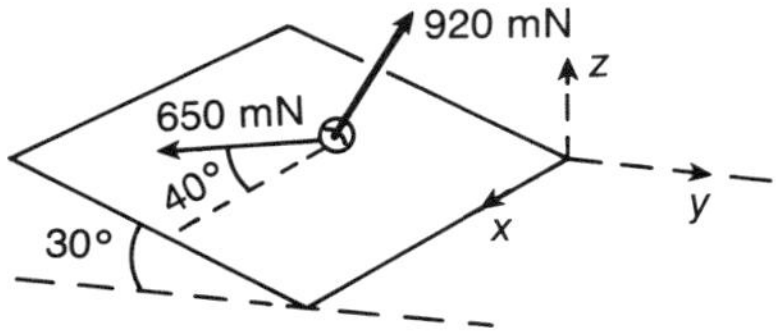

1.40 The tension in the spring shown is 145 N.

(a) Express the force exerted by the spring on the support at C in terms of its *xyz*-components.

(b) Express the force of Problem (a) as the sum of three forces ($\boldsymbol{F}_{BC}$, $\boldsymbol{F}_{EC}$ and $\boldsymbol{F}_{CD}$) directed along the bars BC, EC and CD, respectively. Specify the magnitudes of the three components.

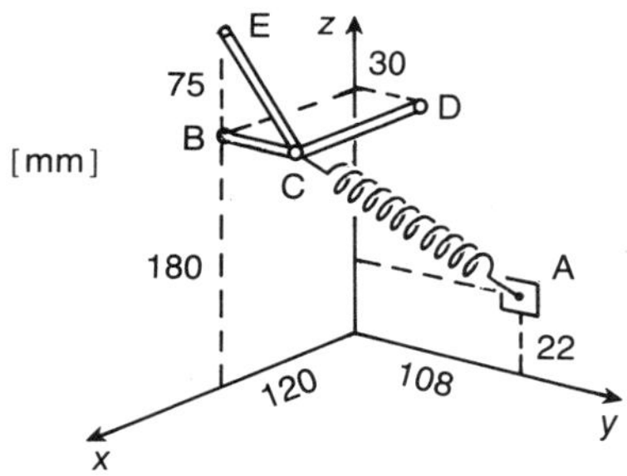

○ **1.41** A force $(0, 0, P)$ is applied at the point $(2a; a; 0)$. Determine the moment about

(a) the origin,
(b) the *x*-axis,
(c) the *y*-axis,
(d) the *z*-axis,
(e) an axis through the origin parallel to $\mathbf{e} = (1/\sqrt{6}, 2/\sqrt{6}, -1/\sqrt{6})$.

○ **1.42** A force $(-1, 1, 2)$ N is applied at the point $(2; 0; 2)$ m. What is the moment of the force about

(a) the origin,
(b) point A of coordinates $(0; 3; 2)$ m,
(c) point B of coordinates $(1; 1; 4)$ m.

1.43 Use the results from the preceding problem to calculate the moment about

(a) the *y*-axis,
(b) an axis passing through A parallel to the *z*-axis,
(c) axis BA.

○ **1.44** The mass of the cube shown is 78 kg. Determine the moment of the gravitational force about corners A, B, C and O, respectively.

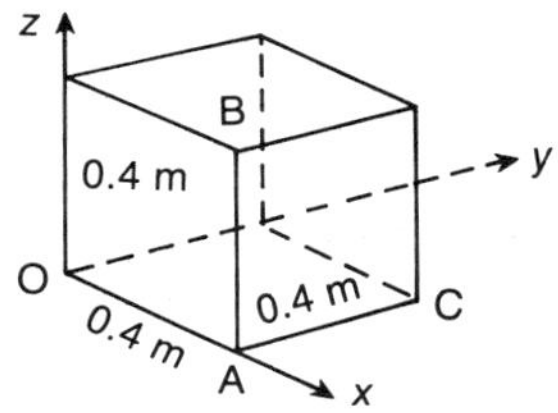

1.45 A force $(2P, P, P)$ is applied at $(3a; 2a; a)$. Determine the moment about

(a) the origin,
(b) $(a; a; a)$,
(c) $(a; a; 0)$,
(d) an axis through the origin parallel to $(1, 1, 1)$,
(e) an axis through the origin parallel to $(2, 1, 1)$,
(f) an axis through $(2a; 2a; a)$ parallel to $(1, 1, 0)$.

○ **1.46** The length of the edges of the cube shown is a. The magnitude of the force is F. Determine the moment about each of the coordinate axes. Use both the alternative methods described below.

(a) Use the (force) × (moment arm) approach. Resolve the force into components in such a manner that the moment arms can easily be found (in this case two components).

(b) Express the force in the form (F_x, F_y, F_z) and the position vector $\boldsymbol{r} = \overrightarrow{OP}$ of the point of application in a similar way. Determine the moment $\boldsymbol{M}_O$ from $\boldsymbol{r} \times \boldsymbol{F}$. The projections of $\boldsymbol{M}_O$ on the axes correspond to the moments about the axes.

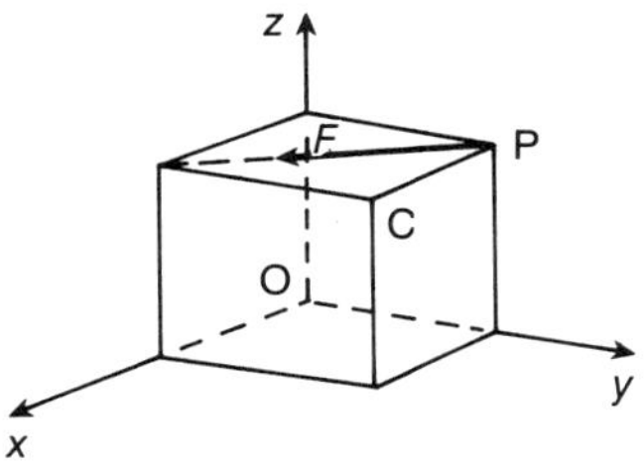

1.47 See the preceding problem. Determine the moment about the diagonal OC. First discuss which one of the two methods (the geometric or the algebraic) is preferable. Calculate the moment $\boldsymbol{M}_{OC}$ using the method chosen. Then repeat the calculation using the other method.

1.48 In problem 1.46 the moment about the origin ($\boldsymbol{M}_O$) was calculated in two ways.
(a) Use method (b) to determine the moment $\boldsymbol{M}_C$.
(b) Determine the moment about OC by projecting $\boldsymbol{M}_C$ on the axis; $M_{OC} = \boldsymbol{M}_C \cdot \mathbf{e}_{OC}$.
(c) Compare M_{OC} calculated in this way with the result from Problem 1.47 which was obtained from $\boldsymbol{M}_O \cdot \mathbf{e}_{OC}$.
(d) In the remark following Worked Example 1.12 a certain property of the moment about an axis was proven. Check that the results found here are in agreement with the content of the remark, especially the last two sentences.

○ **1.49** A load is to be moved from its position on the ground. The tension in the wire of the crane is 17 kN. Consider the force exerted by the wire on the support at D. What is the moment about O if the load is located at A?

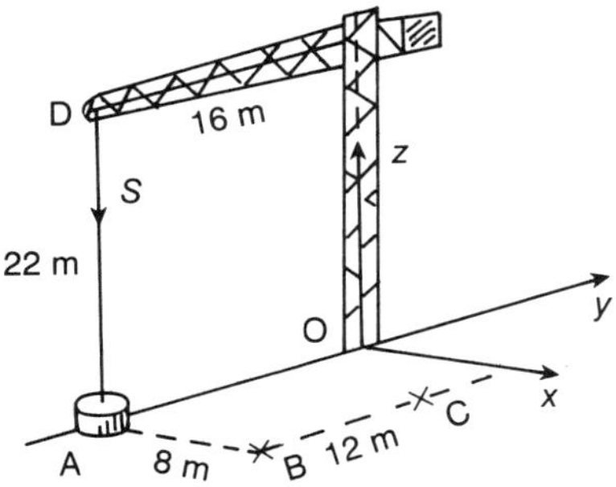

1.50 See Problem 1.49. Assume that the position of the crane and the tension in the wire are unchanged. The load, however, is located at another position. Determine the moment about O in the case where the load is located at
(a) position B,
(b) position C.
(c) What are the moments about the z-axis when the load is in positions A, B and C, respectively?

○ **1.51** A 130 N force is applied to the control lever as shown. Determine the moment of the force about
(a) axis OA,
(b) point O.

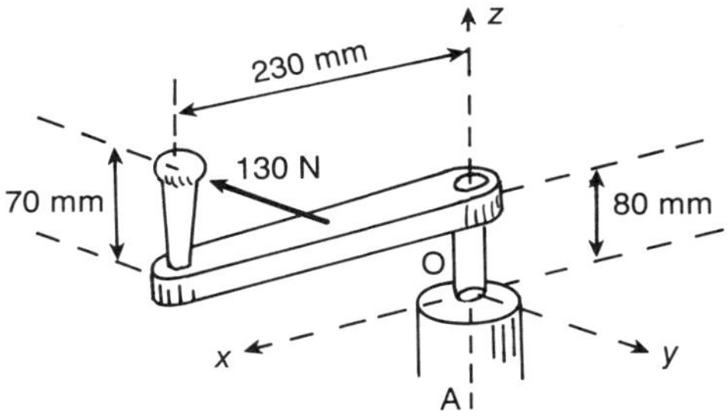

1.52 A horizontal force P is applied to the crank of a hand-driven winch as shown. Determine the magnitude of the moment about O.

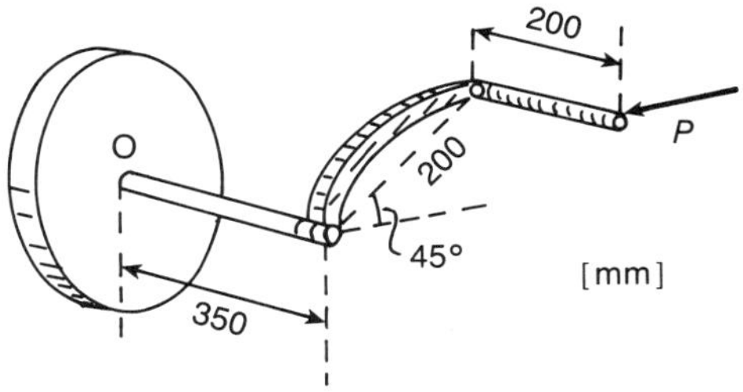

◎ **1.53** A door of mass m in the bulkhead of a sailing boat is shown overleaf. The door may be regarded as thin. In the position shown the boat is tilted through an angle α from the neutral position. Determine the moment of the gravitational force about the axis passing through the hinges.

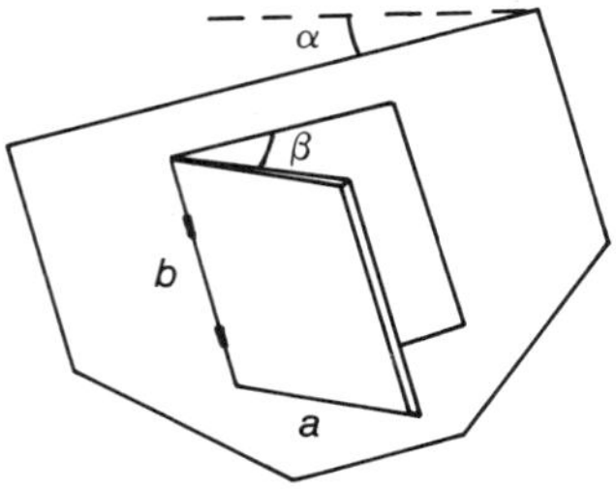

○ **1.54** Three forces $(P, P, -P)$, $(0, 2P, P)$ and $(-P, -P, P)$ are applied at points $(a; a; 0)$, $(-2a; -a; -a)$ and $(0; 2a; 0)$, respectively. Determine the sum of forces.

1.55 Consider the force system of Problem 1.54. Determine the sum of moments about
(a) the origin,
(b) the y-axis,
(c) point C $(3a; 2a; a)$,
(d) an axis through C in the direction $(1, -2, 2)$,
(e) point E $(0; 377a; 188.5a)$.

1.56 A system of three forces is shown here.
(a) Determine the sum of forces.
Determine the sum of moments about
(b) the origin,
(c) point B,
(d) axis OD,
(e) axis BD,
(f) axis BC.

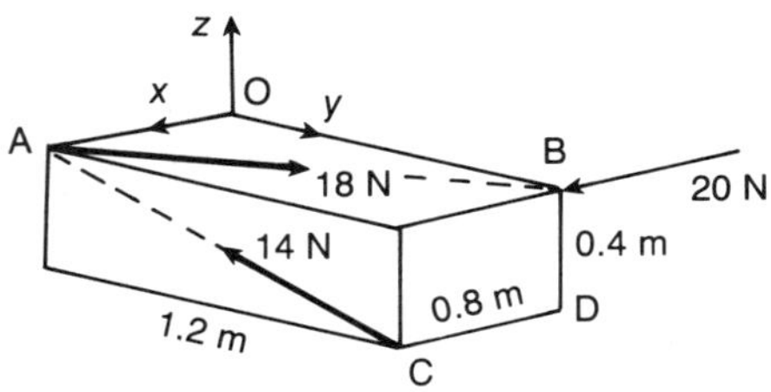

1.57 The sum of forces of a certain system is $(-7, 0, 2)$ N. The sum of moments about the point $(1; 2; 0)$ m is $(3, 7, -3)$ N m. Determine the sum of moments about the point $(0; -1; 1)$ m.

1.58 Three forces are acting on a cube as shown. The magnitude of one of the forces (Q) is known. Determine the magnitudes of the two other forces (P and F) so that the sum of forces is equal to zero. In this case, what is the sum of moments about the origin? What is the sum of moments about point A?

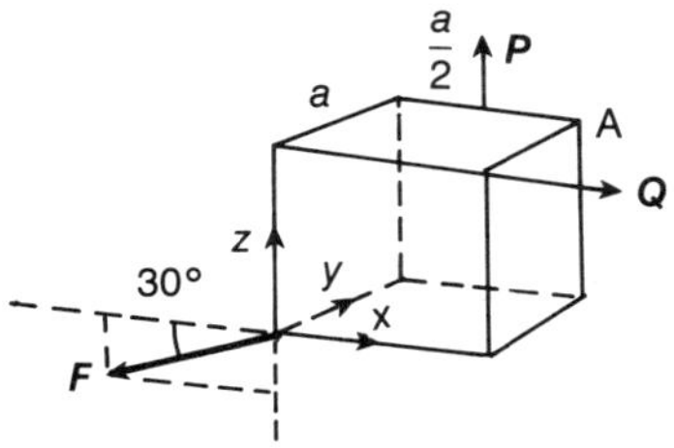

○ **1.59** The edges of the cube shown are all of length a. All forces (1) ... (6) are of magnitude F. Determine the vector expression for the couple consisting of forces
(a) (1) and (2),
(b) (3) and (4),
(c) (5) and (6),
(d) (1) and (3),
(e) (4) and (2).

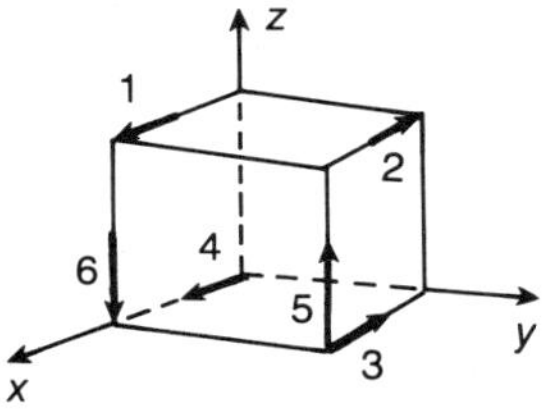

1.60 Determine the vector expression for the couple shown.

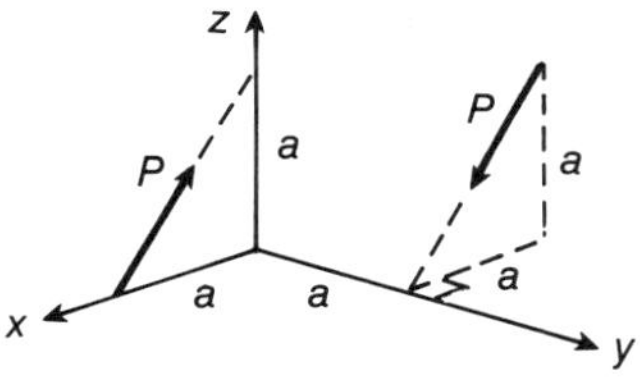

○ **1.61** A horizontal bar AB is screwed into a wall. To loosen it two vertical forces of equal magnitude and opposite direction are applied. Determine the projections on axes BA and AC of the corresponding couple.

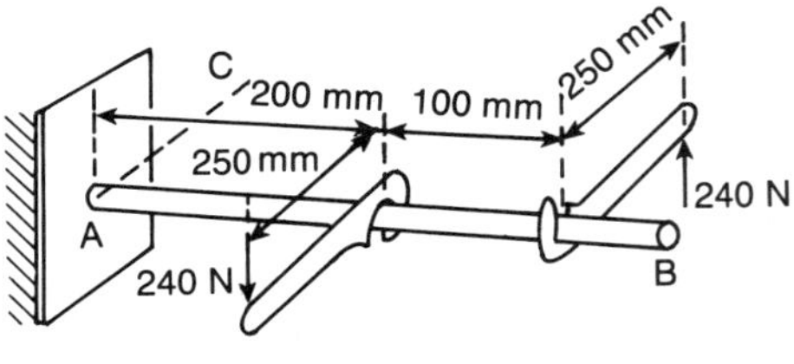

○ **1.62** The length of the edges of the cube shown is 0.80 m. Replace the force system by a force and a couple choosing the following points of reduction:
(a) A
(b) B
(c) the origin
(d) C.

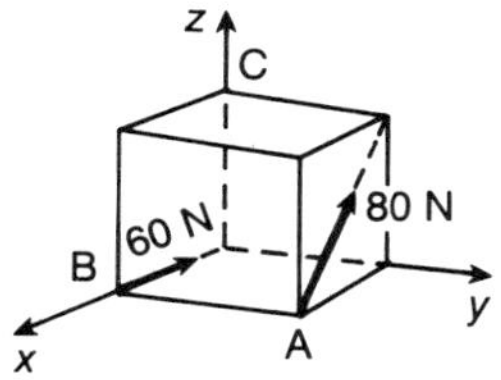

○ **1.63** See the preceding problem. Add the couple 50 N m $\mathbf{e}_z$. Solve problems (a)–(d).

1.64 A system of parallel forces can always be replaced by one single force, a resultant. Perform the reduction for the force system shown here. The resultant may be assumed to act in the *xy*-plane.

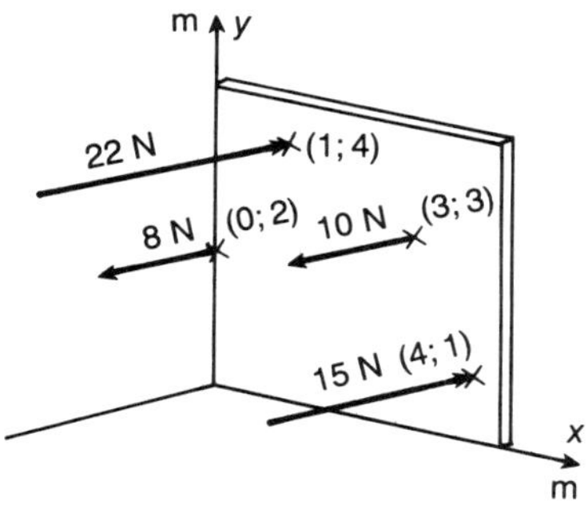

1.65 Replace the forces acting on the bent bar by one single force.

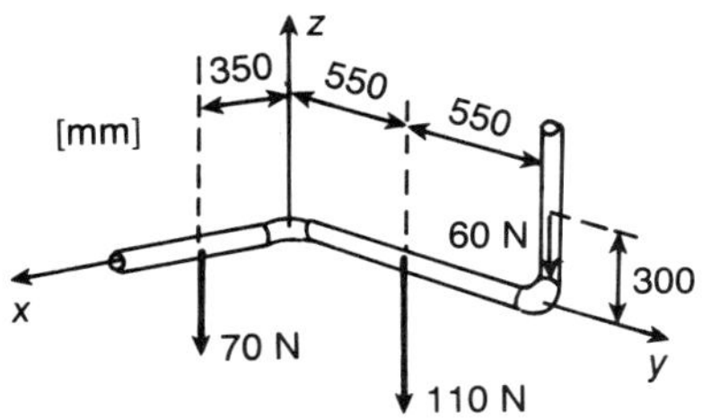

1.66 Replace the force system by a force and a couple. Choose the following points of reduction:
(a) A
(b) O

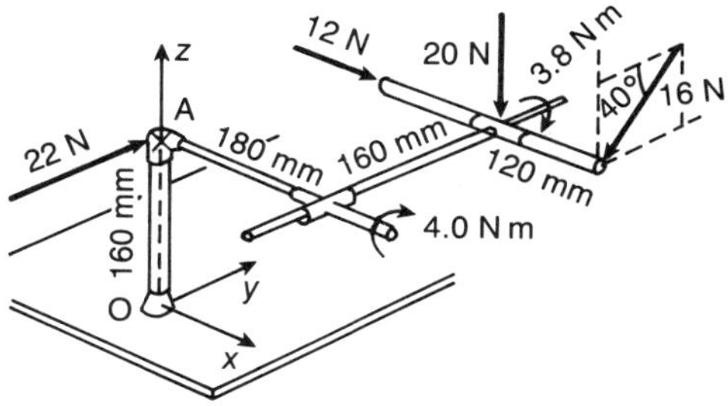

1.67 Replace the 280 N force by a force and a couple. As reduction points choose
(a) A
(b) B
For both cases, also determine the magnitude of that component of the couple that is normal to the corresponding bolt axis.

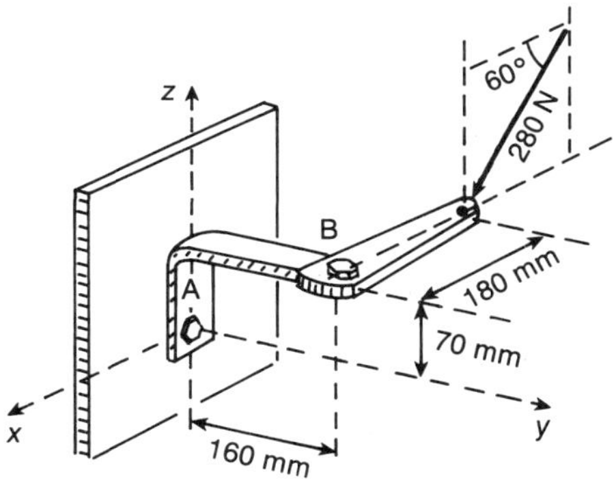

○ **1.68** Replace the system shown by a force at the origin and a couple.

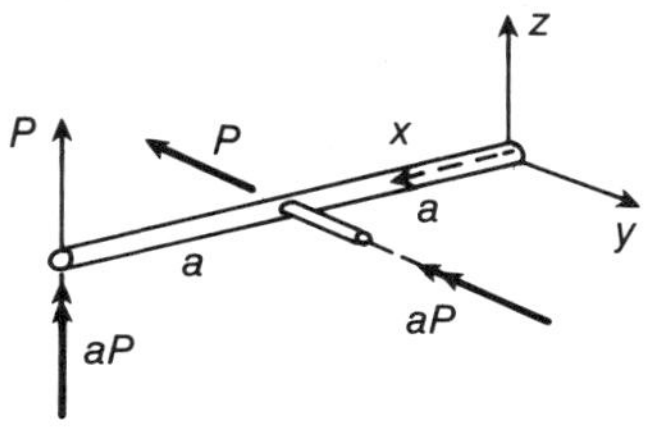

2

Equilibrium

2.1 Conditions of equilibrium, free-body diagrams

2.1.1 Equilibrium, conditions of equilibrium

Frame of reference, the concept of rest

In mechanics, *material systems* or bodies are studied. We assume that the position of such a system is specified by the coordinates of each portion of the system. The coordinate system in its turn is assumed to be connected to a certain *frame of reference*. Among these the *inertial frames of reference* are in a unique position. In most cases we describe the position and motion of the bodies considered, utilizing coordinate systems connected to such frames.

The meaning of the concept of *rest* deserves a comment: a point is at rest with respect to a certain frame of reference if its coordinates related to a system which is rigidly connected to the frame of reference are constant with respect to time.

If the velocity of a point is zero momentarily we do not use the term *rest* to describe the condition at this instant. To clarify the meaning of the term we sometimes use the expression *continuous rest*.

Not only material systems – particles, bodies – can be at rest. Also, a frame of reference may be resting. This may be a practical point of view in many cases, but it is important to understand that there is no frame of reference that can be distinguished from other frames of reference and be treated as *the only one* at absolute rest.

Instead there is an infinite number of frames of reference moving with constant velocities with respect to each other, each of which may be treated as the resting one. These frames, which are all of equal standing and may be defined without any internal contradiction, are called *inertial frames of reference*[1].

1 No single frame of reference may be regarded as being at absolute rest, and no velocity may be referred to as the absolute velocity. Only relative velocities may be measured. This *principle of relativity* will be utilized here within the limits of classical mechanics. In 1905 Albert Einstein formulated his theory of special relativity according to which the principle of relativity is applicable to *all* phenomena. This led to a completely new view of the nature of space and time.

Classical mechanics is based on Newton's three laws, see Section I.1. In the first and second laws the concepts rest, velocity and acceleration are of vital importance. However, these concepts have no well-defined meaning as long as we have not defined a proper frame of reference. Inertial frames distinguish themselves from other frames of reference in the sense that Newton's first and second laws are valid when such a frame is utilized.

Once a certain inertial frame of reference has been chosen, it is customary to refer to this as the *resting frame*. As the frame is being used to describe positions and motions, this approach will facilitate our understanding and discussions.

It is obvious that a frame of reference that is fixed with respect to the earth is not an inertial frame, strictly speaking, as the earth rotates. There is an overwhelming number of applications, however, where it is possible to disregard this fact and treat the earth as resting: the motion of a car along a road, the motion of a swinging pendulum in the laboratory, the motion of a football on the field, etc. All these motions are usually regarded as absolute motions since the rotation of the earth is so slow that it does not affect our everyday observations of motion and rest.

What is the meaning of equilibrium?

The concept of *equilibrium* may be defined as follows. A material system is said to be in equilibrium if every portion of the system is at rest.

In combination with the discussion above this brief formulation is sufficient. The outcome of the discussion may be summarized as:

- By the term 'rest' we understand continuous rest.
- Whenever the concept of equilibrium is used, we understand equilibrium and rest relative to an inertial frame.

Internal and external forces

In order to formulate the conditions that must be satisfied for a body to be in equilibrium we must first introduce the concepts *internal forces* and *external forces*. For this purpose, consider a body which is supported by a horizontal surface and kept in its position by a cord, Figure 2.1.

The *external* forces acting on the body are:

- the gravitational forces on each portion of the body represented by a resultant gravitational force (Q)
- the force (R) exerted by the table
- the tension (S) in the cord

Internal forces are all forces exerted on each other by various portions of the body.

Even when a more general material system than one single body is considered the terms 'external' and 'internal' forces are used. We always define the external forces acting on a certain system as the forces exerted by bodies outside the system. A single force cannot in general be defined to be external or internal; first it has to be related to

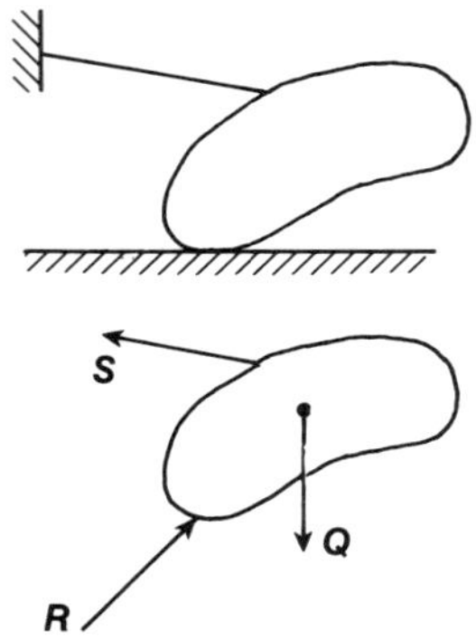

Figure 2.1

the system considered. As an example we consider two blocks placed on a table, see Figure 2.2.

If the two blocks together form our system, then the gravitational forces and the normal forces exerted by the table are the external forces, see Figure 2.3(a). If only one of the blocks is chosen as the system, then the forces exerted by the other block must be regarded as external forces, see Figure 2.3(b).

The forces exerted by the blocks on each other as shown in Figure 2.3(b) are opposite to each other in the sense stated in Newton's third law, the law of action and reaction, Section I.1. When a certain portion of a system is separated or isolated from its surroundings as in Figure 2.3(b), it is absolutely necessary that this law is obeyed. In the following we will be reminded of this on several occasions.

Conditions of equilibrium

A material system will remain in equilibrium if the system of external forces acting on each portion of the system is equivalent to zero.

This condition is *necessary* and *sufficient* for equilibrium. It may be regarded as a postulate that is confirmed by all experimental experience. If the material system is a *rigid body*, i.e. a body that cannot be deformed, it is sufficient that the external forces acting on the body as a whole form a system equivalent to zero.

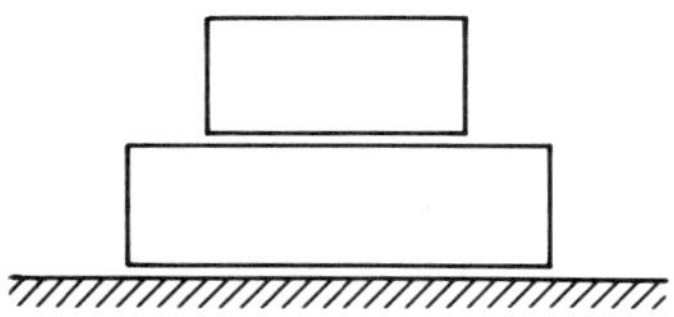

Figure 2.2

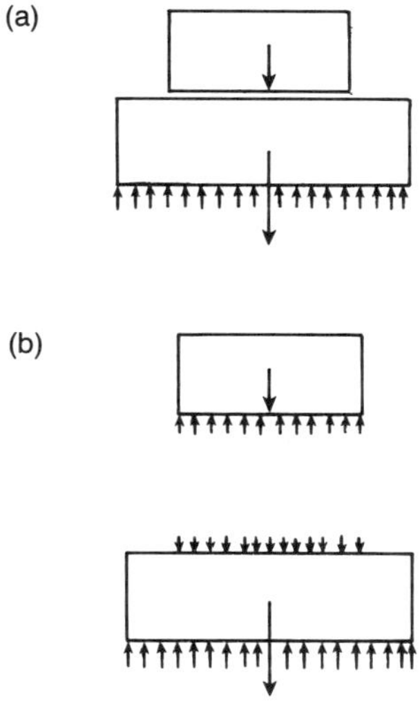

Figure 2.3

Force systems equivalent to zero were considered in Section 1.3.6. Such a system is characterized by the fact that the sum of forces and the sum of moments about an arbitrary point (A) are both equal to zero.

Hence, the *equations of equilibrium*

$$\sum \boldsymbol{F}=0 \qquad \sum \boldsymbol{M}_{\mathrm{A}}=0 \tag{2.1}$$

apply to the material system as a whole, as well as to each portion of the system. The forces involved in Equations (2.1) are the external forces acting on the system chosen.

In Section 2.2 we shall see how the conditions of equilibrium are utilized in specific applications where systematic calculations are performed.

Position of equilibrium

Consider a material system such that the equilibrium conditions, Equations (2.1), are satisfied. The position of the system is then referred to as a *position of equilibrium*.

The pendulum of Figure 2.4(a) is supported at O and is shown in an equilibrium position. If it is at rest we conclude that the pendulum is in equilibrium.

A system may be in an equilibrium position momentarily without being in equilibrium. Assume that the pendulum is swinging. At the instant shown in Figure 2.4(b) the pendulum is in the equilibrium position, but it is not in equilibrium:

- Individual parts of the pendulum are in motion, which is not consistent with the definition of equilibrium.
- The system of external forces is not equivalent to zero – Equations (2.1) are not satisfied.

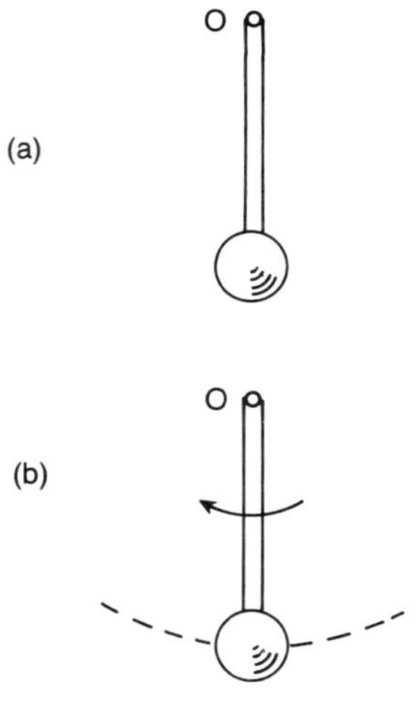

Figure 2.4

'The system is in a position of equilibrium' is not equivalent to 'The system is in equilibrium'.

2.1.2 Free-body diagrams

The concepts of *material system* and *external forces* play important parts in the formulation of the equilibrium conditions. The concept *portion of a system* was also included in the formulation, but this has no absolute meaning of its own; a portion of a material system is also a material system and is treated in the same way.

The first important step when a problem is solved is to choose a material system to which the conditions of equilibrium are to be applied. Once the system has been chosen it is advantageous to isolate the system in a special diagram where the extensions and the limits of the system are clearly identified.

Next, the *external forces* acting on the isolated system are introduced into the diagram. No other forces are allowed here. Suitable symbols are introduced to denote the forces. The resulting diagram is called the *free-body diagram* and is the starting point when the equilibrium conditions are formulated. In the following worked example we shall see how a free-body diagram is drawn in a specific case.

WORKED EXAMPLE 2.1

A boy of weight 700 N is standing on a bridge that may be regarded as a uniform plate. The weight of the bridge is 950 N. It is attached to a hinge at O and is supported by a cable as shown in Figure 2.5. Draw a free-body diagram of the system plate + boy.

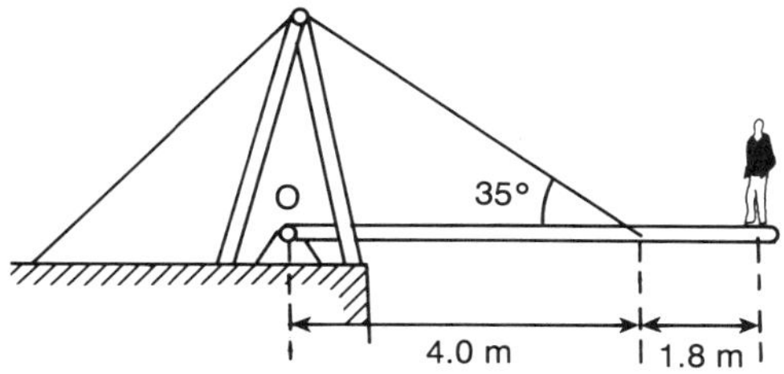

Figure 2.5

SOLUTION

The external forces shown in Figure 2.6 are:

- The gravitational forces of 700 N and 950 N.
- The tension in the cable – the direction is known and the magnitude is denoted by S.
- The reaction (the force exerted by the support) at O – the direction and the magnitude are both unknown; symbols R and α are introduced as shown.

Note that the forces exerted by the feet of the boy on the bridge and vice versa are not included; they are internal forces when the system plate + boy is considered. The free-body diagram is now ready for use. Normally, the next step is to apply the equations of equilibrium, $\Sigma \boldsymbol{F} = 0$ and $\Sigma \boldsymbol{M} = 0$, which are used to determine the unknowns of the problem, e.g. S, R and α.

Centre of gravity

In Section 1.2.5 we have seen how two parallel forces can be replaced by one single force, the resultant. The magnitude and the line of action may be determined uniquely. An arbitrary system of parallel forces may be reduced by repeating this procedure until only one force, the resultant of the system, remains. This is illustrated in Figure 2.7(a). The magnitude of the resultant is equal to the magnitude of the sum of the original forces.

The gravitational forces acting on the various portions or elements of a body form a system of parallel forces. Consequently, they can be replaced by one single force. The magnitude (Q) of the resultant is equal to the sum of the magnitudes of the elementary

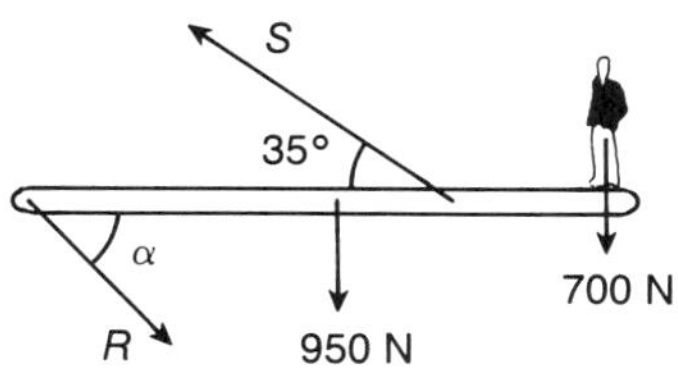

Figure 2.6

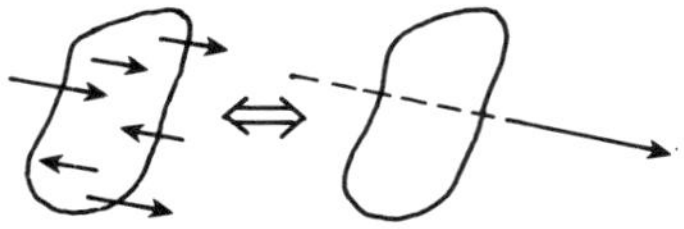

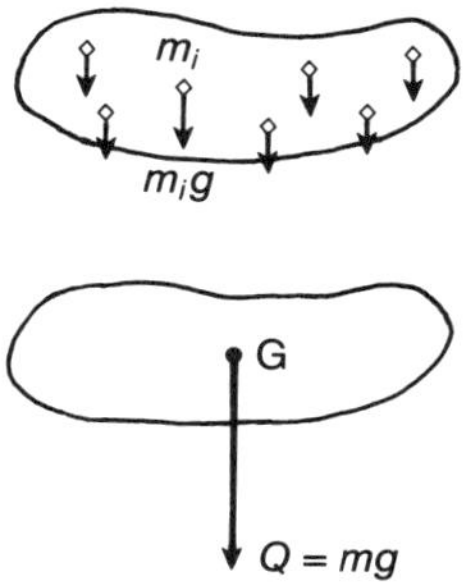

Figure 2.7

gravitational forces. The magnitude of the gravitational force acting on an element of mass m_i is $m_i g$. The resultant Q is given by

$$Q = \sum_i (m_i g) = \left(\sum_i m_i\right) g = mg$$

where $m = \Sigma m_i$ is the total mass of the body. This is shown in Figure 2.7(b).

Thus we know that the total gravitational force can be represented by a single force $Q = mg$, where m is the total mass of the body. The resultant Q is also known as the *weight* of the body. At this stage we do not concern ourselves with the problem of how to determine the line of action of the gravitational force acting on a body of irregular shape. This problem will be treated in Chapter 3. We merely state that the line of action of the gravitational force passes through a special point, called the *centre of gravity*. In some problems in this chapter it is assumed that the location of the centre of gravity is known in advance. If this is not obvious for symmetry reasons, the location will normally be given in a corresponding diagram where it is denoted by the letter G.

Newton's third law in connection with free-body diagrams

The application of the law of action and reaction deserves a special comment. When solving problems where a system has to be divided into several portions, it is extremely important that this law is thoroughly understood. The reader is recommended to work through the following worked example as an exercise.

WORKED EXAMPLE 2.2

The system shown in Figure 2.8 is assumed to be in equilibrium. The weights of the three performers are Q_1, Q_2 and Q_3. The normal and friction forces acting at the contact points at

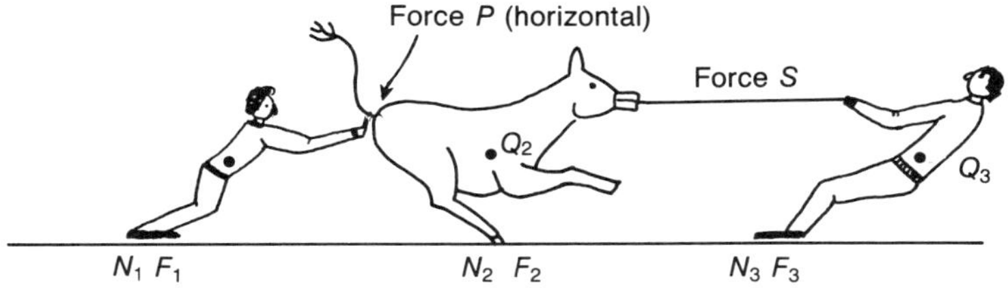

Figure 2.8

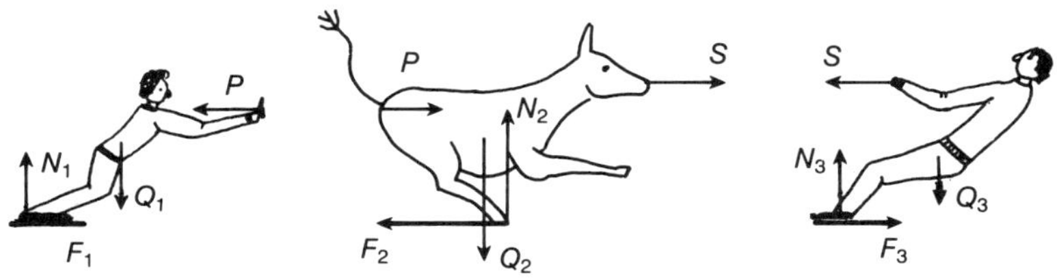

Figure 2.9

the ground are denoted by N_1, F_1, etc. Introduce the external forces acting on each one of the performers in the diagram and label the forces. Make sure that the direction of each force is correct. The centres of gravity are marked with points.

SOLUTION

The solution is shown in Figure 2.9.

REMARK

The two forces of magnitude P must be of opposite direction and equal magnitude according to Newton's third law. The same is true for the two forces of magnitude S.

Let us summarize the procedure for construction of free-body diagrams:

1. Decide which system is to be isolated.
2. Draw a separate diagram showing the chosen system and nothing else.
3. Introduce the forces in the diagram and label them. Remember Newton's third law.

WORKED EXAMPLE 2.3

A sketch of a simple crane is shown in Figure 2.10. AB is a uniform beam of mass 110 kg. Body L is a 280 kg load. The cable FOL runs over a small, frictionless pulley. Determine the tension in the cable CB and the force (the vector expression) exerted on the beam AB by the support at A.

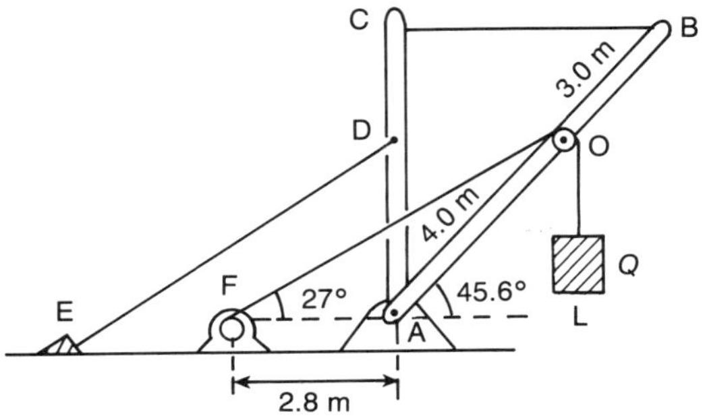

Figure 2.10

SOLUTION

The two forces to be determined are denoted by S and $\mathbf{R}$, respectively. The weight of the load (L) is $Q = 280 \cdot 9.81$ N. As the pulley is frictionless, it can be shown that the tension in the cable FOL is equal to Q everywhere in the cable. We will return to this property in Worked Example 2.7. The weight of the beam AB is $P = 110 \cdot 9.81$ N. The point of application, the centre of gravity, is located at the middle of the beam.

As well as the forces mentioned above, a number of forces are present in the system: internal forces and moments in the engine F, the tension in the cable ED, internal forces in the beam AC, the force exerted on beam AC by the support at A, forces exerted by the pulley O and the beam AB on one another, and so on.

This example shows how important it is to concentrate on *what is known* and *what it is desired to know*. Actually, this information is given in the text of the problem, and we shall now without too many comments present a simple solution. To make the solution simple we shall avoid introducing quantities that are unknown but not requested.

The forces to be determined are S and $\mathbf{R}$. These forces will be external forces if we decide to isolate the system shown in the free-body diagram of Figure 2.11, namely the system consisting of

beam AB + pulley O + load L + portion of the cable

Force $\mathbf{R}$ is unknown with respect to magnitude as well as direction. We resolve it into two components V and H, one of them vertical, the other one horizontal. As we have three unknowns (S, V and H) we need three equations. These are given by the equilibrium conditions, $\Sigma \mathbf{F} = 0$ and $\Sigma M_A = 0$. The first of these gives two scalar equations, $\Sigma F_x = 0$ and $\Sigma F_y = 0$:

$$\sum F_x = 0 \Rightarrow H - S - Q \cos 27^\circ = 0$$

$$\sum F_y = 0 \Rightarrow V - P - Q - Q \sin 27^\circ = 0$$

$$\sum M_A = 0 \Rightarrow P \cdot 3.5 \text{ m} \cdot \cos 45.6^\circ + Q \cdot 4 \text{ m} \cdot \cos 45.6^\circ - S \cdot 7 \text{ m} \cdot \sin 45.6^\circ - Q \cdot 2.8 \text{ m} \cdot \sin 27^\circ = 0$$

Introducing numerical values for P and Q and solving for S, V and H we obtain $S = 1367$ N, $H = 3815$ N, $V = 5072$ N.

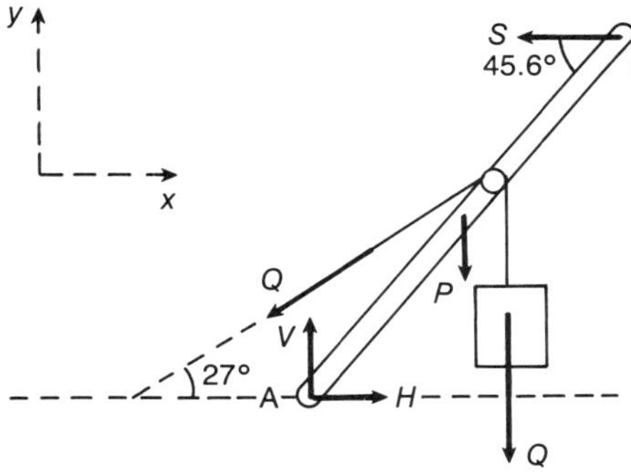

Figure 2.11

ANSWER

The tension in the cable is 1.4 kN. The force supported at A is (3.8, 5.1) kN.

Superposition of force systems equivalent to zero

Let us assume that a certain force system is equivalent to zero. This system is symbolically denoted by S_1. The equilibrium equations $\Sigma \boldsymbol{F}^{(1)} = 0$, $\Sigma \boldsymbol{M}^{(1)} = 0$ are then satisfied for this force system. Let another force system S_2 also be equivalent to zero. This system then satisfies the equations $\Sigma \boldsymbol{F}^{(2)} = 0$, $\Sigma \boldsymbol{M}^{(2)} = 0$.

Assume that S_1 and S_2, one at a time, are external force systems acting on a certain material system, e.g. a rigid body. In both cases the equilibrium conditions are satisfied. What will happen if the two force systems are simultaneously applied to the body? In this case the force systems are said to be *superposed*. If the new force system is denoted by S, then we can express the relation between the force systems in a symbolic way by $S = S_1 + S_2$. The sum of forces and the sum of moments of the system S are, respectively,

$$\sum \boldsymbol{F} = \sum \boldsymbol{F}^{(1)} + \sum \boldsymbol{F}^{(2)}$$
$$\sum \boldsymbol{M} = \sum \boldsymbol{M}^{(1)} + \sum \boldsymbol{M}^{(2)}$$

As the equilibrium conditions are satisfied for S_1 and S_2 it is obvious that the system S is also equivalent to zero. The new situation is thus compatible with equilibrium. If, on the other hand, the systems S and S_1 are known to be equivalent to zero, then we can deduce that the difference between the systems, $S_2 = S - S_1$, is equivalent to zero.

The result of the discussion above can be used to simplify calculations considerably. In a certain situation it may be known in advance that a certain force system (S_1) corresponds to equilibrium, i.e. $S_1 = 0$. However, we may not know exactly what forces are included in the system. If a certain force, e.g. a load, is changed, then other forces of the original force system must be changed too to ensure that the new force system is compatible with equilibrium (meaning $S = 0$). Sometimes these changes are the interesting part of the problem. Together with the change of the load they form an additional system $S_2 = S - S_1$. As $S = 0$ and $S_1 = 0$, it follows that $S_2 = 0$. In other words, it is sufficient to consider the additional system. In the following worked example an application will be shown.

WORKED EXAMPLE 2.4

In the registration certificate of a truck, Figure 2.12, the following information about the axle loads of the unloaded truck (including the driver) is given: 'front 3350 kg, rear 3100 kg'. The truck is used for transporting planks to a construction site on a road where the axle load is limited to '4000 kg'. The length of each plank is 4.80 m and the mass is 14.0 kg. How many planks can be transported at the same time?

SOLUTION

When the unloaded truck is in equilibrium it is subjected to a force system consisting of the gravitational force on the truck and the normal forces exerted on the wheels by the ground, also called the axle loads. When the load is applied the normal forces will be changed. The loaded truck is in equilibrium, meaning that the additional force system must be equivalent to zero.

This force system is shown in Figure 2.13. The forces involved are the total gravitational force Q on the planks and the additional normal forces N_A and N_B. These are not allowed to exceed $650 \cdot 9.81$ N and $900 \cdot 9.81$ N, respectively. The question is: for what value of Q does one of these forces reach its limit?

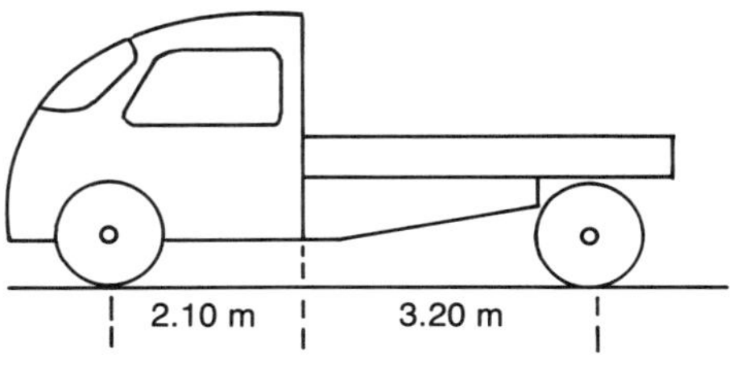

Figure 2.12

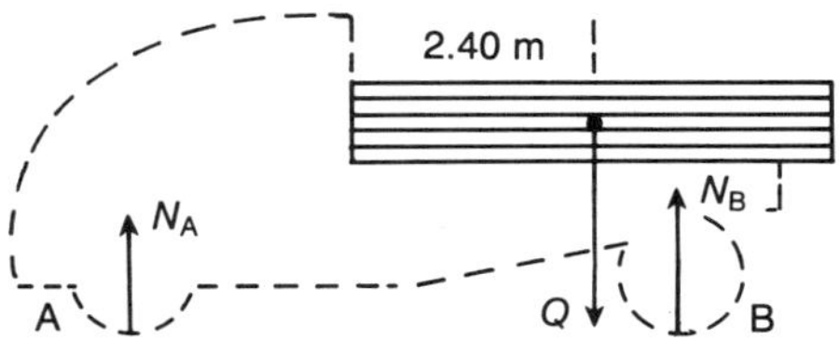

Figure 2.13

The equilibrium equations are:

$$\sum \boldsymbol{F} = 0 \Rightarrow N_A + N_B - Q = 0 \tag{2.2}$$

$$\sum M_A = 0 \Rightarrow Q \cdot 4.50\ \text{m} - N_B \cdot 5.30\ \text{m} = 0 \tag{2.3}$$

$$\sum M_B = 0 \Rightarrow N_A \cdot 5.30\ \text{m} - Q \cdot 0.80\ \text{m} = 0 \tag{2.4}$$

Inspection of Equations (2.3) and (2.4) indicates that N_B is the normal force that first reaches its limit. We thus put $N_B = 900 \cdot 9.81$ N and calculate the corresponding value of Q. Of course, we have to check that N_A really is below the limit of $650 \cdot 9.81$ N.

We find that $Q = 1060 \cdot 9.81$ N and $N_A = 160 \cdot 9.81\ \text{N} < 650 \cdot 9.81$ N. The value of Q corresponds to 75.7 planks. Thus, 75 planks may be transported at the same time.

REMARK

We observe that Equation (2.2) is not used. We can also see that this equation may be obtained by subtracting Equation (2.3) from Equation (2.4). Only two of the three equations are linearly independent. The reason for this is that all forces of the acting force system are parallel. This property will be further explored in Section 2.2.1 where 'Special two-dimensional force systems' are considered

2.2 Equilibrium in two dimensions

2.2.1 Equations of equilibrium

In this section we shall concentrate on material systems that are subjected to two-dimensional force systems according to Section 1.2. In Worked Examples 2.1–2.4 we have seen a few typical examples of this kind.

In fact, it is sufficient that the acting forces can be reduced to a two-dimensional force system. A uniform, rectangular plate is shown in Figure 2.14(a). The plate is supported by two hinges at A and B in such a way that it may be rotated about the horizontal axis AB. The plate is kept in the horizontal position by two cords attached at corners C and D. If the plate is isolated we find that forces according to Figure 2.14(b) are acting, i.e. the gravitational force $\boldsymbol{Q}$, tensions $\boldsymbol{S}_C$ and $\boldsymbol{S}_D$ in the cords and reactions $\boldsymbol{R}_A$ and $\boldsymbol{R}_B$ exerted by the bearings. From symmetry arguments we deduce that $\boldsymbol{S}_C = \boldsymbol{S}_D$ and $\boldsymbol{R}_A = \boldsymbol{R}_B$.

According to Section 1.2.5 the force system can be reduced to a two-dimensional force system according to Figure 2.14(c), where $\boldsymbol{R} = \boldsymbol{R}_A + \boldsymbol{R}_B$ and $\boldsymbol{S} = \boldsymbol{S}_C + \boldsymbol{S}_D$.

If an arbitrary body is in equilibrium it is required that the external force system acting on the body is equivalent to zero. According to Equations (2.1) we then have

$$\sum \boldsymbol{F} = 0$$

$$\sum \boldsymbol{M}_A = 0 \tag{2.5}$$

where the point A may be chosen in an arbitrary manner. Using the coordinate system

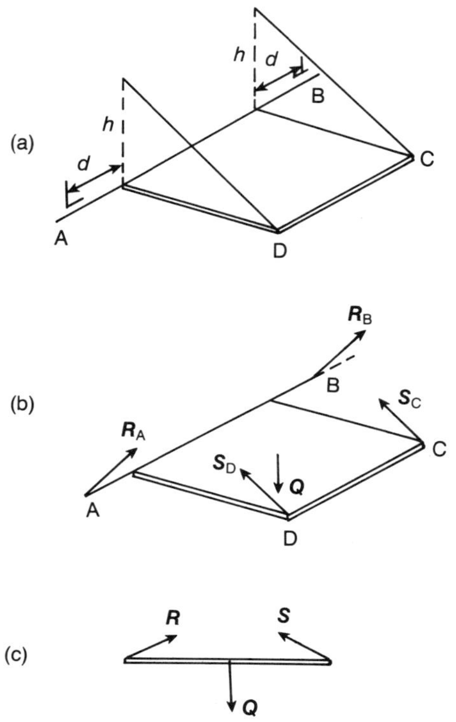

Figure 2.14

of Figure 2.15 we can express these as six scalar equations:

$$\begin{cases} \sum F_x = 0 \\ \sum F_y = 0 \\ \sum F_z = 0 \end{cases} \qquad \begin{cases} \sum M_{Ax} = 0 \\ \sum M_{Ay} = 0 \\ \sum M_{Az} = 0 \end{cases} \tag{2.6}$$

Let us assume that the forces acting on the body form a two-dimensional force system. We introduce a coordinate system such that all forces are located in the *xy*-plane. If couples are included, the corresponding vectors must be perpendicular to the *xy*-plane. Three of the equilibrium equations will be automatically satisfied in this case, namely $\Sigma F_z = 0$, $\Sigma M_{Ax} = 0$ and $\Sigma M_{Ay} = 0$. Consequently, three equations of equilibrium remain:

$$\sum F_x = 0 \qquad \sum F_y = 0 \qquad \sum M_{Az} = 0 \tag{2.7}$$

In this case it is customary to exclude the index z from the last equation.

In a two-dimensional case we thus have three non-trivial equilibrium equations. It is certainly no coincidence that the number of equations is three. This has to do with the fact that three quantities are required to describe uniquely the position of a rigid body that is free to move in a plane.

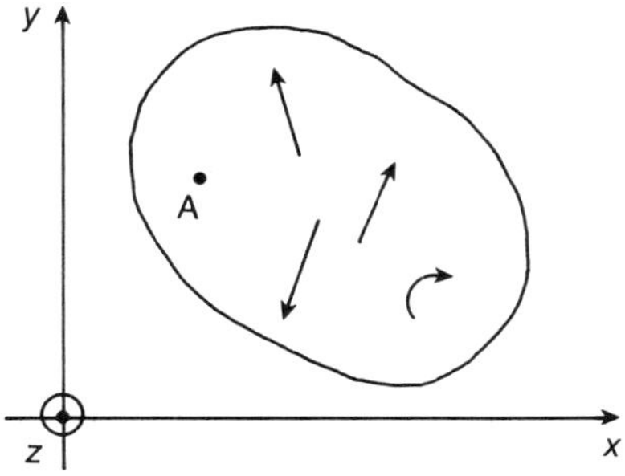

Figure 2.15

As an example we consider a bar in the xy-plane, see Figure 2.16. If we know the coordinates $(x_P; y_P)$ of a point P of the bar and an angle ϕ describing the orientation of the bar, then the position of the bar is uniquely determined. If the forces applied to the bar are either known in advance or given as functions of x_P, y_P and ϕ we can determine the three unknowns x_P, y_P and ϕ from Equations (2.7).

This requires, of course, that the force system applied to the bar is such that equilibrium is possible. Even if a solution exists it is not necessary that the solution is unique.

Alternative equations of equilibrium

As the moment point A of the equilibrium equation $\Sigma M_A = 0$ may be chosen arbitrarily we would perhaps expect that more than three equilibrium equations could be utilized by using other moment points, e.g. point B of Figure 2.17. We then have

$$\begin{cases} \sum F_x = 0 \\ \sum F_x = 0 \end{cases} \quad \begin{cases} \sum M_B = 0 \\ \sum M_A = 0 \end{cases}$$

This would make it possible to solve for four unknowns instead of three. However, a closer look at the equations for a specific case would reveal that the equations are not linearly independent. One of them can always be derived from the other three. No additional information is obtained by adding the fourth equation, since the original

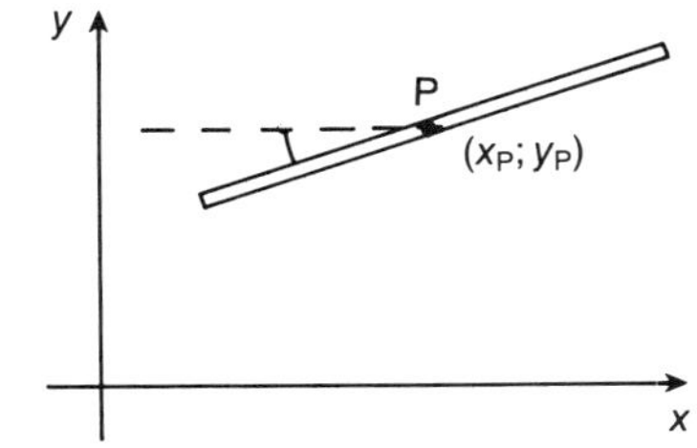

Figure 2.16

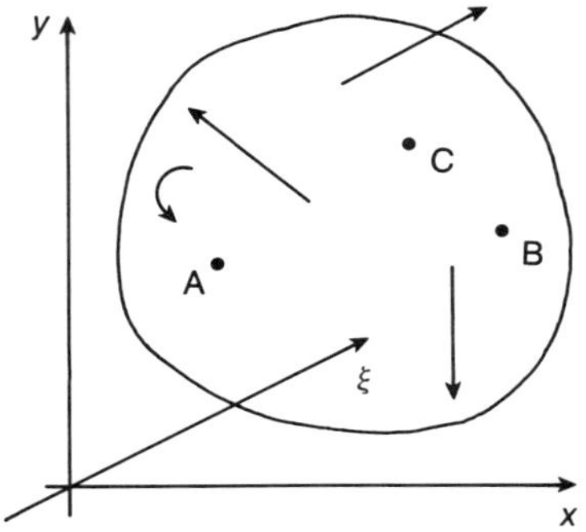

Figure 2.17

three equations already express the fact that the force system is equivalent to zero, i.e. the equilibrium conditions are satisfied. Hence, the fourth equation must be satisfied automatically.

However, it is possible to choose the three equations in other ways than we did in Equations (2.7). One possibility would be to replace one of the force equations by one more moment equation. We then have, for instance,

$$\sum F_x = 0 \qquad \sum M_A = 0 \qquad \sum M_B = 0 \tag{2.8}$$

This requires that the x-direction is not normal to the vector $\overrightarrow{AB}$, because in that case the resultant of the force system could be a force with its line of action passing through A and B. In other words, a force system which satisfies Equations (2.8) is not necessarily equivalent to zero in this case.

Another possibility is to use three moment equations, e.g.

$$\sum M_A = 0 \qquad \sum M_B = 0 \qquad \sum M_C = 0 \tag{2.9}$$

This requires, however, that points A, B and C are not located on a straight line, in which case the resultant of the force system could be a force along that line.

Further possibilities are obtained by choosing two non-perpendicular components of $\Sigma F = 0$. Using denotations according to Figure 2.17 we have, for instance,

$$\sum F_x = 0 \qquad \sum F_\xi = 0 \qquad \sum M_A = 0 \tag{2.10}$$

WORKED EXAMPLE 2.5

A bar of mass m is supported by two smooth rings, A and B, and by a smooth incline as shown in Figure 2.18. Determine the forces acting on the bar at the contact points A, B and C.

SOLUTION

The first step is to draw the free-body diagram of the bar, which is shown in Figure 2.19.

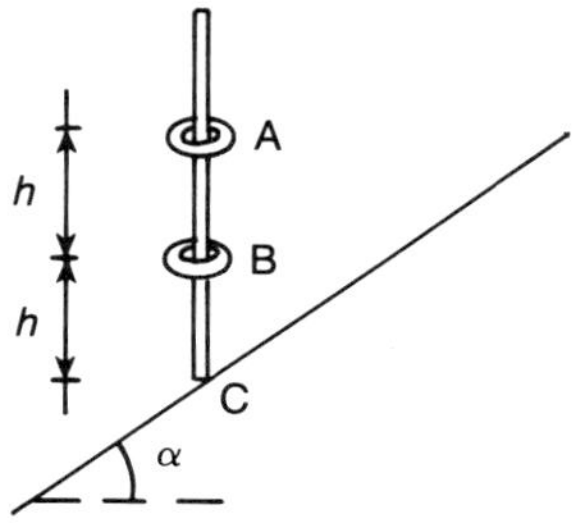

Figure 2.18

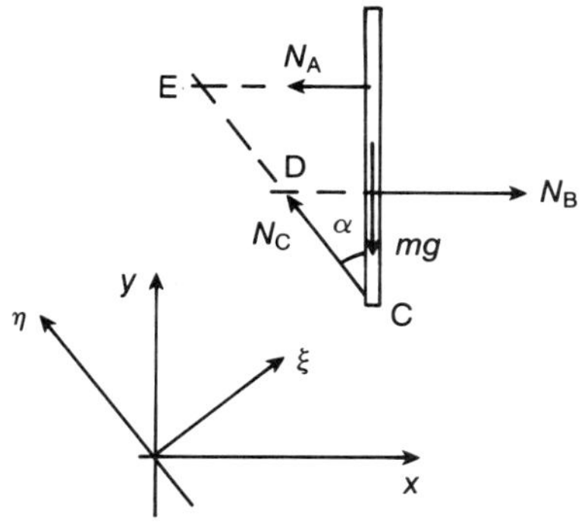

Figure 2.19

As there is no friction the contact forces will be directed as shown (this will be explained in detail in Section 2.2.2). The number of unknowns is three: N_A, N_B and N_C. Using the coordinate system shown, we have

$$\begin{aligned} &\sum F_x = 0 \Rightarrow -N_A + N_B - N_C \sin\alpha = 0 \\ &\sum F_y = 0 \Rightarrow N_C \cos\alpha - mg = 0 \\ &\sum M_C = 0 \Rightarrow N_A \cdot 2h - N_B \cdot h = 0 \end{aligned} \qquad (2.11)$$

The solution is

$$\begin{cases} N_A = mg \tan\alpha \\ N_B = 2mg \tan\alpha \\ N_C = mg/\cos\alpha \end{cases}$$

So far we have used a rather clumsy way to justify our equations of equilibrium. From now on we shall adopt a much simpler convention. We replace the expression $\Sigma F_x = 0$ by an arrow in the *x*-direction. Instead of $\Sigma M_A = 0$ we write $\overset{\frown}{A}$ or $\overset{\frown}{A}$, where the direction of the arrow indicates the positive reference direction.

To illustrate the use of alternative equilibrium equations, we now choose the following set:

$$\begin{aligned} &\uparrow \quad N_C \cos\alpha - mg = 0 \\ &\overset{\frown}{E} \quad N_B h - mg \cdot 2h \tan\alpha = 0 \\ &\overset{\frown}{C} \quad N_A \cdot 2h - N_B h = 0 \end{aligned} \tag{2.12}$$

Here we have used one force equation and two moment equations.

If we choose the moment points D and E instead we obtain

$$\begin{aligned} &\uparrow \quad N_C \cos\alpha - mg = 0 \\ &\overset{\frown}{D} \quad N_A h - mgh \tan\alpha = 0 \\ &\overset{\frown}{E} \quad N_B h - mg \cdot 2h \tan\alpha = 0 \end{aligned} \tag{2.13}$$

Here each equation contains only one of the unknowns. The solution of the simultaneous equations is thus simplified. An alternative choice is

$$\begin{aligned} &\uparrow \quad N_C \cos\alpha - mg = 0 \\ &\nearrow \quad N_B \cos\alpha - N_A \cos\alpha - mg \sin\alpha = 0 \\ &\overset{\frown}{E} \quad N_B h - mg \cdot 2h \tan\alpha = 0 \end{aligned} \tag{2.14}$$

where we have chosen two force equations in non-perpendicular directions.

We can also take three moment equations, for instance

$$\begin{aligned} &\overset{\frown}{A} \quad N_B h - N_C \sin\alpha \cdot 2h = 0 \\ &\overset{\frown}{D} \quad N_A h - mgh \tan\alpha = 0 \\ &\overset{\frown}{E} \quad N_B h - mg \cdot 2h \tan\alpha = 0 \end{aligned} \tag{2.15}$$

On the other hand the following set of equations does not lead to a solution:

$$\begin{aligned} &\overset{\frown}{A} \quad N_B h - N_C \sin\alpha \cdot 2h = 0 \\ &\overset{\frown}{B} \quad N_A h - N_C \sin\alpha \cdot h = 0 \\ &\overset{\frown}{C} \quad N_A \cdot 2h - N_B h = 0 \end{aligned} \tag{2.16}$$

The reason is that points A, B and C lie on a straight line. Hence, Equations (2.16) may be satisfied even if the force system has a resultant along this line.

Neither does the following combination work:

$$\begin{aligned} &\rightarrow \quad -N_A + N_B - N_C \sin\alpha = 0 \\ &\overset{\frown}{B} \quad N_A h - N_C \sin\alpha \cdot h = 0 \\ &\overset{\frown}{C} \quad N_A \cdot \mathbf{2}h - N_B h = 0 \end{aligned} \tag{2.17}$$

This is because the x-axis is perpendicular to line BC.

The reader is urged to check that the alternative sets of equilibrium equations, (2.12)–(2.15), all give the same solution as Equations (2.11) and that the combinations (2.16)–(2.17) do not yield a solution.

We also see that nothing is gained by adding a fourth equation to any of the systems (2.12)–(2.15). If we, for instance, add the last two of Equations (2.12) we essentially obtain the second equation of system (2.13). This means that if this equation is added to system (2.12), no new information is added.

Special two-dimensional force systems

In a few special cases the acting force system is such that not even three independent equations can be utilized. One example is shown in Figure 2.20 where all the forces have a common line of action. Two of the equations of equilibrium are automatically satisfied, namely

$$\sum F_y = 0 \qquad \sum M_A = 0$$

Only one equation remains:

$$\sum F_x = 0$$

If we try to add further equations, e.g. $\Sigma M_B = 0$, no new information is added.

In Figure 2.21 all forces are parallel. The equation $\Sigma F_y = 0$ is then automatically satisfied, and only two independent equations may be used, for instance,

$$\sum F_x = 0 \qquad \sum M_A = 0$$

In Figure 2.22 a force system where the lines of action of all forces intersect at a point A is shown. Thus the equation $\Sigma M_A = 0$ is trivial. Two independent equations remain, for instance

$$\sum F_x = 0 \qquad \sum F_y = 0$$

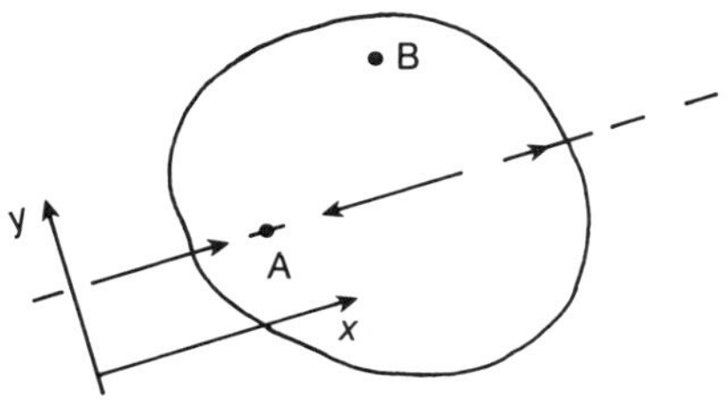

Figure 2.20

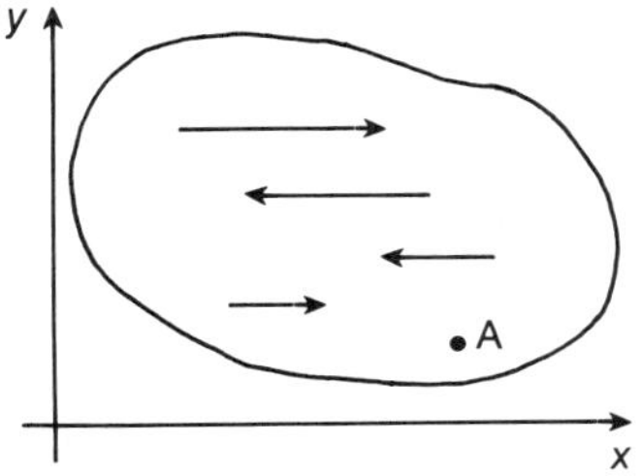

Figure 2.21

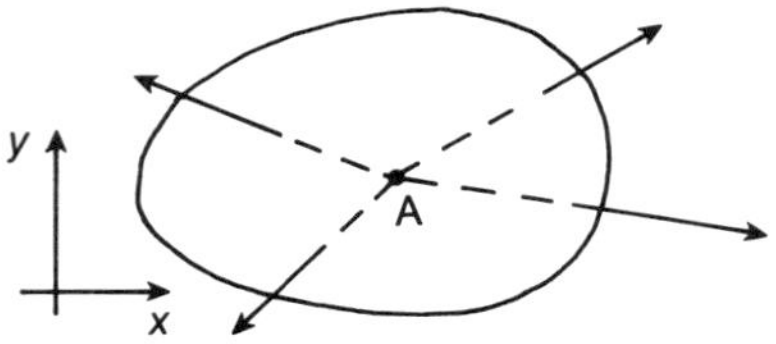

Figure 2.22

Two-force systems

If only two forces are acting on a body some conclusions may be drawn immediately. The body of Figure 2.23(a) is subjected to two forces applied at A and B. The equilibrium condition $\Sigma F = 0$ then implies that

$$F_A + F_B = 0 \Rightarrow F_B = -F_A$$

Hence, the force system must be a couple according to Figure 2.23(b). Since $\Sigma M_B = 0$ at equilibrium the line of action of F_A must pass through B. The force system must thus be any of the two kinds displayed in Figures 2.23(c)–(d).

Three-force systems

If the acting force system consists of three forces, F_A, F_B and F_C, the lines of action must intersect at a point P, see Figure 2.24(a). Should this not be the case, then force F_C alone would have a non-vanishing moment about the point of intersection between

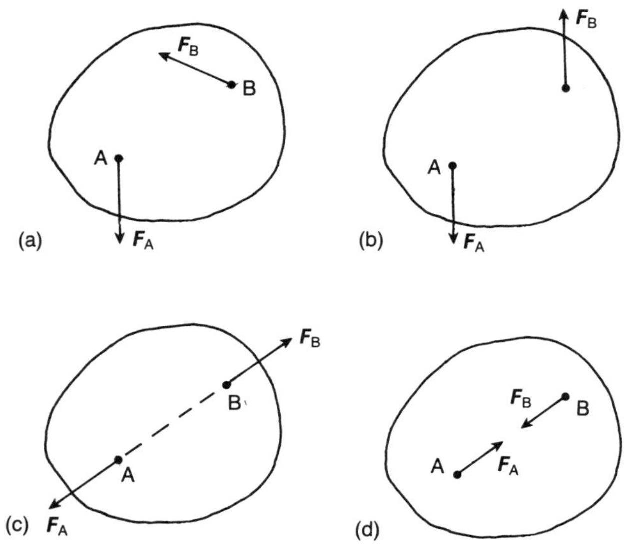

Figure 2.23

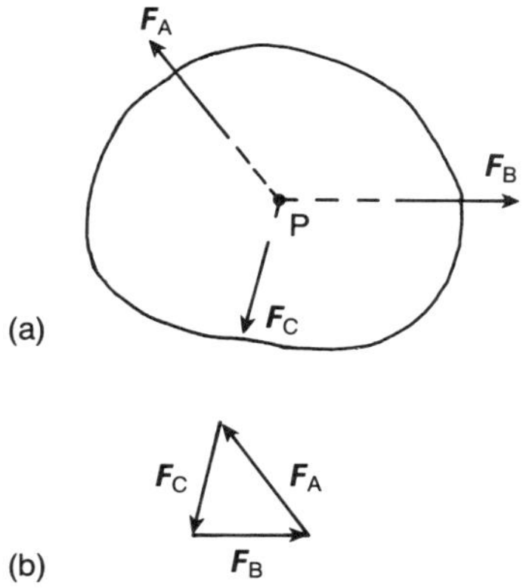

Figure 2.24

the lines of action of F_A and F_B. The sum of moments about this point would then be non-zero, which would be in conflict with the conditions of equilibrium.

Furthermore, the condition $\Sigma F = 0$ implies that the force vectors must form a triangle according to Figure 2.24(b). Sometimes this can be useful when solving problems where the directions of all forces are known.

WORKED EXAMPLE 2.6

A body of mass $m = 2.5$ kg is suspended by two cords as shown in Figure 2.25. The angles made by the cords with the vertical are $\alpha = 30°$ and $\beta = 40°$, respectively. Determine the magnitudes of the tensions in the cords.

SOLUTION

The free-body diagram of the body is given by Figure 2.26(a). Since three forces are acting on the body their lines of action must intersect at one point, namely the point where the cords are attached. This means that the centre of gravity must be located right below this point.

The condition $\Sigma \mathbf{F} = 0$ implies that the forces must form a triangle according to Figure 2.26(b). From the sine law it follows that

$$\frac{mg}{\sin(\pi - \alpha - \beta)} = \frac{S_1}{\sin \beta} = \frac{S_2}{\sin \alpha}$$

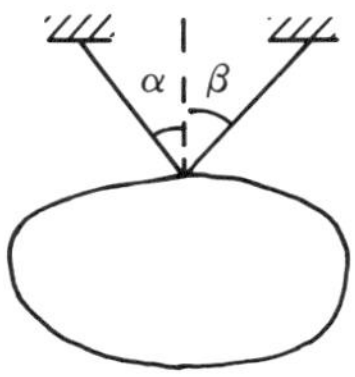

Figure 2.25

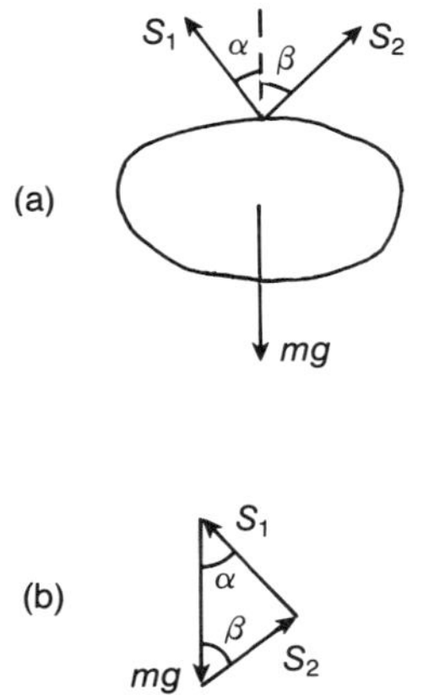

Figure 2.26

Solving for S_1 and S_2 we obtain

$$\begin{cases} S_1 = mg \dfrac{\sin \beta}{\sin(\alpha + \beta)} \approx 16.8 \text{ N} \\ S_2 = mg \dfrac{\sin \alpha}{\sin(\alpha + \beta)} \approx 13.0 \text{ N} \end{cases}$$

REMARK

Using the condition of equilibrium $\Sigma F = 0$ instead, we find that

$$\rightarrow \quad S_2 \sin \beta - S_1 \sin \alpha = 0$$

$$\uparrow \quad S_1 \cos \alpha + S_2 \cos \beta - mg = 0$$

which, of course, leads to the same answer.

2.2.2 Degrees of freedom, constraints, constraining forces

In the preceding section we observed that three variables are needed to describe the position of a rigid body that is free to move in a plane, see Figure 2.16. The number of independent variables required to specify the configuration of a mechanical system is referred to as the number of *degrees of freedom*. A single rigid body that is free to move in two dimensions thus has three degrees of freedom. This is closely related to the fact that there are no more than three independent equations of equilibrium in a two-dimensional case.

In many practical situations, in statics as well as in dynamics, we are concerned with bodies whose abilities to move are restricted in some way owing to contacts with other bodies. Such a body is subjected to a *constraint*. The immediate consequence is that the number of degrees of freedom is reduced, i.e. less than three variables are required to specify the position. As an example we consider the bar AB of Figure 2.27, which is subjected to various types of constraint.

In case (a) point A is restricted to moving in a slot. The position of the bar is determined uniquely if we know the y-coordinate of point A and the angle φ specifying the direction of the bar. In other words there are two degrees of freedom in this case.

In case (b) points A and B are both restricted to moving in slots. The angle φ is sufficient to describe the position of the bar. Thus the bar has one degree of freedom.

In case (c) the bar may rotate freely about an axis A normal to the plane of the figure. In this case too one variable, e.g. the angle φ, is sufficient to specify the location. The number of degrees of freedom is one.

Finally, in case (d) the bar is completely fixed at A, for instance it may be welded at A. The bar is said to be clamped at A. There is no possibility whatsoever for the bar to move. Therefore the number of degrees of freedom is zero.

Let us consider a rigid body which is completely free to move. To specify the location of the body we need to determine three independent variables. Furthermore, we assume that all acting forces are known in advance. In this case it is possible to

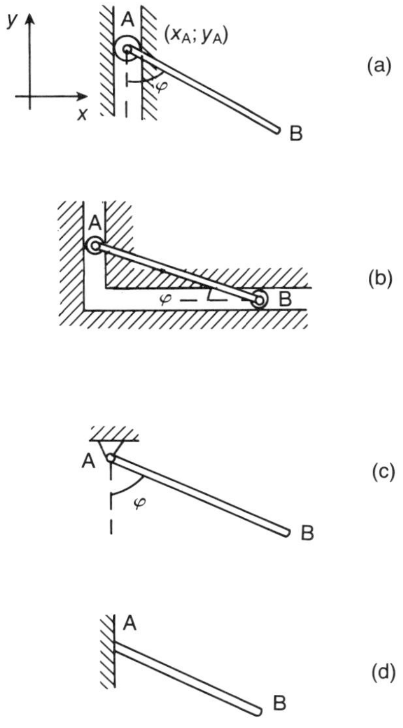

Figure 2.27

determine the three variables from the three equations of equilibrium as was pointed out earlier. If, due to constraints, the number of degrees of freedom is smaller than three, say two, it seems at a first glance as though one of the equations of equilibrium is superfluous. However, this is certainly not the case. If the body is forced into a position determined by the constraint, a corresponding force, which is called a *constraining force* or *reaction*, must act on the body to keep it in place. The third equation may be used to determine the magnitude of this force, which is not known in advance.

In Figure 2.28(a)–(d) the constraining forces acting in cases (a)–(d) of Figure 2.27 are shown. It should be noted that no other forces are shown.

In case (a) point A is prevented from moving perpendicular to the slot. This corresponds to an undetermined constraining force in the opposite direction. This is the force exerted by one of the walls of the slot on the bar at the point of contact A.

In case (b) there are forces acting at both points of contact in an analogous way. These two forces are the unknowns of the problem together with the angle φ.

In general, if a body, due to the action of forces, has a tendency to move in a certain direction at a certain point but is prevented from doing so because of a constraint, then a corresponding constraining force is exerted on the body in the opposite direction at the point of contact.

In case (c) point A of the bar cannot be displaced in any direction. This corresponds to a constraining force whose direction is undetermined. The direction depends on what other forces are acting on the bar. However, the constraining force may always be resolved into two components in an arbitrary way, for instance according to Figure 2.28(c). These two components together with the angle φ are the unknowns of the problem.

The bar of case (d) is completely fixed at A. In a similar manner to case (c) there must be a constraining force of undetermined direction. In this case, however, the bar may not be rotated about A. This additional constraint corresponds to a couple exerted on the bar. In general, if a body is prevented from rotating about a certain axis, this corresponds to what is called a *constraining couple*. In this case the two force components and the couple may be determined from the three equations of equilibrium provided that all other forces are known.

Various types of constraint

In technical applications a number of different types of contacts between a body and

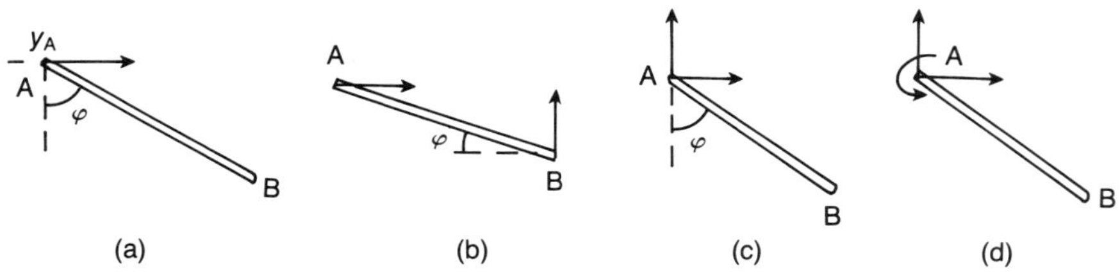

Figure 2.28

its surroundings may occur. In each case constraining forces and couples must act on the body. A review of some important cases is given in Figure 2.29.

In case (a) a body B is supported by another body, e.g. by a fixed surface. The contact surfaces are *smooth* which means that there are absolutely no irregularities that could prevent the bodies from moving sideways with respect to each other. Body B is free to move in any thinkable way except one: it cannot move inwards with respect to the other body, i.e. normal to the tangential plane T at the contact point. This constraint corresponds to a *normal force* N perpendicular to plane T. Furthermore, N must be directed towards body B as the body is not prevented from moving away from the other body. This type of constraint, where the direction of the constraining force is obvious, is called a *one-sided constraint*.

A somewhat more realistic picture of a real situation is shown in case (b). Here we allow the surfaces of the two bodies to be *rough*. This means that if we try to displace the bodies with respect to each other along the tangential plane T, then irregularities of the surfaces will become entangled, thereby preventing the displacement and causing a form of constraint. This corresponds to the existence of a force F in the tangential plane, a *friction force*, in addition to the normal force N. We shall return to the concept of friction in Section 2.5, where a more extensive treatment will be given.

In case (c) a body is attached to a stretched, inextensible cord. Therefore the body is prevented from moving outwards along the direction of the cord, which corresponds to a tensile force or *tension* S exerted by the cord on the body. As the cord cannot resist a displacement of the body such that the cord slackens, the direction of the force must be as shown. Once again we have a case of a one-sided constraint.

A variation on this theme is shown in case (d). Here the cord is replaced by a *light* link, which is assumed to be freely hinged at both ends. Under these conditions only two forces are acting on the link, one at each end. As was shown in Section 2.2.1, where two-force systems were considered, these forces must have a common line of action along the link. Hence, the link exerts an *axial force* N on the body as shown (remember Newton's third law). The only difference compared with case (c) is that the axial force may also be directed inwards towards the body as the link cannot slacken.

In case (e) we show some symbols that are commonly used to indicate that a body is free to move at a point of contact except for displacements normal to the supporting surface. This corresponds to a normal force N as shown. Even though the symbols look the way they do it is customary to use them even in the case of 'two-sided constraints', i.e. N may also be directed outwards from the body.

Some variations are shown in case (f) where a body is free to slide along a rod or a slot. In both cases it is assumed that the body is free to rotate about an axis through point A.

The smooth hinge or pin of case (g) indicates that the body is free to rotate about the axis A but it cannot move away in any direction. This corresponds to a constraining force (or reaction) $\boldsymbol{R}$ with an undetermined line of action. If it is desired $\boldsymbol{R}$ may be resolved into components, for instance as shown. If there is a resistance to rotation about axis A due to friction, we must add a frictional moment on the body. However, if nothing else is stated we will assume that the body is free to rotate whenever this symbol is used.

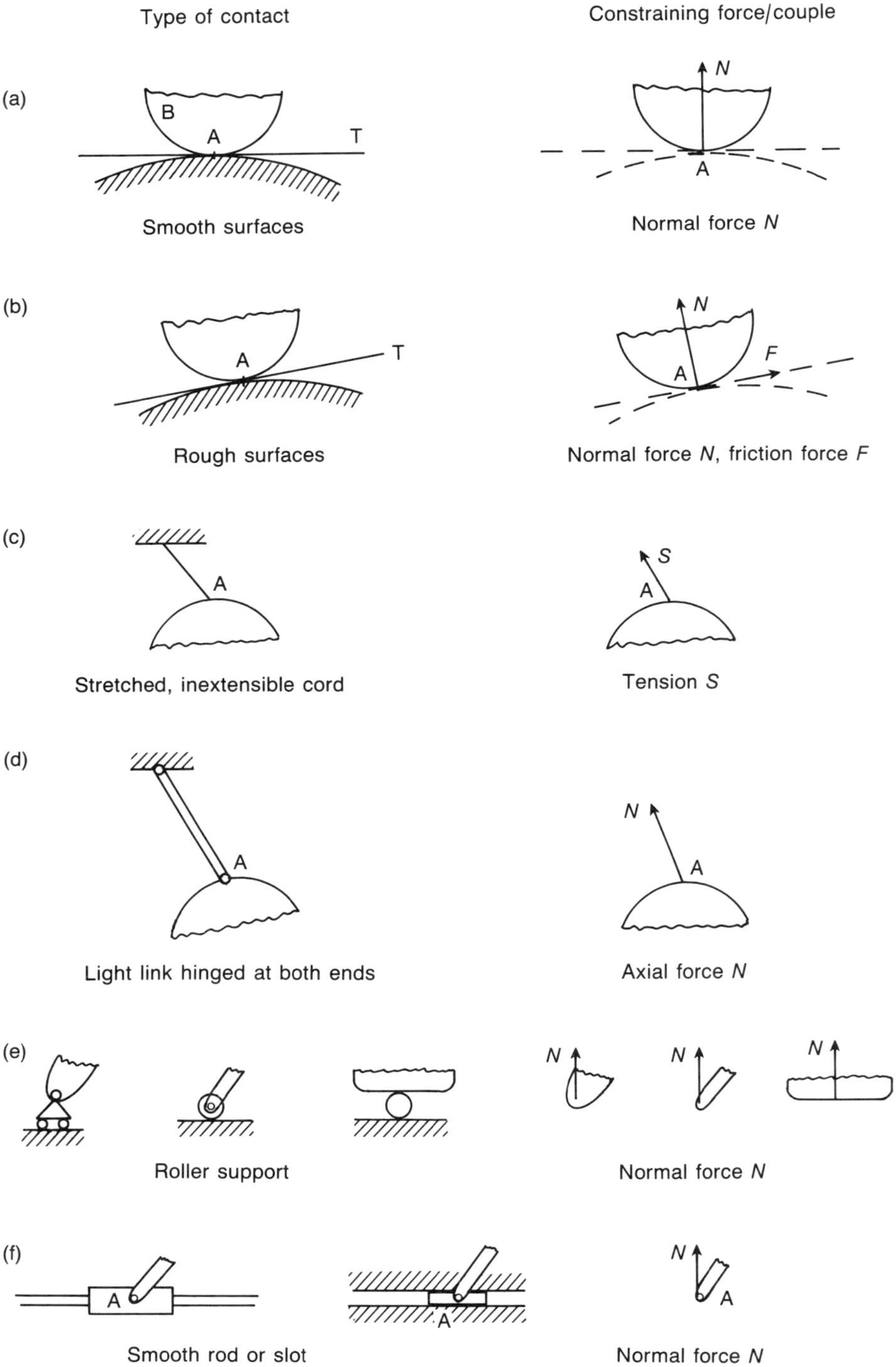
Type of contact
Constraining force/couple
(a)
B
A
T
N
A
Smooth surfaces
Normal force N
(b)
A
T
N
F
A
Rough surfaces
Normal force N, friction force F
(c)
A
S
A
Stretched, inextensible cord
Tension S
(d)
A
N
A
Light link hinged at both ends
Axial force N
(e)
N
N
N
Roller support
Normal force N
(f)
A
A
N
A
Smooth rod or slot
Normal force N

(g)

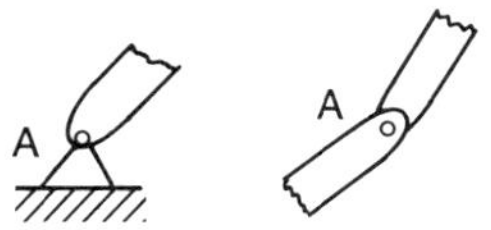

Smooth hinge

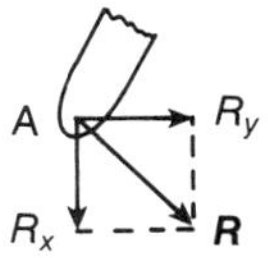

Reaction **R** with components R_x and R_y

(h)

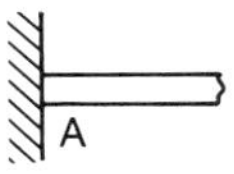

Fixed support

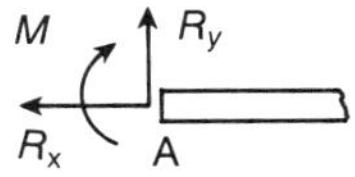

Reaction **R** with components R_x and R_y and couple M

(i)

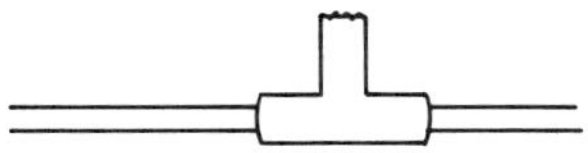

Smooth guide, body not free to turn

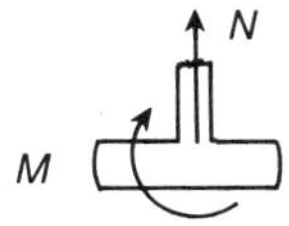

Normal force N, couple M

Figure 2.29

In case (h) a body with a fixed support is shown. The body cannot be displaced in any way. As it cannot be displaced vertically or horizontally a reaction $\boldsymbol{R}$ with components R_x and R_y is assumed to exist. Besides, a constraining couple M must be exerted on the body as it is prevented from rotating about axis A.

Case (i), finally, shows the same situation as case (f) with the distinction that the body is not free to rotate about an axis. In addition to the normal force N a constraining couple M may exist.

At last something should be said about forces that are not constraining forces. These are commonly referred to as *applied forces* and are normally known in advance or at least given as functions of the position of a body. A typical example is the gravitational force on the body; another is the force exerted by an elastic spring.

2.2.3 Methods of problem solving, statical indeterminacy, stability

The student who has a firm grasp of equilibrium equations, free-body diagrams, constraints and constraining forces should now be ready to solve more advanced problems by him- or herself. At this point it should be noted that the problems that follow should be considered as illustrations of the theoretical foundations. It is not until a mastery of various types of problem is achieved that a clear picture of the subject is obtained. On the other hand, solving problems is more or less impossible without a good knowledge of the underlying theory. Theory and problem solving are two disciplines that go hand in hand. The student should try to develop a good understanding of the theory and to work actively in applying the theory to problems. It is not recommended that the student concentrates on studying solutions to a large number of problems in a passive way without attempting to understand the underlying theory. Such a strategy of learning can never give a complete picture of the subject.

From what has been said above it should be evident that in many cases there are several different ways to solve a certain problem. In more complicated problems, e.g. Worked Example 2.3, free-body diagrams may be constructed in more or less suitable ways. Furthermore, alternative choices of equilibrium equations are possible, see Worked Example 2.5. It may be advantageous to try to systematize the methods of solving problems. It may also be worth while to spend some time thinking over each step in the solution. A suitable way may look like this:

1. Choice of system
   Decide which material system (or 'body') is to be considered. In many cases (as in Worked Example 2.5) the choice is natural; in other situations you should think it over carefully. A general piece of advice is that forces to be determined must be external forces to the system under consideration. If a force is not known in advance and/or not of interest you should try to avoid it if possible, see Worked Example 2.3.
2. Free-body diagram
   Isolate the system. It is recommended that you draw a separate diagram where the limits of the system can be clearly indicated. It is important that there is no doubt about what belongs to the system and what belongs to its surroundings.
3. Acting forces
   All *external* forces acting on the system are introduced into the diagram. Here it is important that no forces are left out, for instance because they seem to be uninteresting in this situation. Even more important is not to introduce any non-existing forces. To avoid this it is necessary to have a thorough understanding of the meaning of constraining forces. If a body is connected to a freely hinged pin according to Figure 2.29(g), for instance, then two force components must be introduced but no couple. Of course, the distinction between external and internal forces must be clear.
4. Equations of equilibrium
   In a general case three independent equations of equilibrium may be written down. If any of the special cases of the force systems of Section 2.2.1 is at hand the number of equations is reduced, however. Furthermore, if only two or three forces are acting it is possible to draw some conclusions according to Section 2.2.1 without even utilizing any equations.

   The major problem here is probably to choose the equations in a suitable way. What directions should be chosen for the force components, what moment points are suitable, etc.? It is always advantageous if the simultaneous equations are simple to solve. This is the case if the equations are uncoupled to a large degree. The most elegant solution, that is not always possible to achieve, is when only one unknown is involved in each equation. One way is to choose moment points where the lines of action of two or more forces intersect. On the other hand, difficulties may occur owing to geometrical problems, for instance. If forces that are of no interest in the present situation occur it is recommended that they are kept out of the equations if possible. Worked Example 2.5 shows how different sets of equilibrium equations lead to more or less complicated calculations.

5. Solution of the equations
   Solve the equations of equilibrium for the unknowns to be determined. Here it may be worth while to have some kind of strategy before starting instead of trying to eliminate the unknowns aimlessly.
6. Checks
   Finally the result must be checked:
   Is the physical dimension correct?
   Is the sign of the answer what we expect and, if not, are the expectations or the calculations incorrect?
   Is the order of magnitude reasonable?

Numerical/algebraic calculations

In this context it may be appropriate to say something about numerical calculations. In most cases it is suitable to introduce symbols for various physical quantities involved in the problem (forces, masses, lengths, etc.). When the answer is obtained it is possible to see how the result depends on the values of various quantities. What would have happened if the value of a certain quantity had been twice as large and so on? Unfortunately, this may lead to algebraic expressions that are very difficult to handle. In such cases you may consider introducing numerical values at an earlier stage. However, it is important to avoid rounding off the intermediate results too much as this may lead to an unacceptable inaccuracy in the final result.

The answer to a problem cannot always be expressed explicitly. In practical cases you often find yourself in a situation where equations have to be solved approximately by means of numerical methods.

WORKED EXAMPLE 2.7

A pulley of mass m and radius r may rotate freely about an axis O through the centre of gravity. A cord is running over the pulley as shown in Figure 2.30. Determine the tension S_2 at one end of the cord if the tension at the other end is S_1.

SOLUTION

Isolation of the pulley and the cord gives the free-body diagram of Figure 2.31(a). Here we have assumed that axis O is horizontal. This is of no importance here, however. What is

Figure 2.30

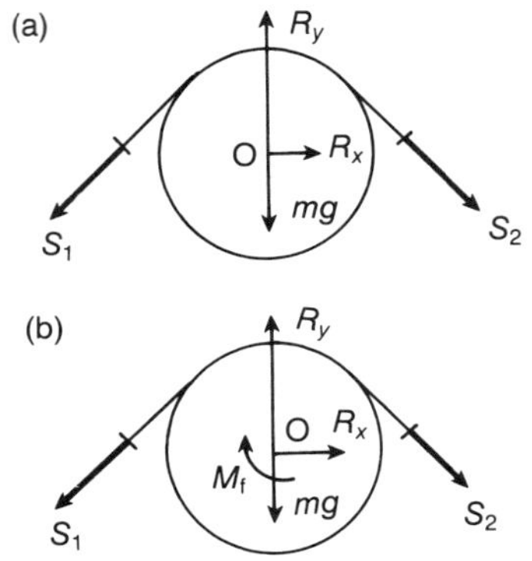

Figure 2.31

important is that the gravitational force mg does not contribute to the sum of moments about O in the following equilibrium equation:

$$\overset{\frown}{O} \quad S_1 r - S_2 r = 0 \Rightarrow S_2 = S_1$$

The tension is obviously the same on both sides of the pulley.

REMARK

If the pulley is not free to rotate owing to friction, a frictional couple M_f must be exerted on the pulley by the axis according to Figure 2.31(b). Hence,

$$\overset{\frown}{O} \quad S_1 r - S_2 r - M_f = 0 \Rightarrow S_2 \neq S_1$$

It is only as long as the pulley is *free to rotate*, i.e. *no friction exists*, that the tension has the same value on both sides of the pulley.

WORKED EXAMPLE 2.8

A circular cylinder of mass m is placed on a rough incline as shown in Figure 2.32. A cord is wound about the cylinder in the mid plane. The cord is attached to the incline so that the straight portion of the cord is horizontal. Determine the tension in the straight portion of the cord and the reaction exerted by the incline on the cylinder.

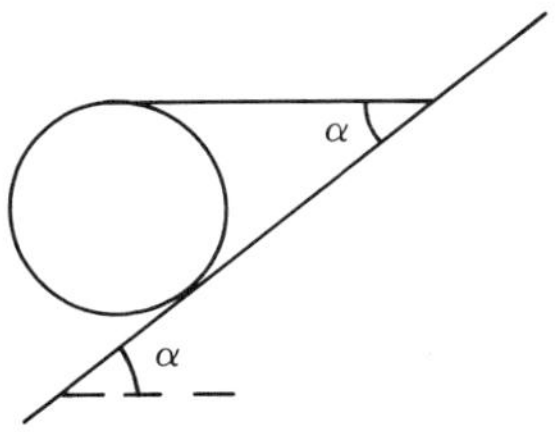

Figure 2.32

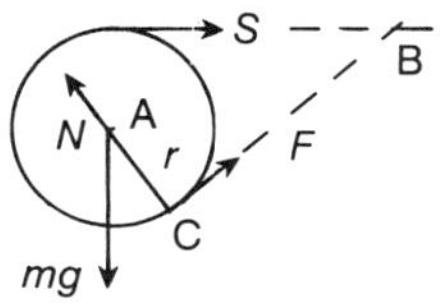

Figure 2.33

SOLUTION

The first step is to draw the free-body diagram of the cylinder, Figure 2.33. The tension is denoted by S. The direction of the reaction exerted by the incline is not known, but it is always possible to resolve the force into two components, the friction force F and the normal force N. These components are parallel to and perpendicular to the incline, respectively. Hence, the number of unknowns is three.

The conditions of equilibrium yield:

$$\nearrow \quad F + S\cos\alpha - mg\sin\alpha = 0$$

$$\nwarrow \quad N - S\sin\alpha - mg\cos\alpha = 0$$

$$\widehat{\mathrm{A}} \quad Fr - Sr = 0$$

Solving the equations we obtain

$$S = F = mg\,\frac{\sin\alpha}{1+\cos\alpha}$$

$$N = mg$$

The magnitude of the reaction is

$$\sqrt{F^2+N^2}$$

$$= mg\sqrt{\frac{2}{1+\cos\alpha}}$$

REMARK

Our choice of equilibrium equations is, of course, not the only one that is possible. As alternatives we could have considered the sum of forces in the horizontal and vertical directions, the sum of moments about points B or C, etc. It is recommended that the student explores and compares the various combinations of equations on his or her own.

WORKED EXAMPLE 2.9

A hemisphere of mass m and radius r is attached to a cord of length $2r$. The cord is attached to a smooth wall at a point A, and the hemisphere is supported by the wall as shown in Figure 2.34. Determine

(a) the position of equilibrium, i.e. the angle φ,
(b) the tension of the cord,
(c) the normal force exerted by the wall on the hemisphere.

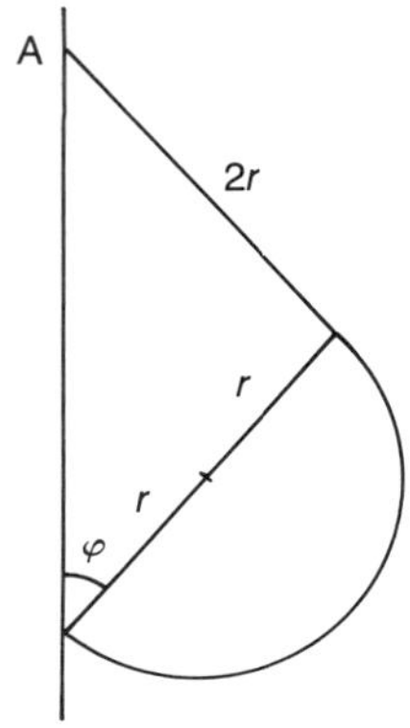

Figure 2.34

SOLUTION

The free-body diagram of the hemisphere is shown in Figure 2.35(a). It may be shown that the centre of gravity of a hemisphere is located on the axis of symmetry at a distance $3r/8$ from point O (methods to show this are described in Section 3.2; see also Appendix 2). Furthermore, we note that triangle ABC is isosceles meaning that the angles at A and C are both φ.

From the equilibrium conditions we find that:

$$\uparrow \quad S\cos\varphi - mg = 0 \tag{2.18}$$

$$\rightarrow \quad N - S\sin\varphi = 0 \tag{2.19}$$

$$\overset{\frown}{A} \quad N\cdot 4r\cos\varphi - mg\left(r\sin\varphi + \frac{3}{8}r\cos\varphi\right) = 0 \tag{2.20}$$

From Equation (2.18) we find that

$$S = \frac{mg}{\cos\varphi} \tag{2.21}$$

Equation (2.20) yields

$$N = \frac{1}{4}mg\left(\tan\varphi + \frac{3}{8}\right) \tag{2.22}$$

Combining Equations (2.22), (2.21) and (2.19) we obtain

$$\frac{1}{4}\left(\tan\varphi + \frac{3}{8}\right) - \tan\varphi = 0 \Rightarrow \tan\varphi = \frac{1}{8} \tag{2.23}$$

$$\therefore \varphi = \arctan\frac{1}{8} \approx 7.13^\circ$$

Substituting into Equation (2.21) we have

$$S = \frac{\sqrt{65}}{8}mg \approx 1.01\,mg$$

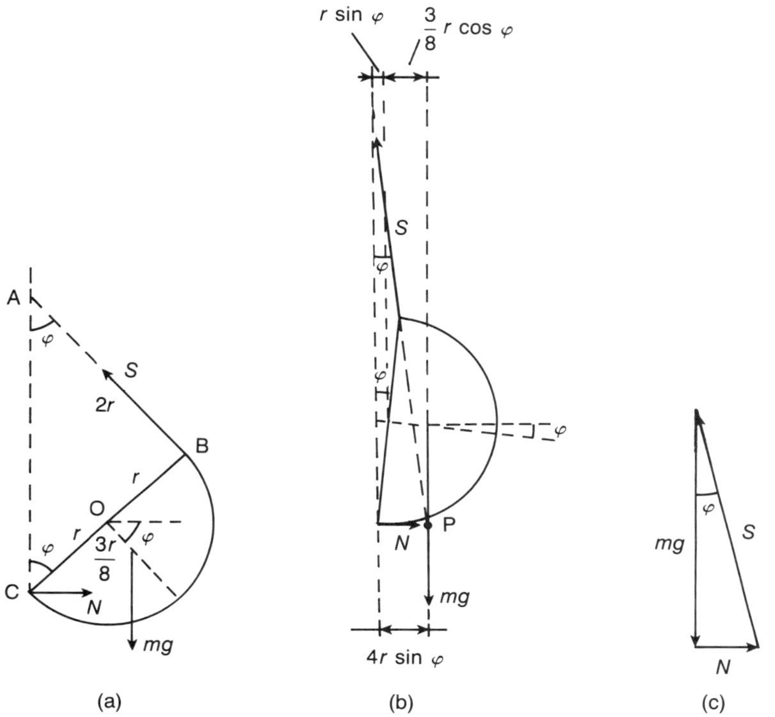

Figure 2.35

Combining Equations (2.22) and (2.23) we have

$$N = \frac{1}{8}\, mg$$

The results are summarized as follows:

$$\text{(a) } \arctan \frac{1}{8} \qquad \text{(b) } \frac{\sqrt{65}}{8}\, mg \qquad \text{(c) } \frac{1}{8}\, mg$$

REMARK

The lines of action of the three forces must intersect at a point P. This can be utilized to solve the problem geometrically. From Figure 2.35(b) it is evident that

$$r \sin \varphi + \frac{3}{8} r \cos \varphi = 4r \sin \varphi \Rightarrow \tan \varphi = \frac{1}{8}$$

Furthermore, the forces must form a right-angled triangle according to Figure 2.35(c). It then follows immediately that

$$S = \frac{mg}{\cos \varphi} = \frac{\sqrt{65}}{8}\, mg \quad \text{and} \quad N = mg \tan \varphi = \frac{1}{8}\, mg$$

Statical indeterminacy

So far we have assumed that the problems considered lead to three simultaneous equations with three unknowns. However, this is not always true.

Let us consider what is called a simply supported beam according to Figure 2.36. From the free-body diagram we see that three unknown forces, V_A, H_A and V_B, are acting. It is an easy matter to determine these from the equations of equilibrium:

$$\begin{aligned} &\leftarrow \quad H_A = 0 \\ &\uparrow \quad V_A + V_B - F = 0 \Rightarrow \\ &\widehat{A} \quad V_B L - F\frac{L}{3} = 0 \end{aligned} \quad \left\{ \begin{aligned} &H_A = 0 \\ &V_A = \frac{2}{3}F \\ &V_B = \frac{1}{3}F \end{aligned} \right.$$

A problem of this type, where all the unknowns may be determined from the equations of equilibrium, is called *statically determinate*.

Now consider the same beam supported at three points as in Figure 2.37. From the free-body diagram of the beam we note that four undetermined forces are involved since a force V_C has been added. The equations of equilibrium are:

$$\begin{aligned} &\leftarrow \quad H_A = 0 \\ &\uparrow \quad V_A + V_B + V_C - F = 0 \\ &\widehat{A} \quad V_C\frac{L}{4} + V_B L - F\frac{L}{3} = 0 \end{aligned}$$

If we try to solve the equations we find as before that $H_A = 0$. Then two equations and three unknowns remain. Obviously, all of the unknowns cannot be determined from the equations of equilibrium. Such a problem is said to be *statically indeterminate*.

You may ask if anything really has changed just because the 'unnecessary' support at C is added. The beam was already in a state of equilibrium, and nothing new has

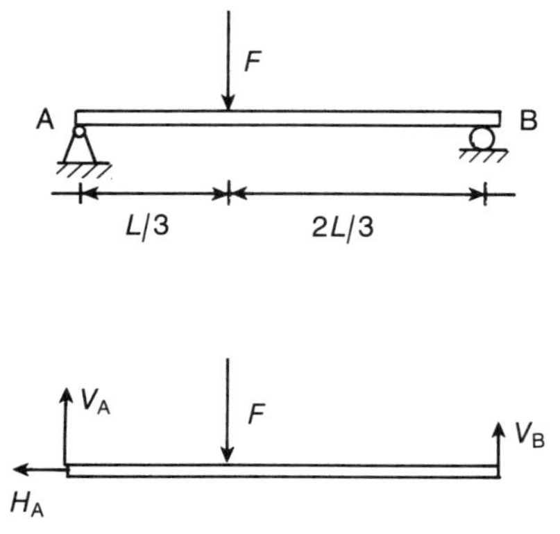

Figure 2.36

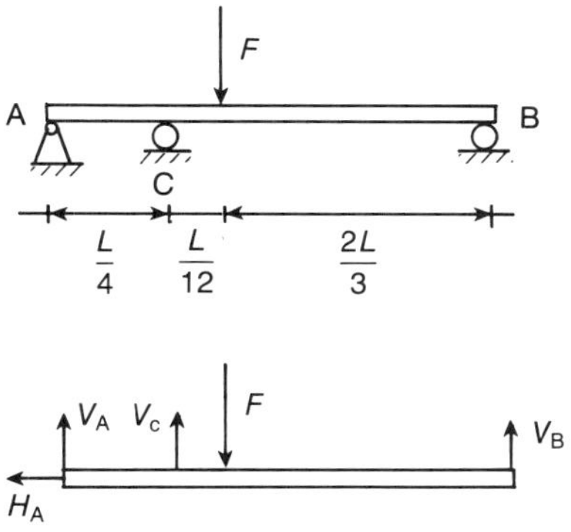

Figure 2.37

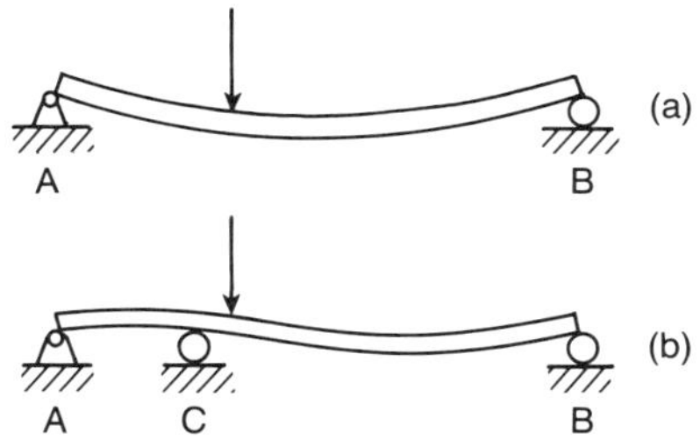

Figure 2.38

happened to the beam – or has it? To understand what takes place here we have to consider the fact that a real beam is not a rigid body.

If the beam is supported at two points it will in a real case be deformed according to Figure 2.38(a). The deformations have, of course, been exaggerated here. If a support C is added the deformation will be approximately as in Figure 2.38(b). Evidently, the addition of the support at C must require that a force V_C is added at the same time as the supports at A and B are partially unloaded. The equilibrium equations, however, are not sufficient to determine the reactions V_A, V_B and V_C, i.e. how the load F is distributed on the three supports.

In a problem that is statically indeterminate it is necessary to take the deformation of the body and its connection to the acting forces into account. This, in turn, requires a knowledge of certain material properties of the body. This type of problem is studied within the subject of strength of materials and is outside the scope of this book.

WORKED EXAMPLE 2.10

Figures 2.39(a) and (b) show two beams, each of mass m, that are loaded by gravitational forces and supported in different ways.

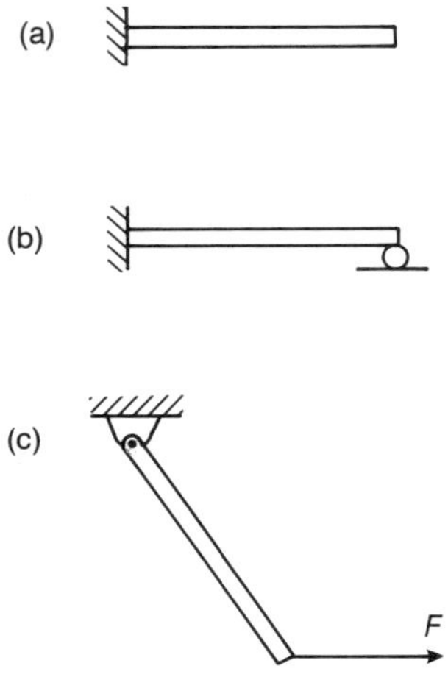

Figure 2.39

Figure 2.39(c) shows a bar of mass m that is free to rotate about a horizontal axis. As well as the gravitational force a horizontal force of magnitude F is applied to the bar. In this case the position of equilibrium is not known in advance.

Investigate for each of the three cases if the problem is statically determinate, in the sense that all external forces and couples acting on the bodies and the position of equilibrium may be determined.

SOLUTION

(a) From the free-body diagram of Figure 2.40 we deduce that the beam is subject to two force components and one couple. It is easy to show that there are three independent equilibrium equations. Hence, the problem is *statically determinate*. It is left to the reader to show that the solution is $H_A = 0$, $V_A = mg$, $M_A = mgL/2$.

(b) We see from the free-body diagram of Figure 2.41 that three undetermined forces and one couple are acting at the supports. From the three equations of equilibrium we find that $H_A = 0$. The other unknowns cannot be determined, however. Thus, the problem is *statically indeterminate*.

(c) The free-body diagram of Figure 2.42 shows that two unknown force components, V and H, and the angle φ are to be determined. From the condition $\Sigma F = 0$ it follows that $V = mg$ and $H = F$. From the moment equation $\Sigma M_A = 0$ we find that $\tan \varphi = 2F/mg$, which leads

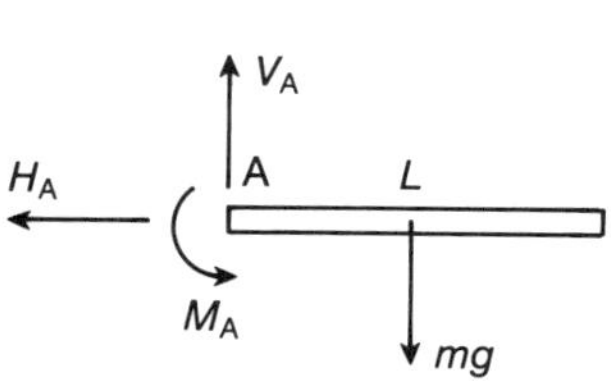

Figure 2.40

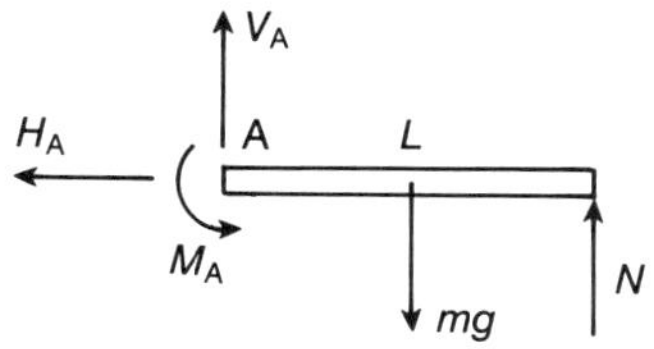

Figure 2.41

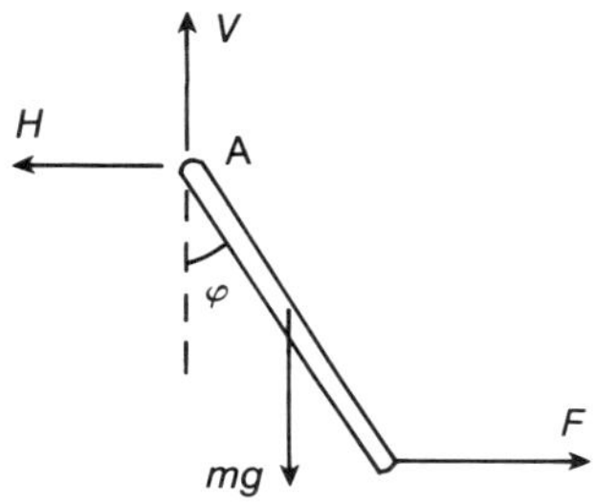

Figure 2.42

to the solutions $\varphi = \varphi_1 = \arctan(2F/mg)$ and $\varphi = \varphi_1 + \pi$. Hence, the problem is *statically determinate*.

Stability

In Worked Example 2.10(c) above we found that two different positions of equilibrium exist. In the special case $F = 0$, Figure 2.43, these are given by $\varphi = 0$ and $\varphi = \pi$. If we place the bar in the equilibrium position $\varphi = 0$ and give it a small disturbance it is reasonable to assume that it will start to oscillate about the position of equilibrium. It will stay in the neighbourhood of this position and the velocity of each point will be small. Such a position of equilibrium is said to be *stable*.

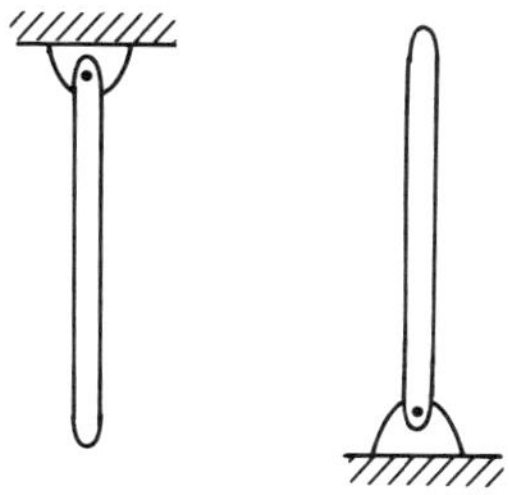

Figure 2.43

If we place the bar in the position $\varphi=\pi$ instead and disturb it slightly it will move away from the equilibrium position in a dramatic way. This type of equilibrium position is called *unstable*. In reality the bar cannot be placed in this position and remain in equilibrium.

We may formulate a general definition of *stability* as follows. If, after a small disturbance, a mechanical system in a position of equilibrium is moving in the neighbourhood of this with small velocities, then the equilibrium position is *stable*. In the opposite case the position of equilibrium is *unstable*.

As an additional illustration of the concept of stability we consider a wire that is bent to the shape shown in Figure 2.44. The wire is located in a vertical plane. The tangent of the curve described by the wire is horizontal at $x=x_1$, $x=x_2$, $x=x_3$ and for all values of x such that $x_4 \leqslant x \leqslant x_5$. A small pearl is threaded on the wire and is free to move along it without friction.

The pearl is only affected by the gravitational force and a normal force exerted by the wire. The equations of equilibrium are satisfied if and only if the normal force is vertical, i.e. when the pearl is located at any of these points where the tangent of the wire is horizontal.

Let us assume that we try to place the pearl at point $x=x_1$ and that we disturb it slightly, e.g. by displacing it sideways. We realize intuitively that the pearl will swing back and forth in the vicinity of the equilibrium position with a small velocity. The equilibrium position is thus *stable*.

If we place the pearl at point $x=x_2$ instead and disturb it, it will move far away from the equilibrium position. The same is true for the inflection point $x=x_3$. The equilibrium positions $x=x_2$ and $x=x_3$ are thus *unstable*.

In the interval $x_4 < x < x_5$ the wire is horizontal. The pearl can be placed anywhere, but the equilibrium is nowhere stable. If the pearl is given a small initial velocity it will move far away, albeit with a small velocity. Hence, the equilibrium position is *unstable*. In a practical case, however, it will stay in the neighbourhood because the wire is never perfectly smooth since we cannot avoid friction completely. This type of equilibrium is called *neutral*.

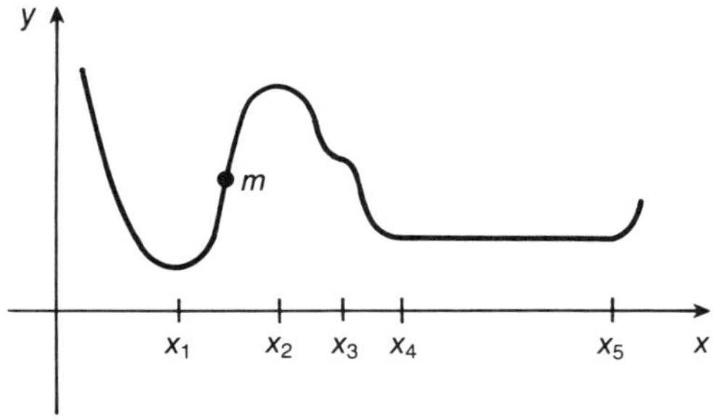

Figure 2.44

2.2.4 Systems of rigid bodies

So far in Section 2.2 the conditions of equilibrium have been applied to material systems consisting of one single rigid body. In many cases, however, we are interested in systems consisting of two or more rigid bodies which are connected to each other in some way. In some cases it may be sufficient to consider the entire system as one unit. In other cases it may be necessary to study each different portion of the system, either because we are interested in the forces exerted by the different portions on each other, or because an analysis of the entire system will lead to a statically indeterminate problem. To illustrate this we consider the following example.

WORKED EXAMPLE 2.11

A system consists of two bars, AB and BC, of equal length, Figure 2.45. The masses of the bars are m_1 and m_2, respectively. The system is located in the vertical plane. Determine all reactions exerted on the bars at the smooth joints A, B and C.

SOLUTION

First we try to isolate the entire system as shown by the free-body diagram in Figure 2.46. We then have the following equations of equilibrium:

$$\overset{\frown}{A} \quad V_C \cdot 2L \cos\alpha - m_1 g \frac{L}{2} \cos\alpha - m_2 g \frac{3L}{2} \cos\alpha = 0 \tag{2.24}$$

$$\uparrow \quad V_A + V_C - m_1 g - m_2 g = 0 \tag{2.25}$$

$$\rightarrow \quad H_A - H_C = 0 \tag{2.26}$$

From Equation (2.24) we find that $V_C = m_1 g/4 + 3m_2 g/4$ which, when inserted into Equation (2.25), yields $V_A = 3m_1 g/4 + m_2 g/4$.

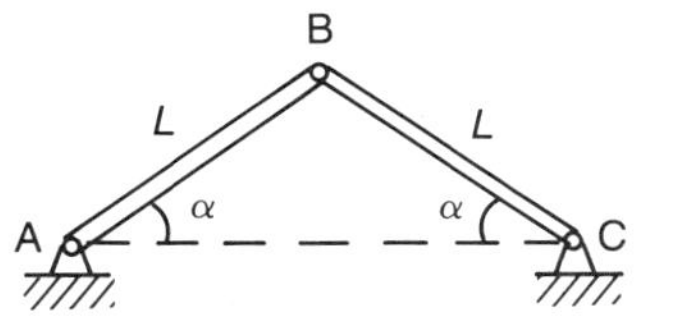

Figure 2.45

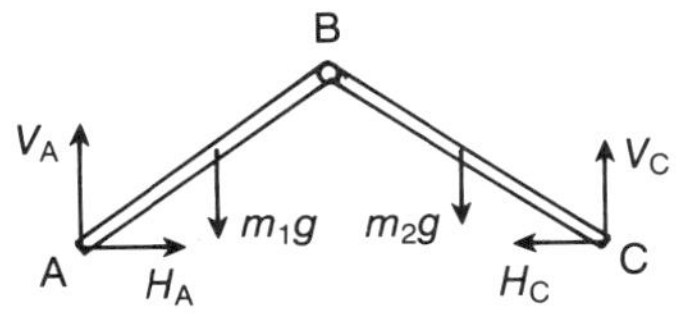

Figure 2.46

If we only consider both bars as one system we are not able to determine

- the forces H_A and H_C (the problem is statically indeterminate), and
- the forces exerted by the bars on each other at B (internal forces to the system as a whole).

To proceed we divide the system into two portions, i.e. we isolate one bar at a time, Figure 2.47. As well as the forces acting on the composite system we must now introduce the forces exerted by the bars on each other at joint B. These are external forces when the separate bars are considered. It is important that we are careful in applying Newton's third law. This means that forces V_B and H_B are directed in the opposite sense when they are applied to bar AB compared with the case where they are applied to bar BC.

For each one of the bars we have three equilibrium equations, which together with Equations (2.24)–(2.26) above give a total of nine equations while the number of unknowns is six. The first three equations, however, may be derived from the latter six. Thus we only have six *independent* equations, and the problem *may* be statically determinate.

Considering bar AB we have the following equilibrium equations:

$$\uparrow \quad V_A - m_1 g - V_B = 0 \tag{2.27}$$

$$\rightarrow \quad H_A - H_B = 0 \tag{2.28}$$

$$\overset{\frown}{A} \quad H_B L \sin\alpha - V_B L \cos\alpha - m_1 g \frac{L}{2} \cos\alpha = 0 \tag{2.29}$$

These equations together with the results above give the answer to the problem:

$$H_A = H_B = H_C = \frac{1}{4}(m_1 + m_2) g \cot\alpha$$

$$V_A = \frac{3}{4} m_1 g + \frac{1}{4} m_2 g$$

$$V_B = \frac{1}{4}(m_2 - m_1) g$$

$$V_C = \frac{1}{4} m_1 g + \frac{3}{4} m_2 g$$

It is recommended as a check that the student verifies that the equations of equilibrium of bar BC are satisfied by the solution.

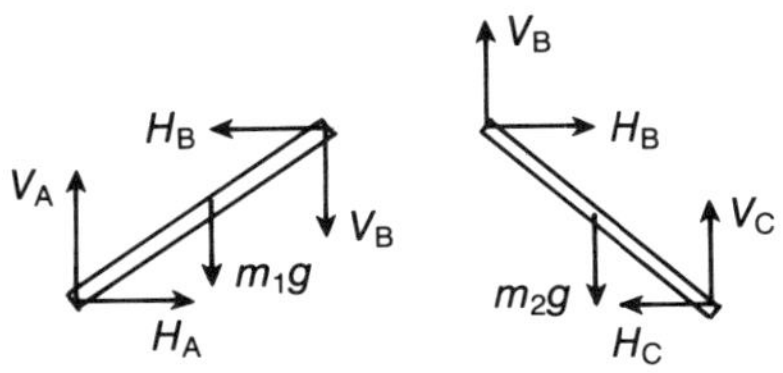

Figure 2.47

As we have seen above, solving a problem where a more complex system of bodies is involved is relatively straightforward. No additional theoretical background is required by the student. All there is to do is to isolate each separate body and draw the corresponding free-body diagram. However, it is absolutely necessary to have a complete understanding of which forces are external to the body considered. Furthermore, care when using Newton's third law is vitally important. Once the correct free-body diagram is obtained all that is left is to write down the equations of equilibrium and solve them. However, it should be pointed out that the equations may normally be chosen in a large number of different ways.

WORKED EXAMPLE 2.12

Three identical smooth cylinders, each of mass m, are held together by an elastic cord. The whole system is placed on a smooth table, Figure 2.48. What is the minimum value of the tension required in the cord?

SOLUTION

If the tension is not sufficiently large the two lower cylinders will slide apart, thereby changing the geometry of the problem. As long as the cylinders are in contact, however, the free-body diagrams of the cylinders shown in Figure 2.49 are relevant to the problem. Here all external forces acting on each one of the cylinders have been introduced. Besides, we have utilized the symmetry of the problem.

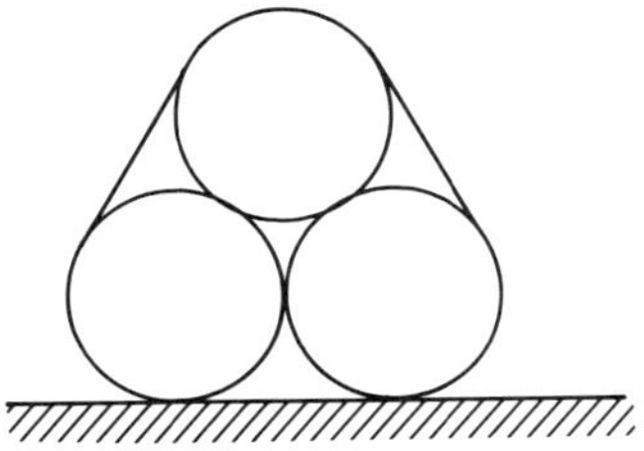

Figure 2.48

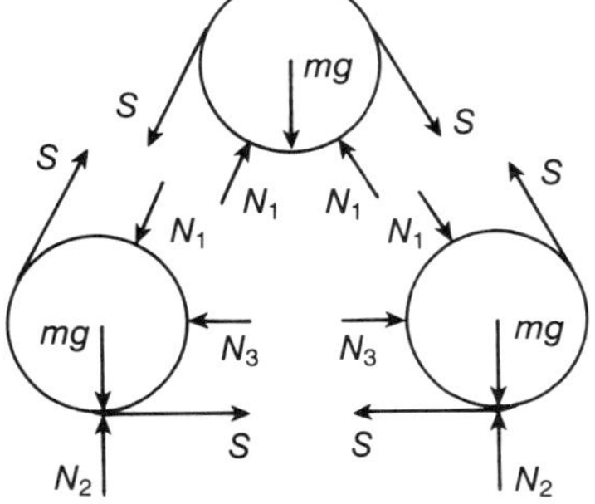

Figure 2.49

To understand why the tension has the same value (S) everywhere we can study the equilibrium equation $\Sigma M = 0$ with respect to the symmetry axis of each one of the cylinders.

As we have assumed that the cylinders are in contact, all normal forces (N_1, N_2, N_3) must be positive. The limiting case where the two lower cylinders are barely in contact corresponds to $N_3 = 0$. Hence, the condition that has to be satisfied is $N_3 \geqslant 0$.

For the top cylinder we now have:

$$\uparrow \quad 2N_1 \cos 30^\circ - 2S \cos 30^\circ - mg = 0 \tag{2.30}$$

For the lower left cylinder we find that

$$\leftarrow \quad N_3 - S \cos 60^\circ - S + N_1 \cos 60^\circ = 0 \tag{2.31}$$

Eliminating N_1 from Equations (2.30) and (2.31) we obtain

$$S = N_3 + \frac{mg}{2\sqrt{3}}$$

From the condition $N_3 \geqslant 0$ it follows that $S \geqslant mg/2\sqrt{3}$.

The minimum allowable value of the tension is thus $mg/2\sqrt{3}$.

WORKED EXAMPLE 2.13

Determine the cutting forces applied to the bolt K shown in Figure 2.50 if two forces, each of magnitude F, are applied to the handles of the shears. Neglect the influence of gravity.

SOLUTION

Isolating certain portions of the shears we obtain the free-body diagrams shown in Figure 2.51.

Our purpose is to determine force P. A moment equation for AB yields

$$\overset{\frown}{A} \quad Pa - Rb = 0$$

In the same way for BD

$$\overset{\frown}{C} \quad Fd - Rc = 0$$

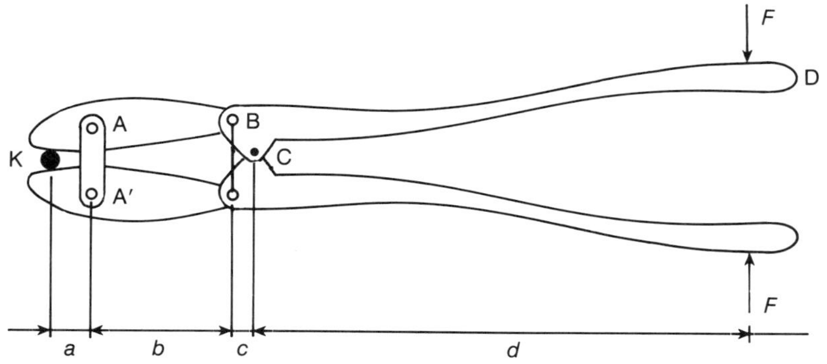

Figure 2.50

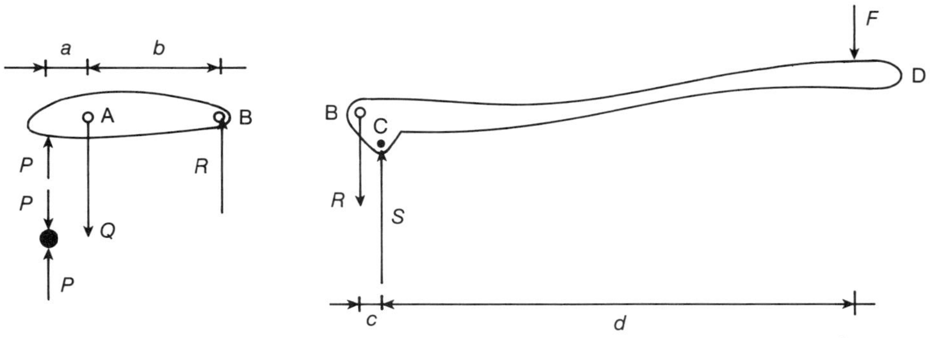

Figure 2.51

Eliminating R we obtain the answer:

$$P = \frac{bd}{ac} F$$

REMARK

It might not be intuitively clear that the directions of the forces are those shown in the free-body diagrams. This can be understood in the following way.

Only two forces are exerted on the bolt. These must then be of equal magnitude and opposite sense according to the discussion on two-force systems in Section 2.2.1. This is also true for AA′. The directions of forces P and Q are thereby determined. If AB is in equilibrium force R must have the direction shown. Finally, equilibrium of BD requires that the direction of force S is as shown. This also follows from the symmetry of the shears.

PROBLEMS

○ **2.1** Determine the force exerted on the pulley by its support at the axis.

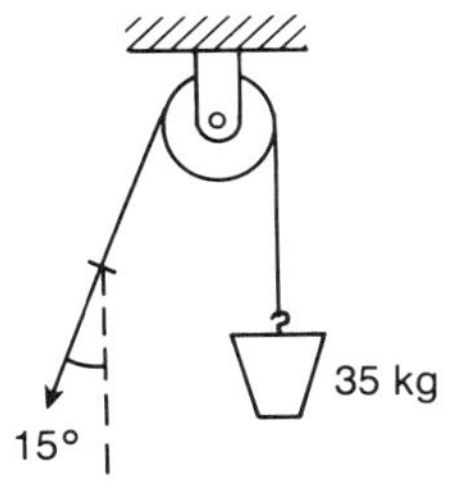

○ **2.2** A ball of mass m is suspended by a cord and supported by a smooth wall. Determine
(a) the angle φ,
(b) the tension in the cord,
(c) the normal force acting at the wall.

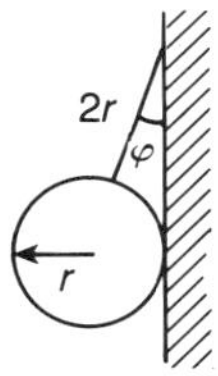

2.3 In the figure overleaf a painter who weighs 70 kg is standing on a 10 kg ladder. The ladder is supported by a smooth wall and a rough floor.
(a) Determine the contact forces at A and B.
(b) What are the values of these forces if one more person, who weighs 55 kg, is standing half-way up the ladder?

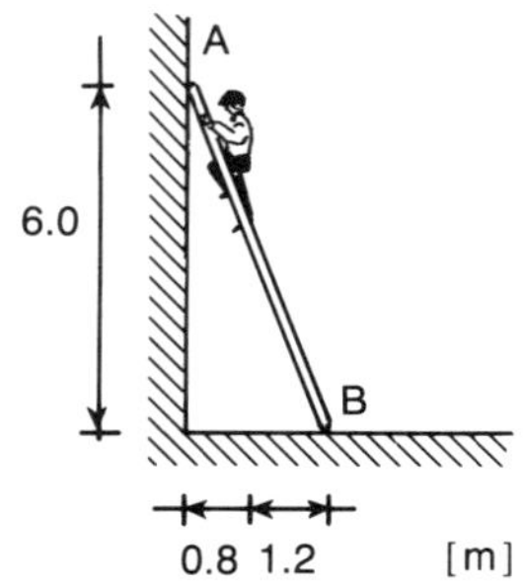

○ **2.4** The mass of the painting shown is 4.0 kg. The cord can withstand a tension of 35 N. What is the minimum allowable value of the distance h?

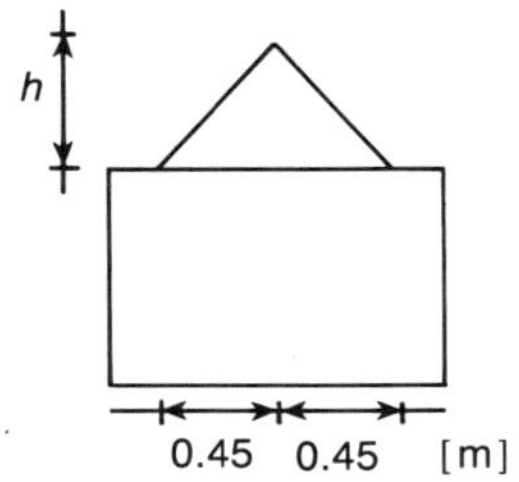

2.5 Determine the tensions in all the cables in both cases assuming that $m_1 = 1.0$ kg and $m_2 = 1.2$ kg.

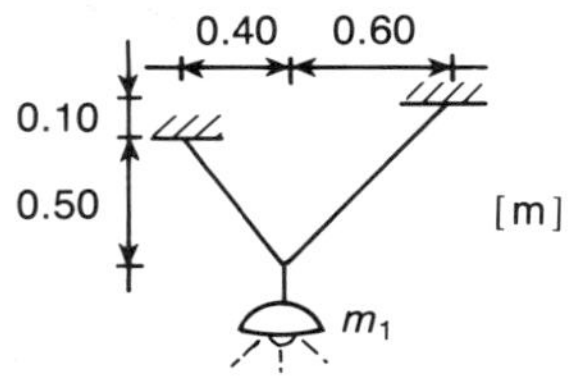

(a)

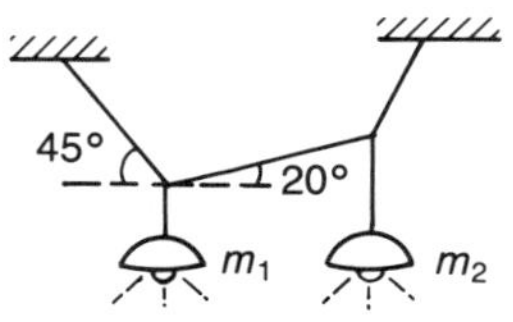

(b)

○ **2.6** A cylinder of mass m is supported by a smooth incline and a smooth wall.

(a) Determine the reactions at A and B.

(b) Determine the angle α knowing that the reaction at B is twice the one at A.

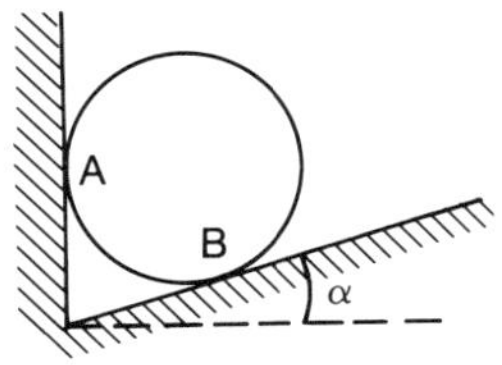

○ **2.7** Determine all the reactions on the beams shown in cases (a)–(d).

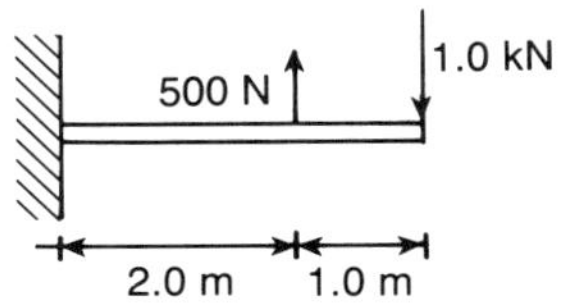

(a)

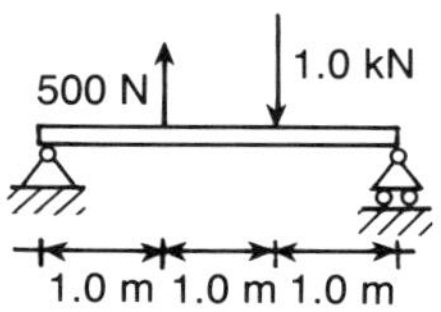

(b)

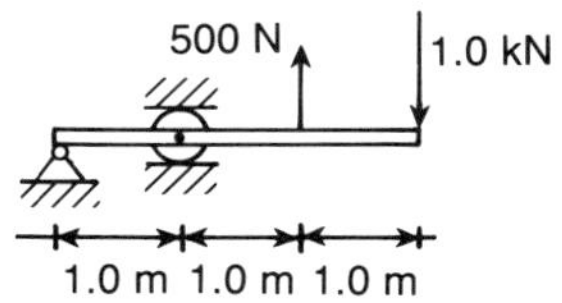

(c)

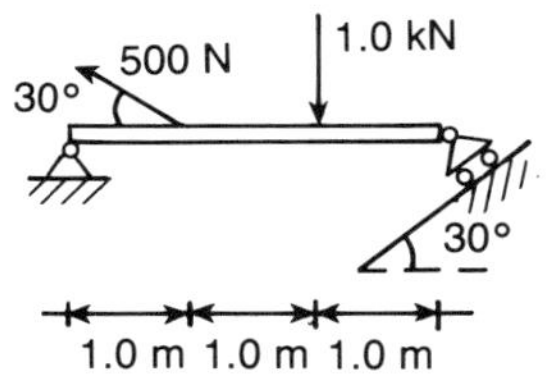

(d)

○ **2.8** A gear box which is mounted on a beam is subjected to two couples as shown. Determine the reactions at A and B.

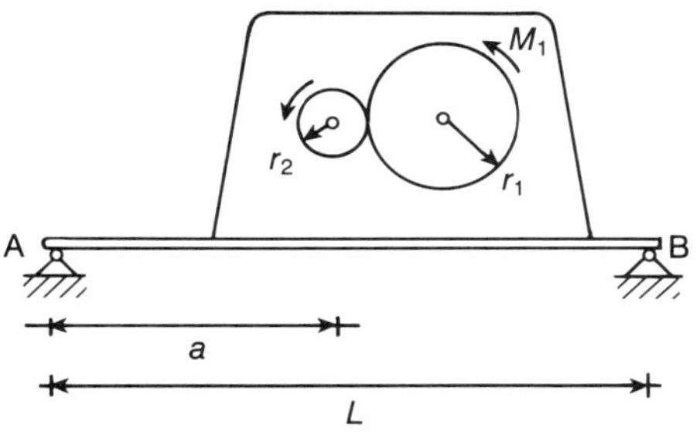

2.9 A truck platform of mass 1200 kg (including the load) is supported by two hydraulic cylinders (only one of them is visible here). Determine the force exerted by each one of the hydraulic cylinders on the platform.

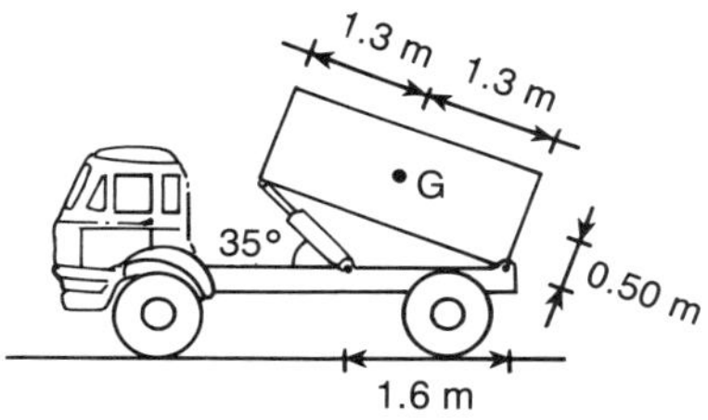

2.10 Determine the tension in the cord and the reaction at point O for the three cases (a)–(c). The weights of the beams and the pulleys may be neglected.

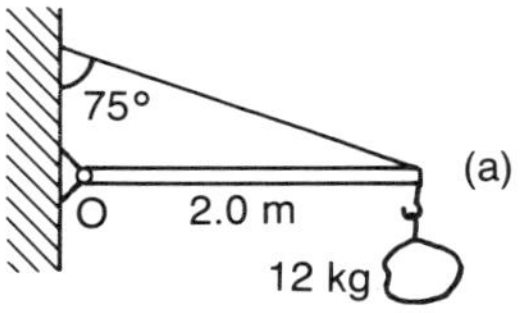

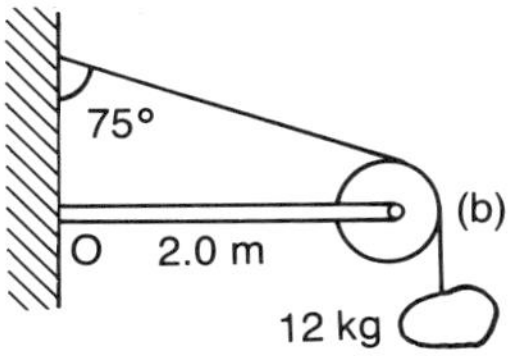

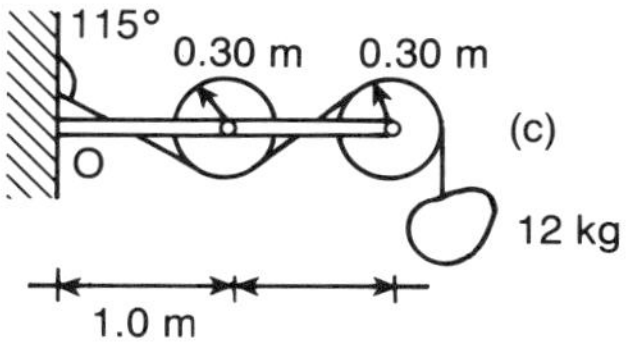

2.11 A block of mass m is placed on a smooth horizontal surface. Two springs of spring constants k and $2k$ and unstretched lengths b_1 and b_2, respectively, are attached to the block in three different ways as shown. The block is in equilibrium. If a force of magnitude F is applied to the block in the direction of the springs the equilibrium position is displaced a distance Δ in the direction of the force. Determine Δ for cases (a)–(c).

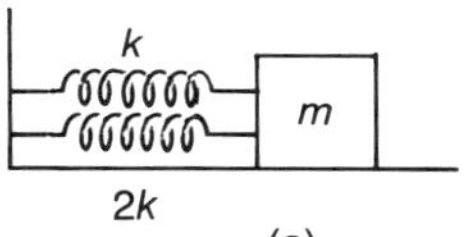

(a)

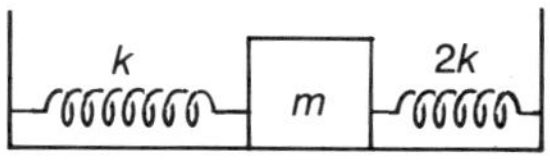

(b)

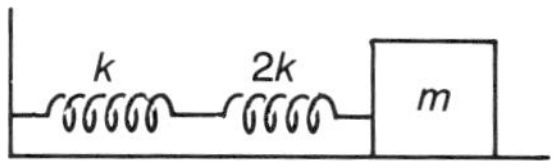

(c)

2.12 A belt is running over two wheels that are mounted on a horizontal disk as shown overleaf. The tension in the tape is 10 N. Determine

(a) the force exerted by the spring at A,
(b) the reaction at B.

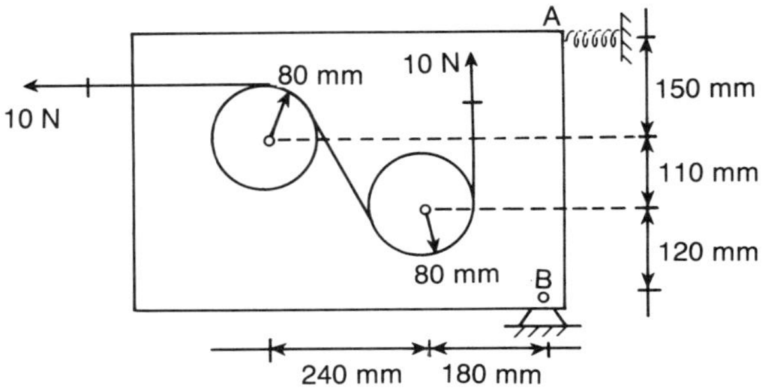

2.13 A crate of mass 400 kg is suspended by two cables. Assuming that the crate is in equilibrium, determine
(a) the angle α,
(b) the tensions S_1 and S_2.

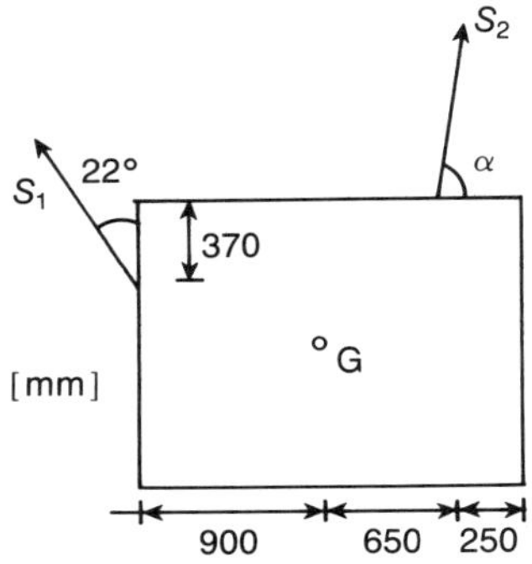

2.14 The jib of a crane is supported by a hinge at A and a cable BC. A load of mass 2.0 tonnes is supported by another cable running over a pulley at B. The weights of all portions of the crane may be neglected compared with the weight of the load. Determine
(a) the tension in cable BC,
(b) the magnitude of the reaction at A.

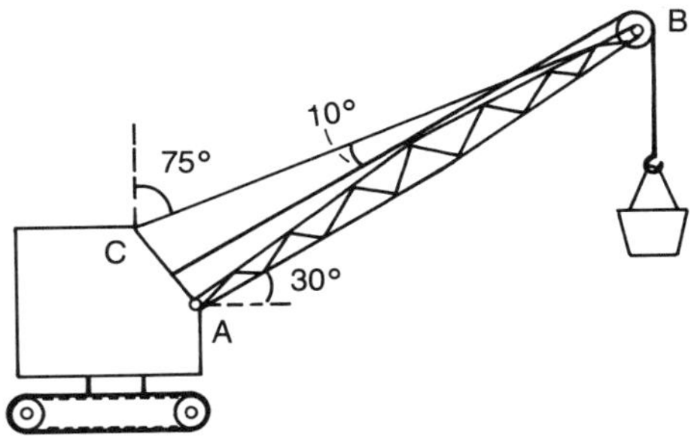

○ **2.15** A car of mass 1300 kg is being unloaded from a ship. A man is pulling the rope AC in order to adjust the position of the car. Determine the force that must be exerted by the man and the tension in the wire AB.

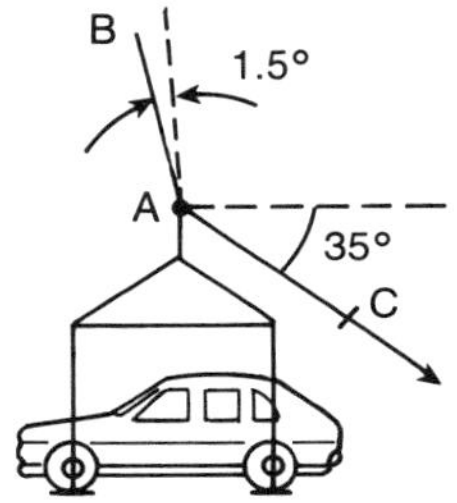

2.16 An electric motor of mass 90 kg is used to drive a belt pulley as shown. When the motor is not operated the tensions in the belt, S_1 and S_2, are both equal to 1.6 kN. The motor is mounted on a platform of negligible weight. Determine the force and the couple that must be exerted on the platform at the support A.

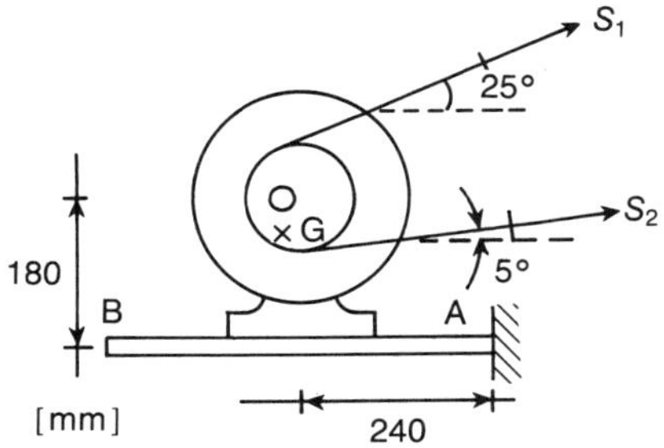

○ **2.17** In cases (a)–(g) a load of mass 200 kg is supported by a cable and a number of pulleys, all of which are of negligible weight. Determine the tension in the cable in each case. Compare case (b) with case (f) and case (d) with case (g).

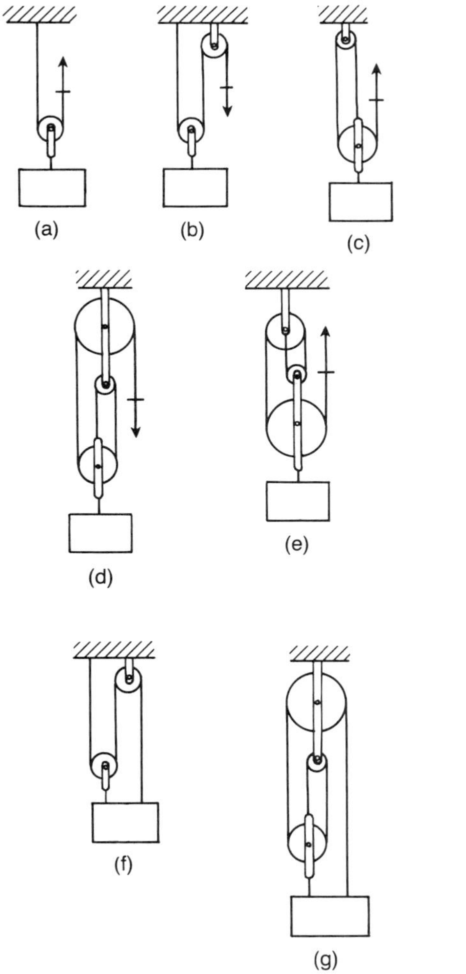

2.18 Determine the force F (or the couple M) required to make the wheel (mass 14.0 kg, radius 0.32 m) roll over the kerb in cases (a)–(c).

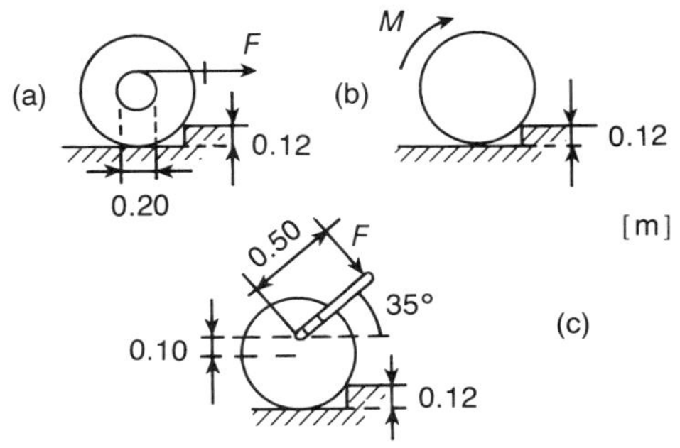

2.19 Determine all the tensions and reactions exerted on the bent beam AB in the four cases shown.

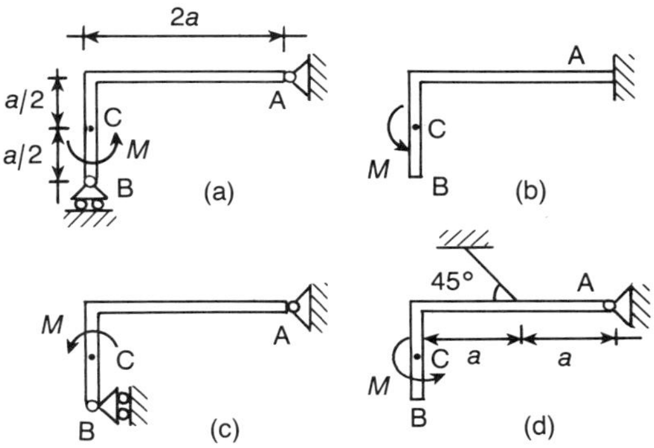

2.20 Solve Problem 2.19 replacing the couple M by a force F applied at point C directed to the right.

2.21 Determine the reactions at A and B. All bars are of negligible weight; the effects of friction may be neglected.

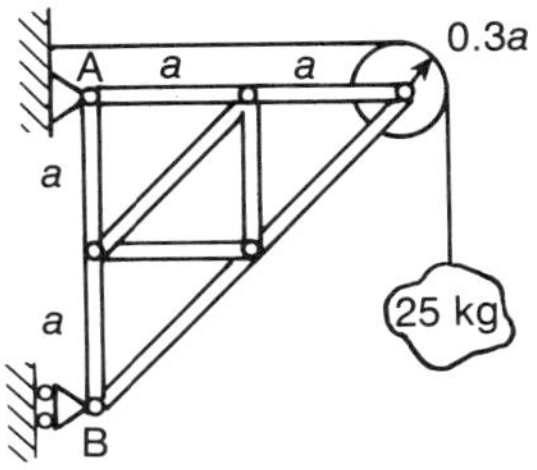

(a)

continued overleaf

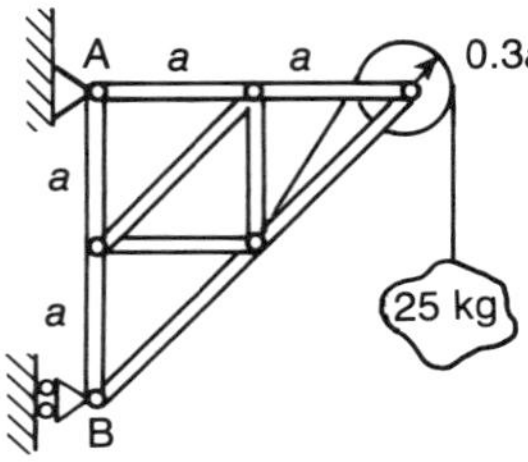

(b)

2.22 Determine the angle φ at equilibrium for cases (a)–(e) assuming that $0 < \varphi < \pi/2$. The effects of friction may be neglected.

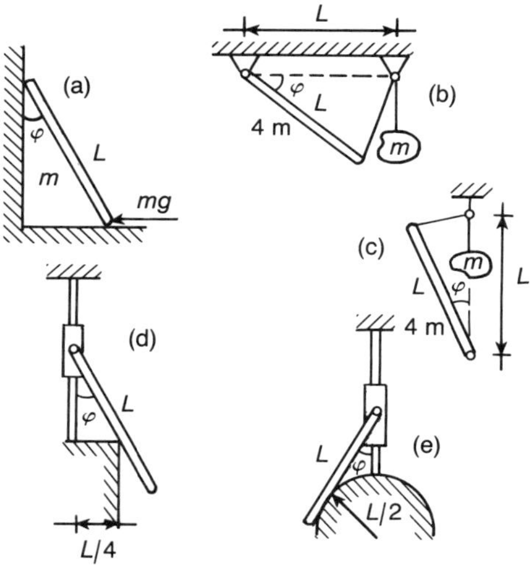

◎ **2.23** The bar AD of mass 10 kg is lifted from its supports using the cables AC and BC of lengths 300 mm and 500 mm, respectively. Determine the tensions in AC and BC.

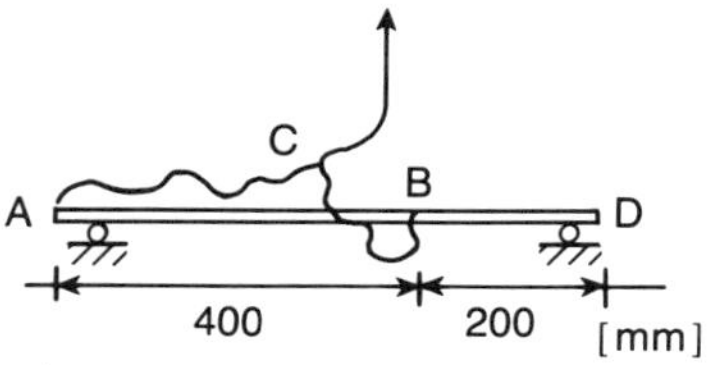

○ **2.24** Two identical smooth cylinders, each of weight Q, are supported by two 45° inclines as shown. Determine the normal forces at A, B, C and D.

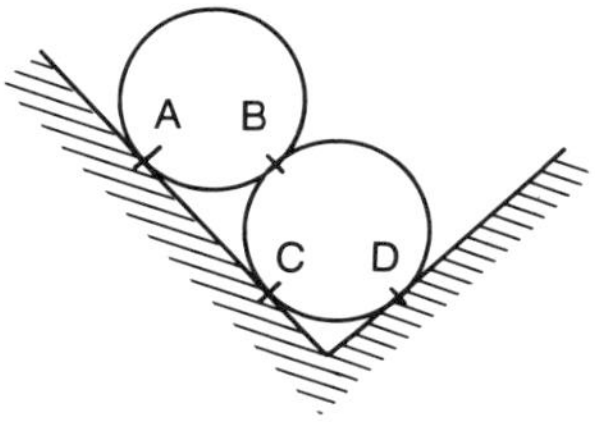

2.25 Determine the force F that must be applied to maintain equilibrium.

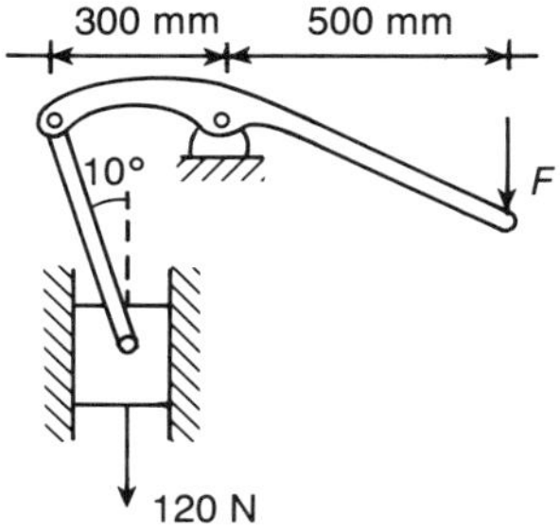

2.26 A circular disk of mass m and radius r and a bar of mass m are connected as shown. Determine the horizontal and vertical components of the reactions at joints A, B and C.

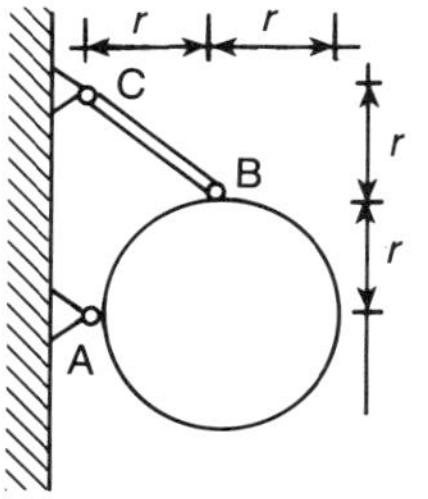

2.27 Determine the reactions at A and B in cases (a)–(e). The weights of all bars may be neglected.

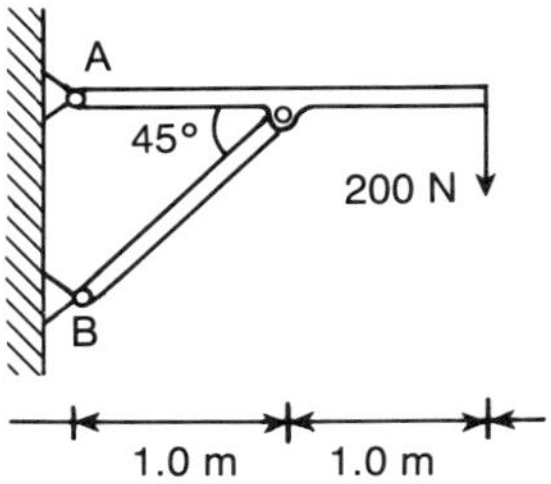

(a)

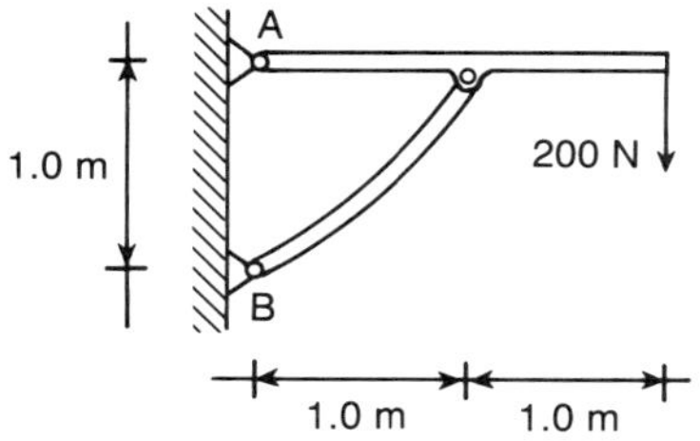

(b)

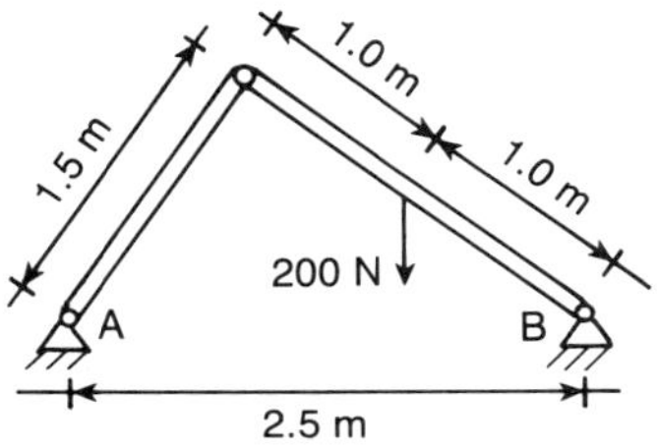

(c)

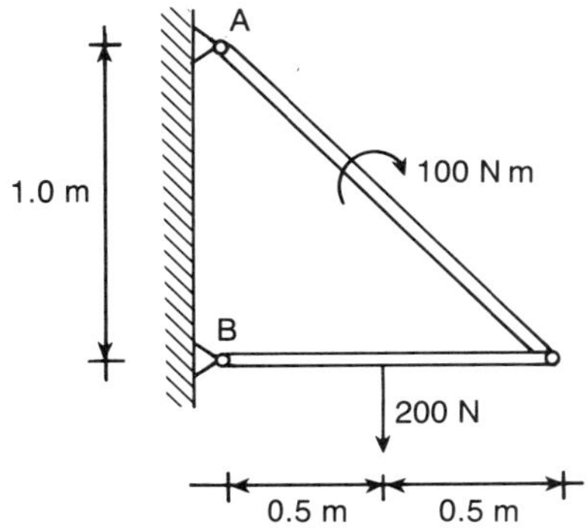

(d)

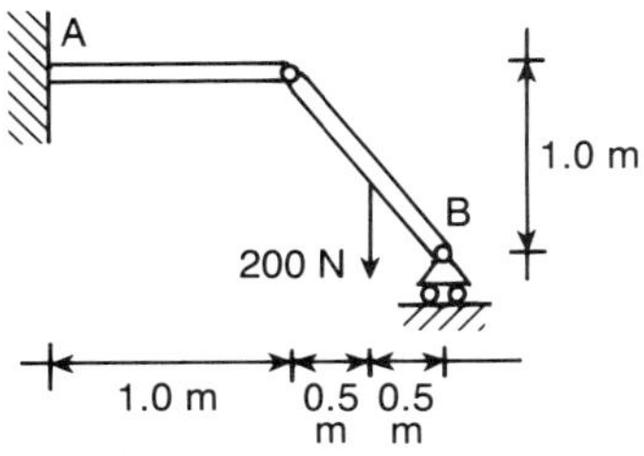

(e)

2.28 Determine the reactions exerted by the supports on the beams in cases (a)–(d).

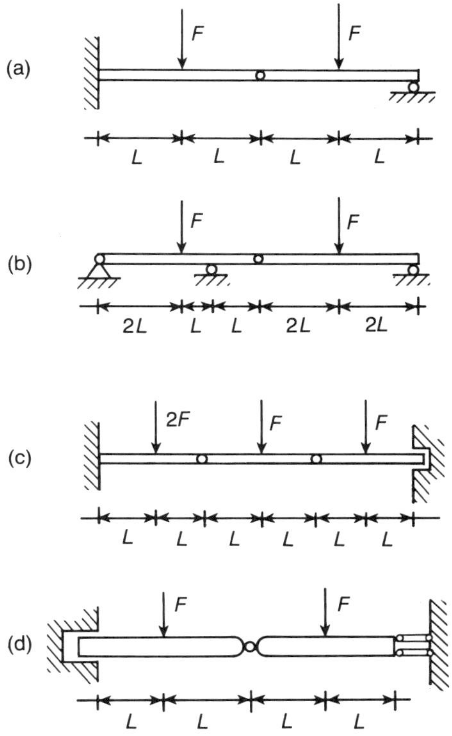

2.29 Determine the reactions at A and the force supported by the pin at B in cases (a)–(c).

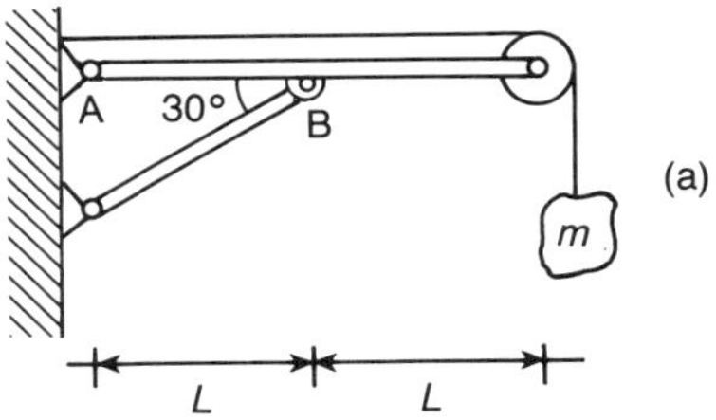

continued overleaf

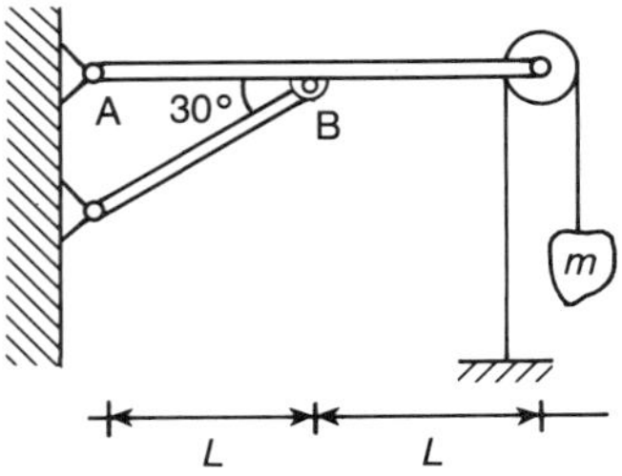

(b)

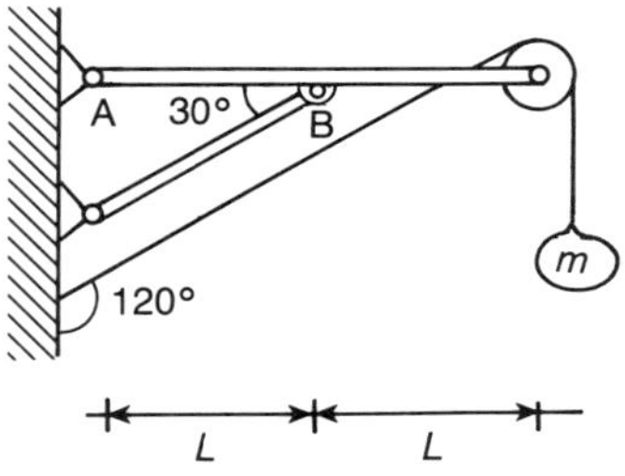

(c)

2.30 The piston shown is subjected to a force $P = 5.7$ kN. Determine the couple M required to maintain equilibrium.

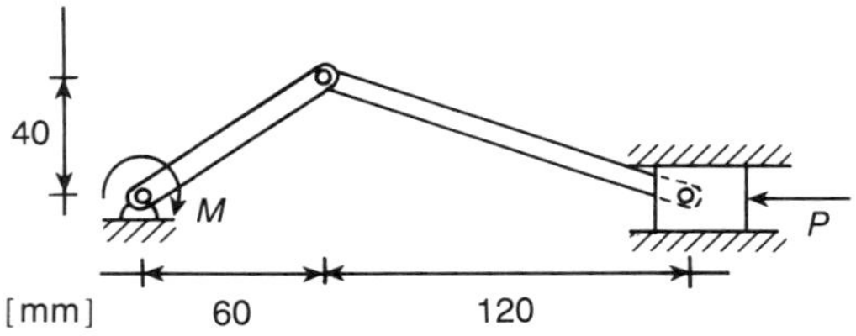

2.31 The mass of the boy shown is 70 kg. The mass of the seat is 6 kg. The masses of all other components may be neglected. Determine

(a) the tensions S_1, S_2 and S_3,

(b) the normal force exerted by the seat on the boy.

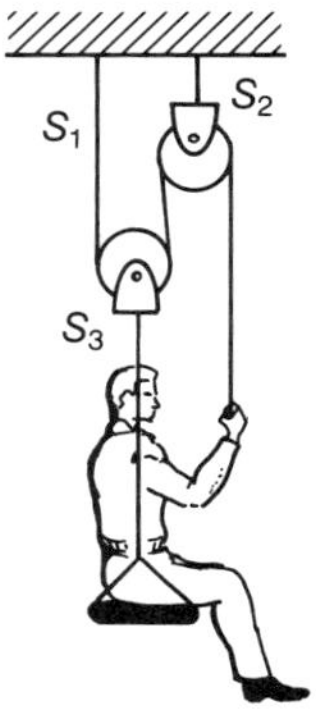

◎ **2.32** Two different devices for compressing aluminium cans are shown here. In both cases AB = 1.5 m, BC = 0.5 m and CD = 0.4 m. The can collapses when it is subjected to a horizontal force of 500 N. What is the vertical force that has to be applied at A if $\varphi = 30°$?

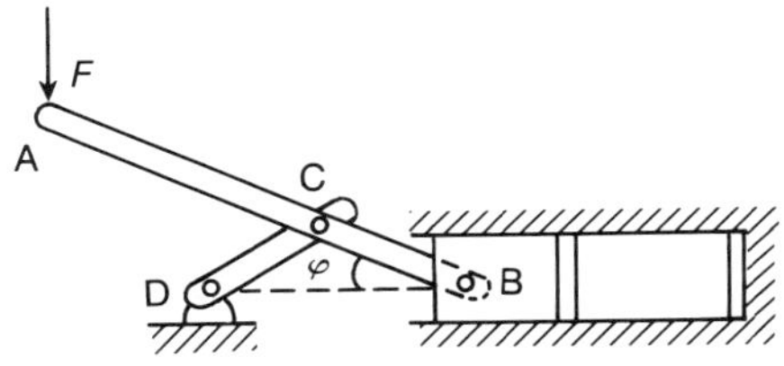

(a)

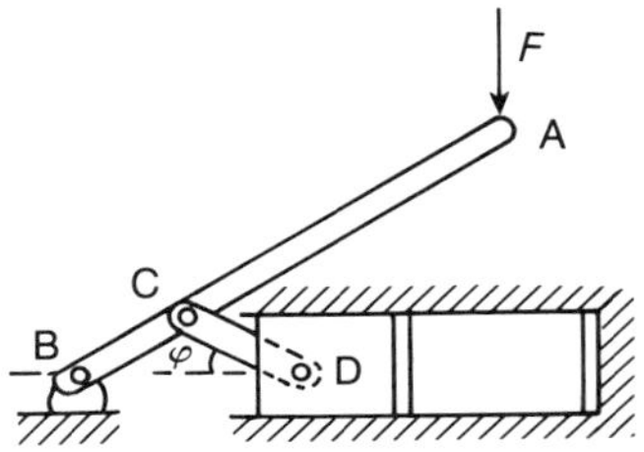

(b)

2.33 Determine the horizontal and vertical components of the forces exerted on the bars at A and C.

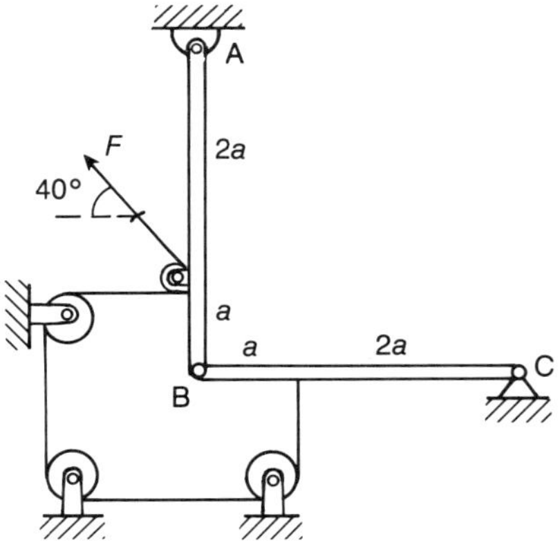

2.34 The device shown is used to raise a load C of mass 450 kg. Determine
(a) the force F required,
(b) the reactions supported at A and B.

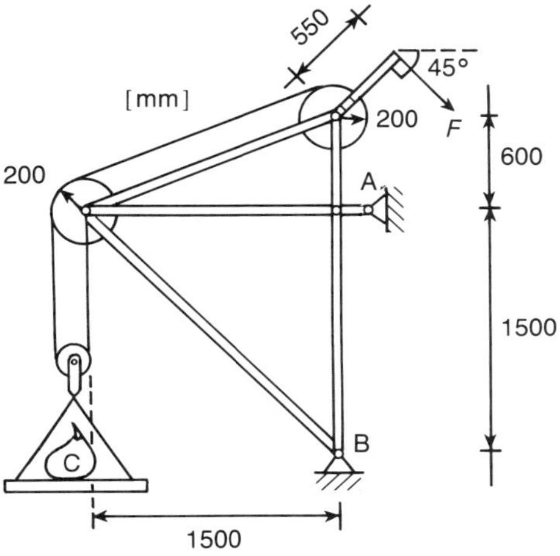

2.35 Determine the reactions at A and B.

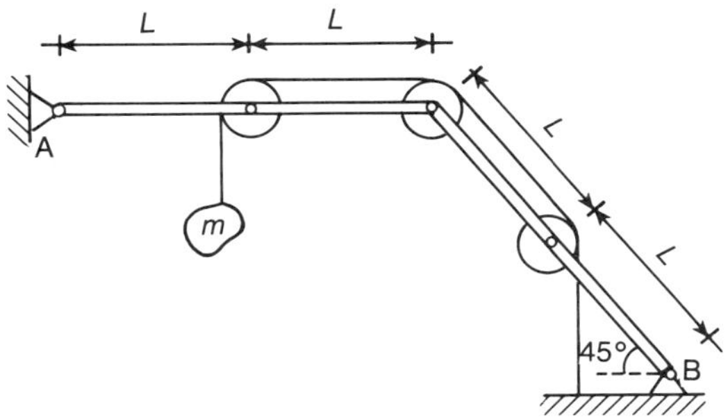

○ **2.36** Determine the couple M required to raise the 450 kg load.

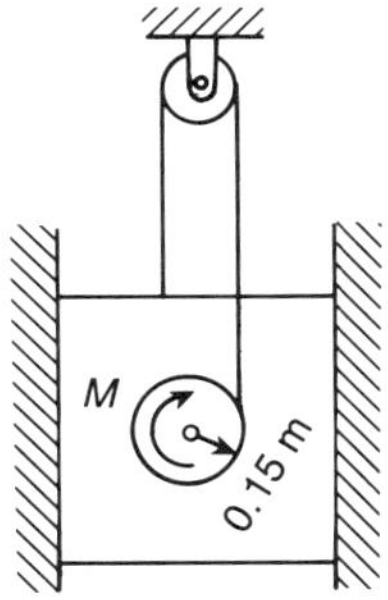

2.37 Determine the ratio of the forces acting on the bolt and the force P. Also determine the reaction at A in terms of P.

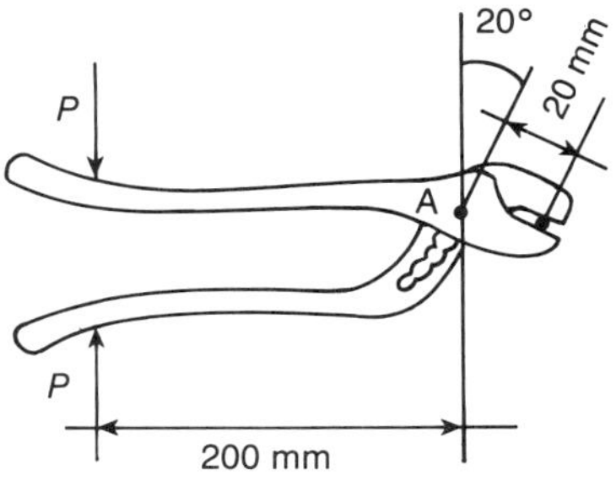

2.38 Determine the cutting forces exerted on the wire A (the handles are *not* joined at B).

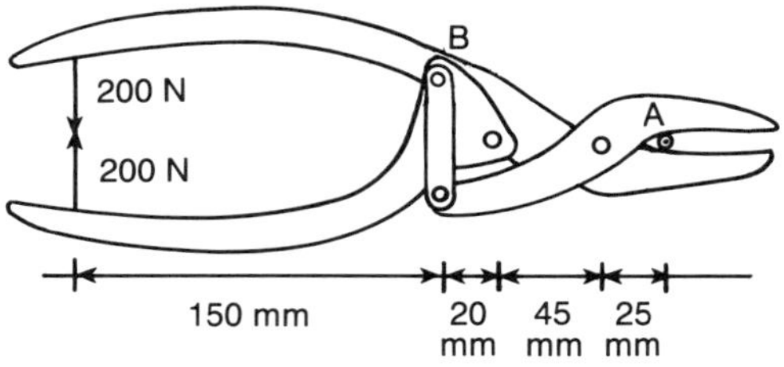

2.39 A pipe wrench of negligible weight consists of two rigid portions connected by a pin at B. Assuming that the wrench does not slip with respect to the pipe, determine the force supported by the pin at B.

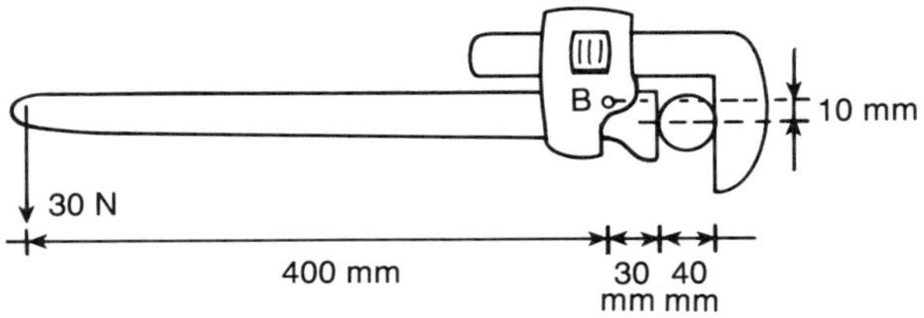

◎ **2.40** A 4.0 kN force is transmitted from the ground to the chassis of a car by the suspension shown here. Determine
(a) the spring force,
(b) the reactions at A and B.

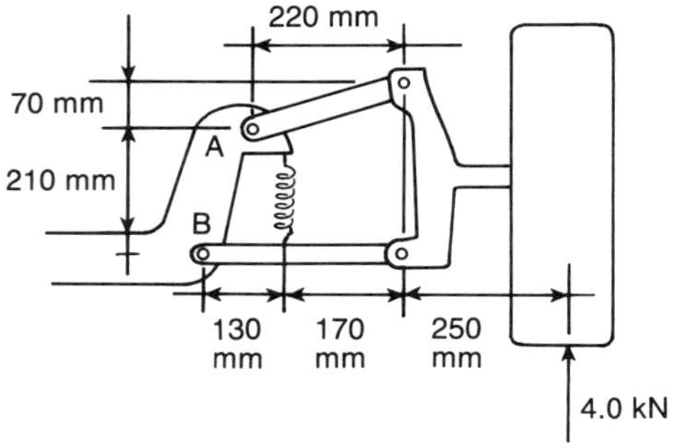

○ **2.41** A car of mass 1400 kg is parked on an uphill road. The brakes are applied on all wheels.
(a) Determine the normal force exerted on each of the wheels.
(b) Is it possible to determine the corresponding friction forces?
(c) What happens if the brakes are applied on the front wheels only?
(d) What happens if the car is parked on a downhill road of the same slope with the brakes applied to the front wheels only?

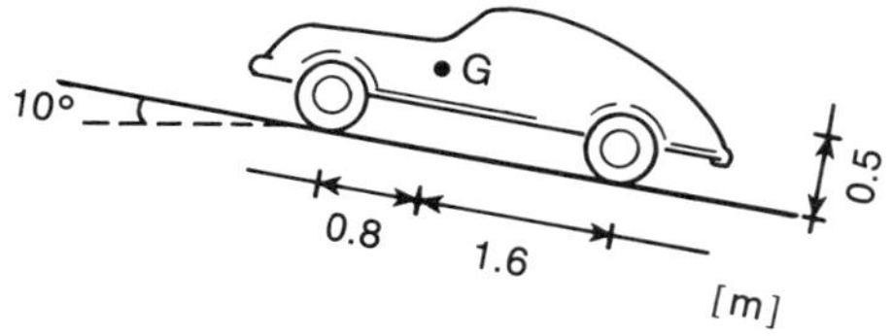

◎ **2.42** The scoop of a tractor is supported by two identical mechanisms; only one is visible here. The scoop carries a load of 700 kg with its centre of gravity at G. The weights of all other parts may be neglected. Determine the reactions at joints A, C, D, E and F.

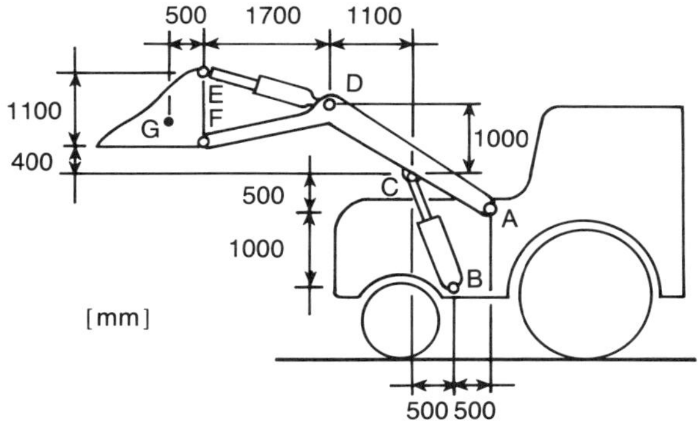

2.43 A planetary gear system is shown opposite. The radius of the central gear A is 35 mm, the radii of the planetary gears B, C and D are 40 mm each and the radius of the outer gear E is 115 mm. The gear teeth are not shown here. A couple $M_A = 200$ N m clockwise is applied to gear A. Assuming equilibrium, determine
(a) the couple that must be applied to the spider,
(b) the couple that must be applied to the outer gear.

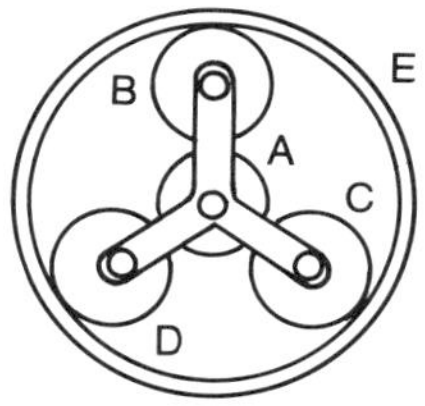

◎ **2.44** Three bars of the same weight q per unit length are connected by smooth joints to form a right-angled triangle ABC, which is suspended by a cord. A load is applied at point B in order to make bar BC horizontal. Determine the horizontal and vertical components of the force exerted on bar AC by bar BC.

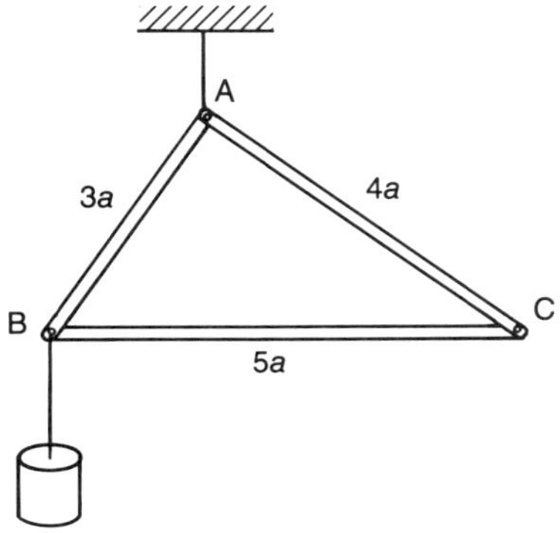

2.45 A cylindrical pipe of radius r is placed on a table. Two smooth balls of weights Q_1 and Q_2 and radii r_1 and r_2, respectively, are placed inside the pipe as shown. What is the minimum value of the weight of the pipe to prevent tipping?

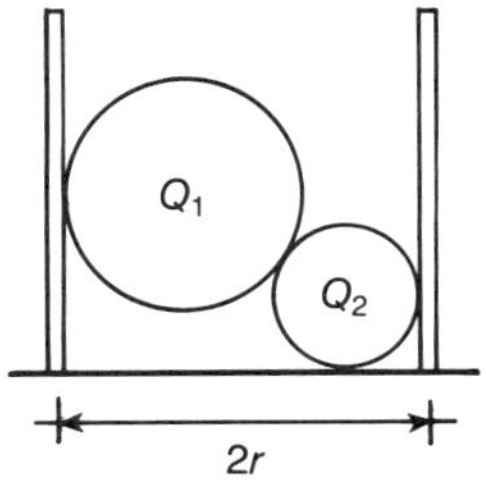

◎ **2.46** Four identical bricks are piled up so that each one sticks out as far as possible without falling.
(a) Determine the distances x_1, x_2 and x_3.
(b) What is the distance x_n if $(n+1)$ bricks are piled in the same way?
(c) If bricks are piled up in this way from two opposite directions, we obtain what is considered to be the prototype of the Gothic arch. What is the maximum width of such an arch if there is an unlimited supply of bricks? For simplicity you may neglect the fact that the two halves support each other at the top.

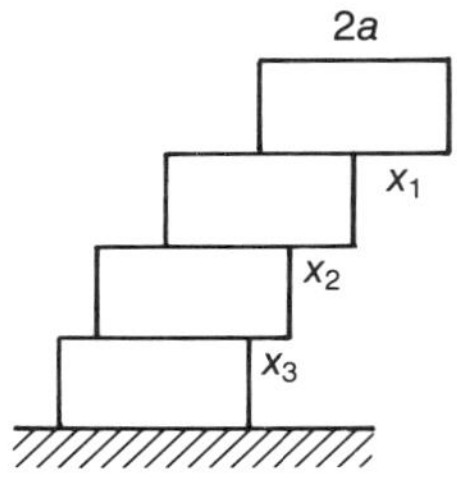

2.3 Equilibrium in three dimensions

2.3.1 Conditions of equilibrium

The conditions of equilibrium, Equations (2.1), state that for every material system in equilibrium the external force system must be equivalent to zero. The relations

$$\sum \boldsymbol{F} = 0 \qquad \sum \boldsymbol{M}_{\mathrm{A}} = 0$$

will then contain six scalar equations. If a coordinate system Axyz is introduced with the origin at the moment point A the relations may be expressed as

$$\begin{aligned} \sum F_x &= 0 \qquad \sum M_x = 0 \\ \sum F_y &= 0 \qquad \sum M_y = 0 \\ \sum F_z &= 0 \qquad \sum M_z = 0 \end{aligned} \tag{2.32}$$

Note that the directions of the coordinate axes as well as the moment point may be chosen arbitrarily. Let us assume that a set of equilibrium equations have been formulated according to Equations (2.32). We are now free to choose another coordinate system A$'x'y'z'$. From the equilibrium conditions we obtain six new equations. This is possible and permitted but these new equations will not give any additional information compared with the first six.

The new equations and the six original equations will be *linearly dependent*. On the other hand, it may be useful to replace one or more of the equations of the Axyz-system by equations obtained from another system. The reason may be that the solution to the problem will be simplified, when it comes to the geometrical or the algebraic part of the problem or, in many cases, both.

The total number of independent equations, however, cannot be increased. The number will remain six as long as we consider the same material system. The corresponding problem in the two-dimensional case was extensively investigated and illustrated in Section 2.2.1.

When the conditions of equilibrium are applied to a rigid body there is a relation between the number of degrees of freedom and the number of independent equations. In Section 2.2.1 this was shown for the case of a two-dimensional force system, see Figure 2.52, where the number is three:

$$\sum \boldsymbol{F}=0 \quad \Rightarrow \begin{cases} \sum F_x=0 \\ \sum F_y=0 \end{cases}$$

$$\sum \boldsymbol{M}_{\mathrm{O}}=0 \Rightarrow \sum M_z=0$$

In a similar way we find for the general three-dimensional case that the body has six degrees of freedom. This corresponds to the fact that six independent variables are needed to specify the location of the body. These variables may be the coordinates x, y and z of a certain point P of the body and three angles. Two of these (θ and φ in Figure 2.53) are used to specify the location of a certain axis PQ of the body, and the third angle (ψ) represents the rotation of the body about axis PQ. These angles (θ, φ and ψ) are called *Euler*'s[2] angles.

Specified values of x, y, z, θ, φ and ψ correspond to a unique position of the body. Let us consider the case where all external forces and couples are given as functions

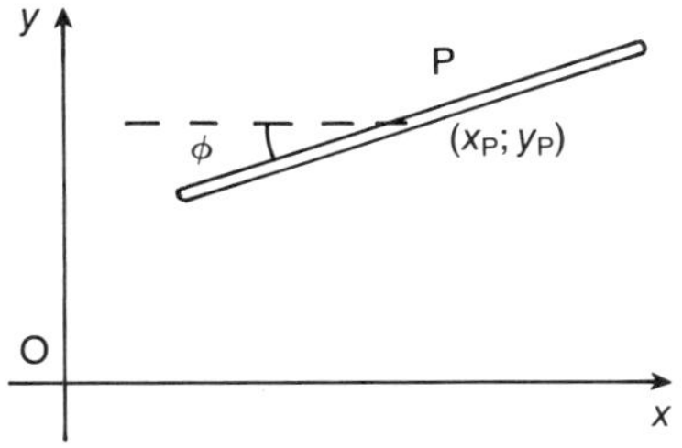

Figure 2.52

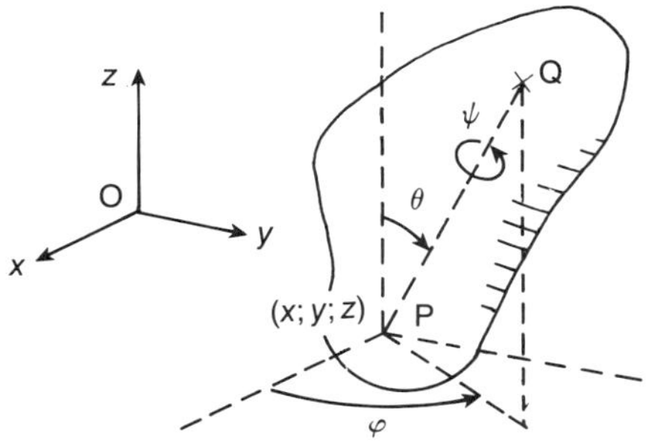

Figure 2.53

2 Leonhard *Euler*, a Swiss mathematician, lived during the eighteenth century. He made fundamental contributions to the development within various areas of pure and applied mathematics.

of the position of the body. Furthermore, we assume that we are interested in determining what position or positions of the body are positions of equilibrium. Accordingly, we wish to find a set of variables corresponding to such forces that Equations (2.32) are satisfied. These equations will involve x, y, z, θ, φ and ψ as unknowns. As there are six equations we normally expect that the problem may be solved.

Special three-dimensional force systems

As we saw in Section 2.2, where two-dimensional problems were considered, the number of equations of interest can in many cases be less than six. The maximum number of independent equations that can be obtained from the equilibrium equations $\Sigma \boldsymbol{F}=0$ and $\Sigma \boldsymbol{M}_{\mathrm{A}}=0$ is three in the two-dimensional case.

There are other situations where the force systems are three dimensional, but the number of independent scalar equations nevertheless is less than six. We give two examples in Figure 2.54. It should be noted that the missing equations still exist in the sense that they are trivially satisfied; normally they are of the type $0 \equiv 0$. The reader is advised to check that this is true.

2.3.2 Constraints, constraining forces

In the previous section we considered the hypothetical case where all six variables describing the position of the body were free to vary while all forces acting were known

(a)

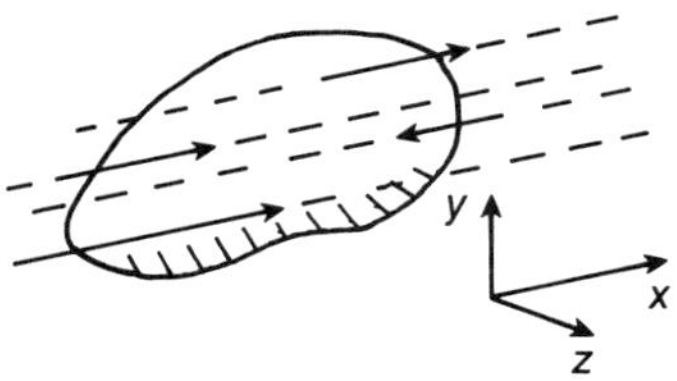

Types of force system — Check

Parallel forces

$\Sigma F_x = 0$	$\Sigma F_y \equiv 0$
$\Sigma M_y = 0$	$\Sigma F_z \equiv 0$
$\Sigma M_z = 0$	$\Sigma M_x \equiv 0$

(b)

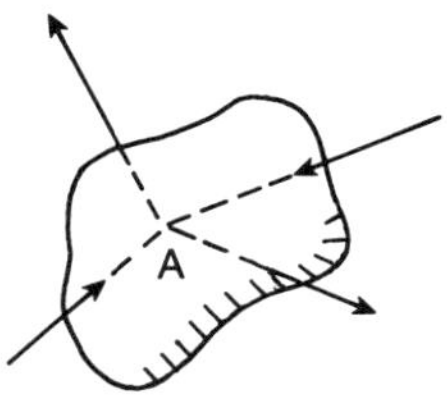

The lines of action intersect at a point A. Choosing the origin at A:

$\Sigma F_x = 0$	$\Sigma M_x \equiv 0$
$\Sigma F_y = 0$	$\Sigma M_y \equiv 0$
$\Sigma F_z = 0$	$\Sigma M_z \equiv 0$

Figure 2.54

as functions of these variables. In practical applications this situation is not very common. It does occur, however. As an example we may take a body that is suspended by three elastic springs, see Figure 2.55. For each position of the body each spring force is uniquely determined, i.e. the magnitude, the sense and the line of action may all be determined. There is at least one position where the force system, consisting of the three spring forces and the weight, is equivalent to zero.

Normally the material system that is considered is subjected to some kind of restriction regarding its ability to move, i.e. *constraints* may exist. To keep the system in a position compatible with the constraints, *constraining forces* or *reactions* must act. These are in general not known when we start solving the problem. On the other hand, we know more about the position of the system than we did in the above case where a body was suspended by springs, and no constraints existed.

Constraints and *constraining forces* were described in Section 2.2.2 in connection with two-dimensional problems. Here we shall only give some additional examples that reflect the more general three-dimensional situation. Constraints occur when there is a contact between two bodies. Usually one of them is fixed. It will then provide a fixed support for the other body at the contact point.

Some common examples of contacts in three-dimensional cases are shown in Figure 2.56. The reader is recommended to return to the corresponding two-dimensional survey (Figure 2.29) in Section 2.2.2 for comparison.

Those three-dimensional cases that distinguish themselves from the corresponding two-dimensional cases in an interesting way have been included here. To make it easier to compare the two- and three-dimensional cases we have kept the denotation (a, b ...) that was used in Figure 2.29.

A few comments with regard to Figure 2.56 are given here. More extensive comments, for instance concerning the nature of various contacts, are given in connection with Figure 2.29. The reader is recommended to return to these if necessary.

In case (a), *smooth* surfaces or ball support, only a *normal force* N perpendicular to the tangential plane T will occur. If friction is considered, as in case (b) where the

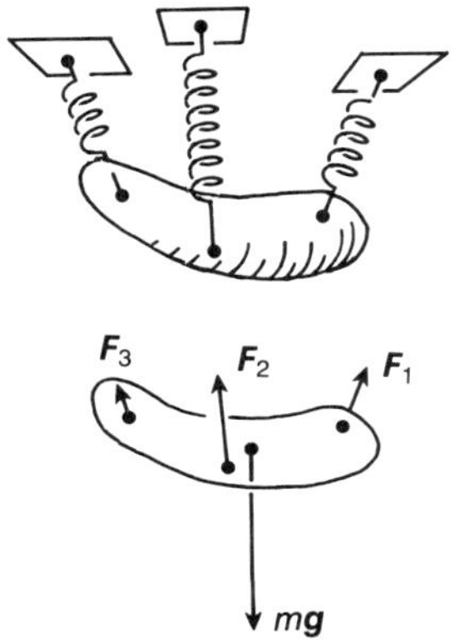

Figure 2.55

Type of contact	Constraining force/couple

(a)

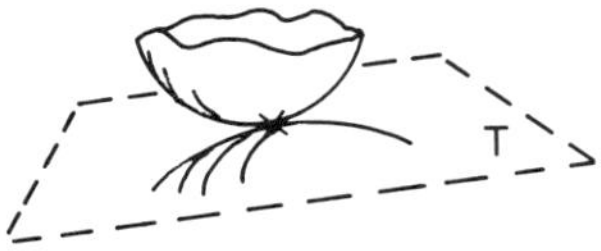

Smooth surfaces

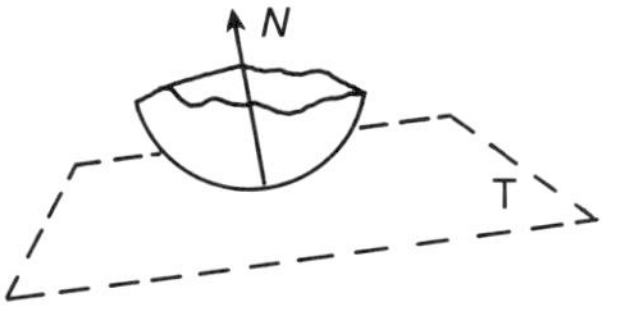

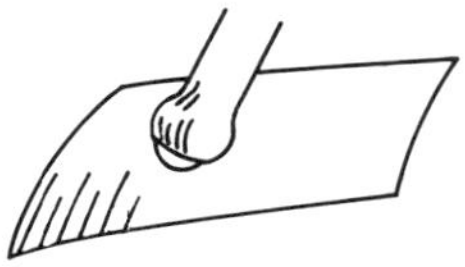

Ball support

Normal force N

(b)

Rough surfaces

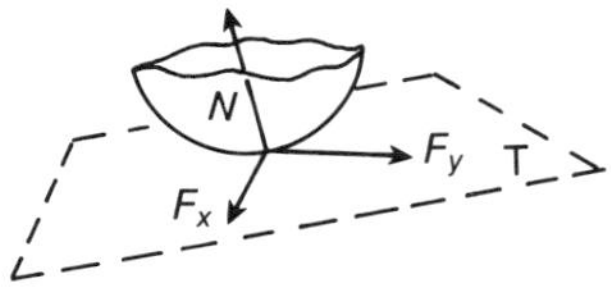

Normal force N, friction force **F** with components F_x *and* F_y

(e)

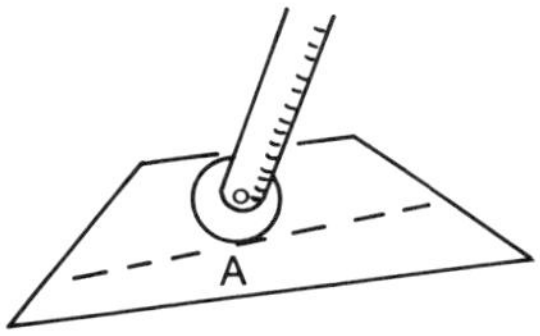

Wheel support

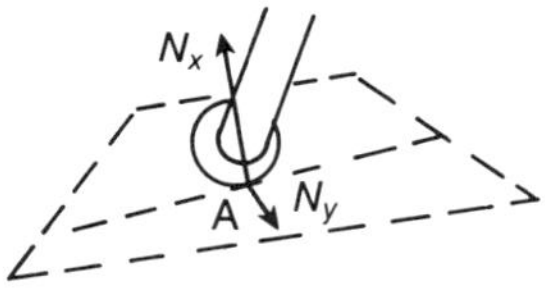

Constraining force with components N_x and N_y

(f)

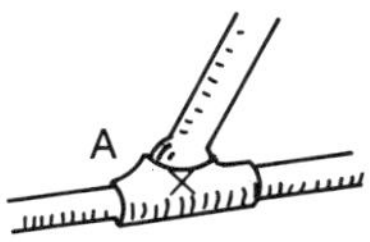

Smooth guide

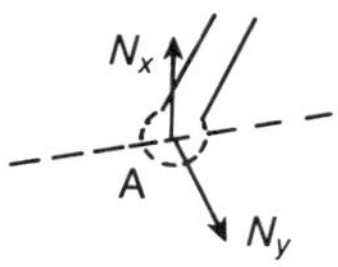

Constraining force with components N_x and N_y

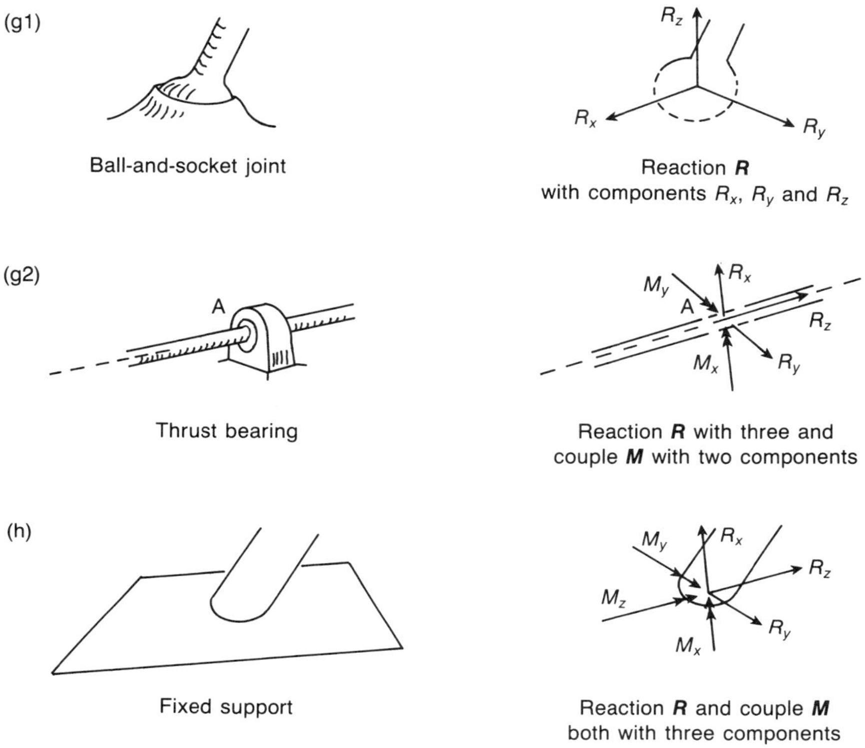

Figure 2.56

surfaces are *rough*, the situation is more complicated. The *friction force* (***F***) may be directed in any sense in the plane T. Hence, two components (F_x, F_y) must be introduced in addition to the normal force.

In cases (e) and (f) we have a couple of symbols indicating that a certain point A of the body is restricted to moving along a line. Otherwise the body is entirely free to move. The constraining force may then have an arbitrary direction in the plane perpendicular to the line and may thus be described by two components (N_x, N_y).

The two-dimensional smooth joint (g) corresponds to the two cases (g1) and (g2) shown here. The *ball-and-socket joint* of case (g1) can support an arbitrary force but no couple. Three force components (R_x, R_y, R_z) are needed. The *thrust bearing* support of case (g2) implies that a certain line (λ) which is attached to the body is fixed. The bearing A must then be able to support the couple components (M_x and M_y) in addition to the force components R_x, R_y and R_z. Otherwise an external force could cause line (λ) of the body to deviate from its fixed location.

A body with a fixed connection, case (h), is completely fixed. This means that the directions of the force and couple vectors (***R*** and ***M***) are completely arbitrary. Each of these is then represented by three components as shown.

WORKED EXAMPLE 2.14

A door of mass 52 kg is supported by two hinges A and B as shown in Figure 2.57. It is kept in the horizontal position by a cable CD. Determine the tension in the cable and the forces exerted by the hinges on the door assuming that the hinges may be treated as ball-and-socket joints.

SOLUTION

It is advisable to introduce symbols for the calculations instead of numerical values. Hence, put $a = 400$ mm and $Q = 52 \cdot 9.81$ N, i.e. the weight of the door. The tension in the cable is denoted by S. Furthermore, the reactions supported by the hinges A and B are introduced. These are so far unknown. We choose to resolve them into components, denoted by R_{Ax}, R_{Ay}, R_{Az}, R_{Bx}, R_{By} and R_{Bz}, as shown in the free-body diagram of Figure 2.58.

The force system which is acting on the door is three dimensional. We expect that six independent equations of equilibrium can be written down:

$$\sum \mathbf{F} = 0$$

$$\sum F_x = 0 \Rightarrow R_{Ax} + R_{Bx} = 0$$

$$\sum F_y = 0 \Rightarrow R_{Ay} + R_{By} + S\cos\alpha = 0$$

$$\sum F_z = 0 \Rightarrow R_{Az} + R_{Bz} - Q + S\sin\alpha = 0$$

Here $\alpha = 30^\circ$.

$$\sum \mathbf{M} = 0:$$

It may be a matter of discussion which moment point is the most suitable – O, A, B or perhaps some other point. In this solution we choose O. One reason is that the line of action of the tension S intersects one of the axes through O. The scalar equations thus obtained are

$$\sum M_x = 0 \Rightarrow R_{Az}a + R_{Bz} \cdot 3a - Q \cdot 2a = 0$$

$$\sum M_y = 0 \Rightarrow Qa - S\sin\alpha \cdot 2a = 0$$

$$\sum M_z = 0 \Rightarrow S\cos\alpha \cdot 2a - R_{Ax}a - R_{Bx} \cdot 3a = 0$$

Combining the six equations of equilibrium we obtain the following results:

$$R_{Ax} = -\frac{\sqrt{3}}{2}Q \qquad R_{Bx} = \frac{\sqrt{3}}{2}Q \qquad S = Q$$

$$R_{Az} = -\frac{1}{4}Q \qquad R_{Bz} = \frac{3}{4}Q \tag{2.33}$$

$$R_{Ay} + R_{By} = -\frac{\sqrt{3}}{2}Q$$

The last relation, Equation (2.33), reflects the statical indeterminacy caused by the fact that the number of unknown quantities (seven) exceeds the number of equations (six). If one of the hinges, say A, is replaced by a sliding contact allowing point A to move along line AB, then R_{Ay} will be zero owing to the absence of the corresponding constraint. In this case Equation (2.33) is replaced by the result $R_{By} = -(\sqrt{3}/2)Q$ and the mathematical problem of solving six simultaneous equations has a unique solution as shown in Figure 2.59.

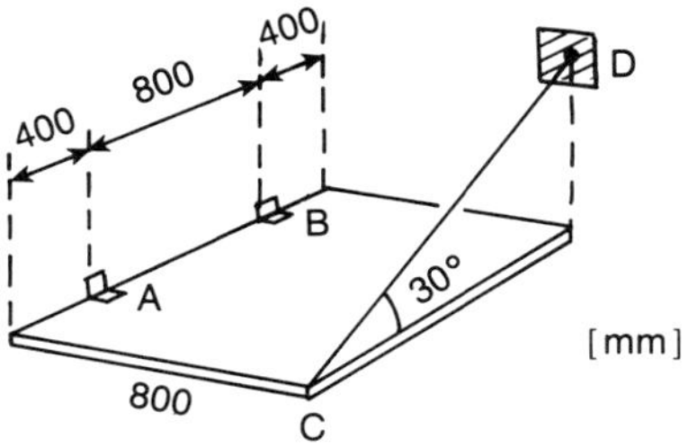

Figure 2.57

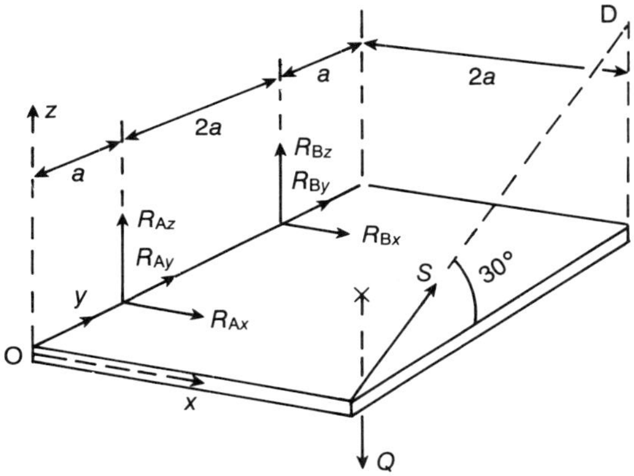

Figure 2.58

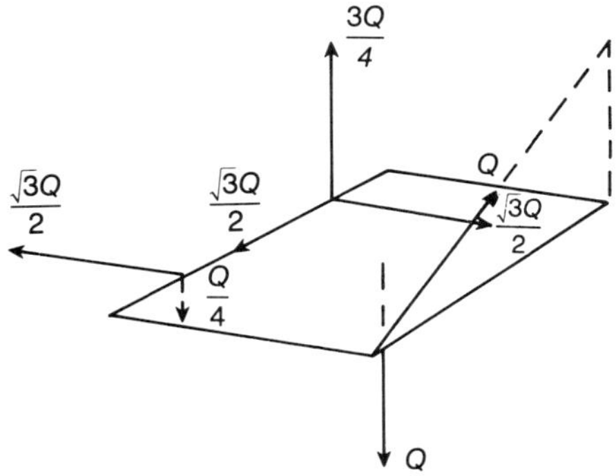

Figure 2.59

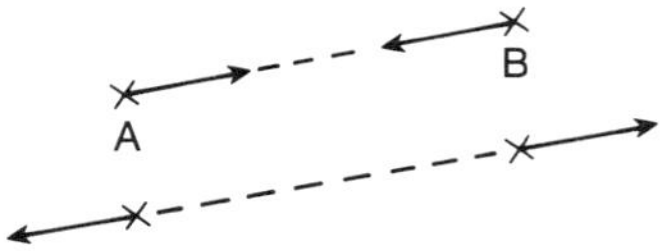

Figure 2.60

Statical indeterminacy

Once again we consider Worked Example 2.14 and Equation (2.33) above. We observe that the latter corresponds to an indeterminacy concerning the components of the reactions along the axes of the hinges.

This *statical indeterminacy* is directly connected with the fact that the door has been secured twice from moving along the y-axis. Hence, we have assumed the existence of force components parallel to the y-axis at both points of contact, A and B. The result is that only the sum of these forces, $R_{Ay} + R_{By}$, may be determined from the equilibrium equations. The individual forces, however, are undetermined. We can only conclude that these two force components together must compensate the y-component $(\sqrt{3}/2)Q$ of the tension in the cable.

In other words, to an arbitrary force system applied to the door we can always add a pair of forces, having the same magnitude, the same line of action and the opposite sense, without affecting the equations of equilibrium, see Figure 2.60.

Such a pair of forces is related to certain stresses in the door. These may have arisen when the door was mounted or later due to temperature changes. Temperature changes are related to deformations of a body. Our assumption that the body is rigid is no longer valid, and the whole truth cannot be obtained from the restricted model originating from the assumption of a rigid body.

As we mentioned above in Section 2.2.3, deformations of bodies and their relations to the acting forces are normally considered within the subject of strength of materials.

WORKED EXAMPLE 2.15

A 75 kg person is standing at a point E on a rectangular table of mass 25 kg. It is assumed that the legs can only support vertical forces. Determine these forces.

SOLUTION

We isolate the table as shown in the free-body diagram of Figure 2.61. The external forces are the weight P of the table, the force Q exerted by the person and the reactions (normal forces) N_A, N_B, N_C and N_D exerted by the floor on the legs of the table.

The forces are all parallel. From Figure 2.54 we conclude that no more than three independent equations of equilibrium are to be expected. We then have, for instance,

$$\sum F_z = 0 \Rightarrow N_A + N_B + N_C + N_D - P - Q = 0 \quad (2.34)$$

$$\sum M_x = 0 \Rightarrow N_A(92) + N_D(92) - Q(42) - P(46) = 0 \quad (2.35)$$

$$\sum M_y = 0 \Rightarrow Q(25) + P(61) - N_C(122) - N_D(122) = 0 \quad (2.36)$$

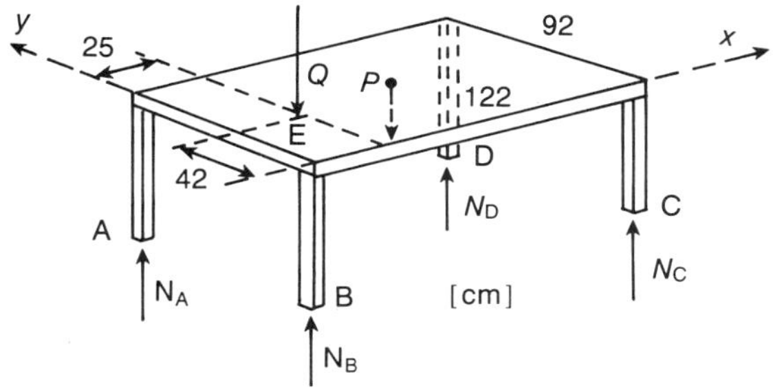

Figure 2.61

The four reactions are to be determined, while the gravitational forces are known in advance. This is one more example of a problem that is statically indeterminate. In this case none of the unknowns may be solved for.

REMARK

As an exercise write down one more equation of equilibrium, for instance by using the condition that the sum of moments about axis CD is zero. Then prove that the equation thus obtained could also have been obtained by a suitably chosen linear combination of Equations (2.34), (2.35) and (2.36).

Obviously, an ordinary table with four legs is an example of a statically indeterminate system. For comparison, the reader is referred to Worked Example 2.17 where a statically determinate system consisting of a loaded table with three legs is considered.

2.3.3 Applications

In order to analyse a statics problem in a successful way it is necessary to have a good knowledge and a firm understanding of the foundations of the subject. A considerable ability to utilize mathematical tools is also required here. To understand the foundations of mechanics it is important, not to say necessary, to put their applications in practice by solving problems. This is the primary purpose when problems are solved in a mechanics course.

The secondary purpose of importance is to gain experience from certain types of problem, to learn to recognize various situations and simply to acquire a practical knowledge of the treatment of problems. This purpose is in many cases the most obvious and the most immediate one from the student's point of view.

To prepare the student for further work in this area we shall present some additional examples and comments. We shall demonstrate a few different types of problems in the form of worked examples.

The position of equilibrium is not known in advance

This situation is perhaps not the most common one in the applications we are concerned with here. As well as the position of equilibrium one or more of the acting forces may be unknown. Some forces may be completely known in advance or may be known as functions of the coordinates describing the position of the system. In a typical case we are interested in determining the position of equilibrium and certain forces.

WORKED EXAMPLE 2.16

The end points of a 6.0 kg bar are connected to two small collars by ball-and-socket joints. The collars may move freely along two fixed bars; one of these is horizontal and the other one is vertical. A body of mass 3.0 kg is suspended by a cable which is connected to collar A by a pulley as shown in Figure 2.62. Determine the distance x at equilibrium.

SOLUTION

To avoid numerical values we introduce the following symbols:

$$L = 1.40 \text{ m}, \quad a = 0.60 \text{ m}, \quad Q = 6 \cdot 9.81 \text{ N}, \quad P = 3 \cdot 9.81 \text{ N}$$

We also observe that the distance b is a function of x given by the relation

$$L^2 = b^2 + a^2 + x^2 \tag{2.37}$$

Isolating the bar and the collars we obtain the free-body diagram shown in Figure 2.63.

The following external forces are applied to the system: H, V, R, S, Q and P. Of these, P and Q are known in advance. The force system is three dimensional, and we have the following equations of equilibrium:

$$\sum \mathbf{F} = 0 \qquad \sum \mathbf{M}_O = 0$$

$\Sigma F_x = 0$, $\Sigma F_y = 0$ and $\Sigma F_z = 0$ yield

$$\begin{aligned} P - S &= 0 \\ V - Q &= 0 \\ R - H &= 0 \end{aligned} \tag{2.38}$$

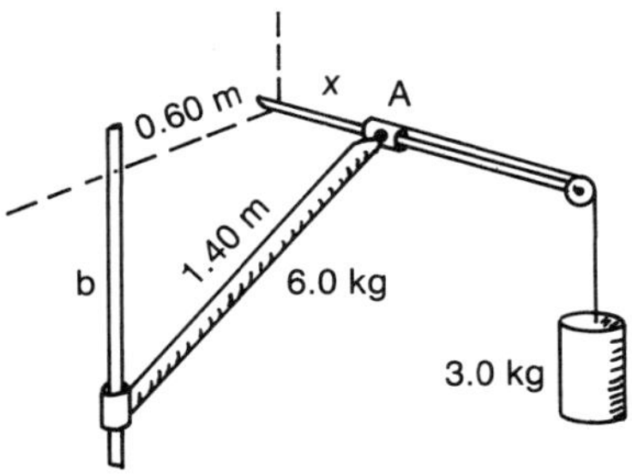

Figure 2.62

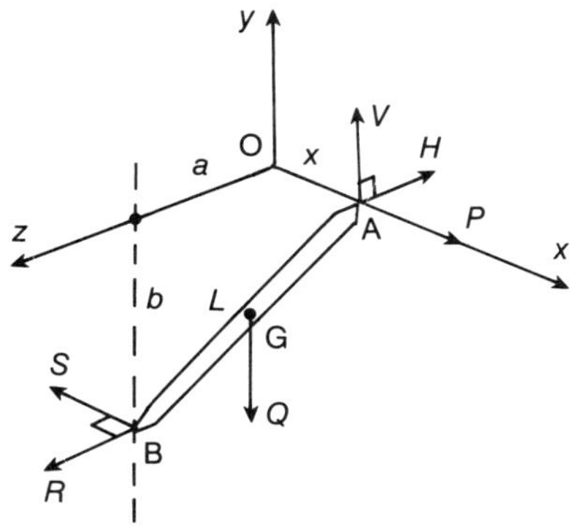

Figure 2.63

The three scalar equations included in $\Sigma \boldsymbol{M}_O = 0$ are obtained by calculating the vector

$$\sum \boldsymbol{M}_O = \vec{OA} \times \boldsymbol{F}_A + \vec{OB} \times \boldsymbol{F}_B + \vec{OG} \times \boldsymbol{Q}$$

$$\begin{vmatrix} \mathbf{e}_x & \mathbf{e}_y & \mathbf{e}_z \\ x & 0 & 0 \\ P & V & -H \end{vmatrix} + \begin{vmatrix} \mathbf{e}_x & \mathbf{e}_y & \mathbf{e}_z \\ 0 & -b & a \\ -S & 0 & R \end{vmatrix} + \begin{vmatrix} \mathbf{e}_x & \mathbf{e}_y & \mathbf{e}_z \\ 0.5x & -0.5b & 0.5a \\ 0 & -Q & 0 \end{vmatrix}$$

$$= (0, Hx, Vx) + (-Rb, \ -Sa, \ -Sb) + 0.5(Qa, 0, \ -Qx) = 0$$

Adding the vectors we obtain three scalar equations corresponding to $\Sigma M_x = 0$, $\Sigma M_y = 0$ and $\Sigma M_z = 0$. Using Equations (2.38) we obtain

$$\begin{aligned} &-2Hb + Qa = 0 \\ &Hx - Pa = 0 \\ &0.5Qx - Pb = 0 \end{aligned} \tag{2.39}$$

Combining the third of Equations (2.39) with Equation (2.37) and utilizing the fact that $Q = 2P$ we obtain

$$x = \sqrt{\frac{L^2 - a^2}{2}} \approx 0.89 \text{ m}$$

REMARK

In this problem there are five unknowns, namely the forces H, V, R and S and the distance x. However, there are six equations given by (2.38) and (2.39). Obviously there is no sixth unknown to determine. This is due to the fact that the number of independent equations reduces to five when the lines of action of all forces acting on the system intersect a certain line, in this case the line coinciding with the bar. The reader is recommended to return to the survey of Figure 2.54. This survey may be supplemented by the case where the lines of action of all forces intersect a certain line; obviously the sum of moments about this line is identically equal to zero. Hence, one equation of equilibrium becomes superfluous, cf. Problem 2.58. With our choice of axes this means that Equations (2.38) and (2.39) are not linearly independent; if the first of Equations (2.39) is multiplied by $0.5x$ and the second one by b and the resulting equations are added, we obtain the third equation.

Hence, we have five linearly independent equations and five unknowns may be determined.

The position of equilibrium is known in advance

In most applications the position of equilibrium is known in advance. Some forces are known, others are not. The unknown forces or some of their components are requested. Sometimes we want to determine whether certain conditions must be fulfilled for a state of equilibrium to exist at all.

WORKED EXAMPLE 2.17

A small box of mass 12 kg is placed on a round table with three legs, Figure 2.64. How is the weight of the box distributed on the three legs?

SOLUTION

This is a case of the type considered in Section 2.1.2.

The weight of the unloaded table is not known. Consequently, we do not know the corresponding contributions to the forces acting on the legs. A vertical force of magnitude $Q = 12 \cdot 9.81$ N is added when the table is loaded by the box. This will cause changes in the normal forces exerted on the legs by the floor. These changes are denoted by N_A, N_B and N_C, Figure 2.65. Together with Q they form a superposed force system which is equivalent to zero. As the question was put this is the only force system we have to consider.

The force system is a parallel force system. Hence, we know that only three independent equations of equilibrium can be obtained. In this case it will be practical to choose three moment equations:

$$\sum M_{AB} = 0 \Rightarrow N_C(600 + 600 \cos 60^\circ) - Q(600 \cos 60^\circ) = 0$$

$$\sum M_{AC} = 0 \Rightarrow -N_B(600 + 600 \cos 60^\circ) + Q(600 \cos 60^\circ - 400 \cos 30^\circ) = 0$$

$$\sum M_{BC} = 0 \Rightarrow N_A(600 + 600 \cos 60^\circ) - Q(600 \cos 60^\circ + 400 \cos 30^\circ) = 0$$

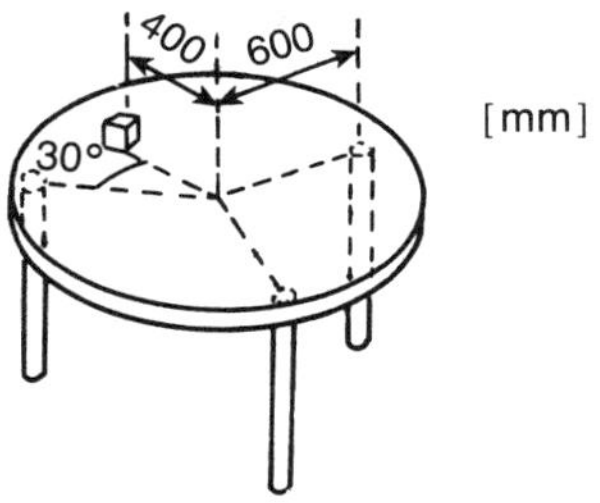

Figure 2.64

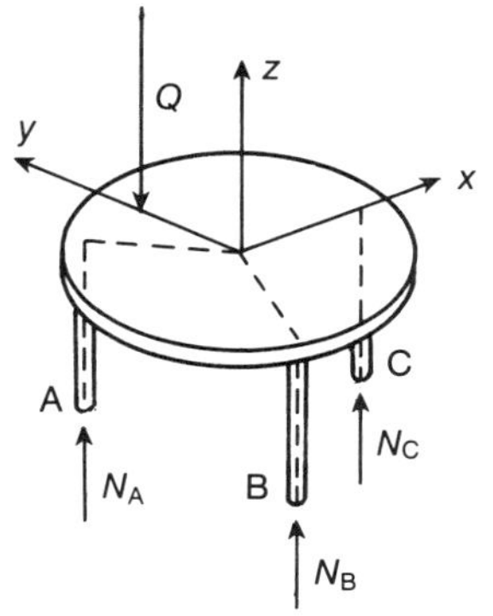

Figure 2.65

Knowing that $Q = 117.7$ N we easily determine the three normal forces:

$N_A = 39.2$ N $\quad N_B = -6.1$ N $\quad N_C = 84.6$ N

CHECK

Is the sum of these forces equal to the weight of the box? Adding them we obtain 117.7 N, meaning that the equilibrium equation $\Sigma F_z = 0$ is satisfied.

PROBLEMS

○ **2.47** A body of mass *m* is attached to a small hook A which is supported by three stretched cords AB, AC and AD. The coordinates of points A, B, C and D are $(a; 2a; 0)$, $(3a; 0; 0)$, $(0; 4a; 0)$ and $(0; 0; 4a)$, respectively. The *z*-axis is directed vertically upwards. Determine the tensions in the cords.

○ **2.48** A small body of weight 150 N is suspended by two cords. It is in equilibrium at a point with coordinates (1; 2; 0)m. Here the *z*-axis is in the vertical upward direction. One of the cords is attached to the point with coordinates (2; 0; 5) m and the other to a point on the *y*-axis. Determine the location of this point and the tensions in the cords.

2.49 A bar AB is free to rotate in the horizontal *xy*-plane. A body of weight *Q* is suspended by a cord. The cord runs through a smooth loop and is attached to the bar at B. It is desired to keep the bar in equilibrium in the position shown by applying a force at point A. What is the minimum magnitude of this force and the corresponding direction?

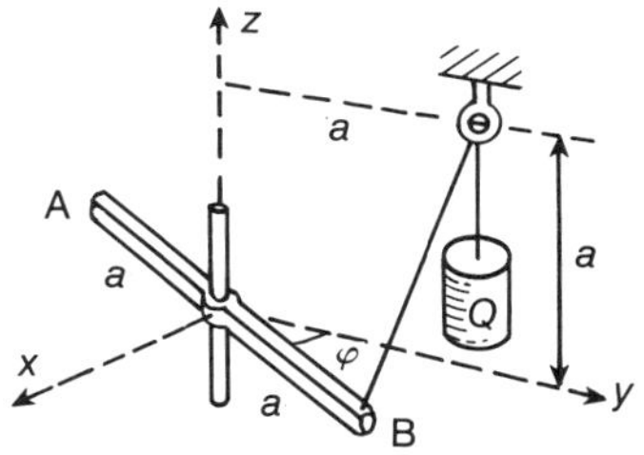

◎ **2.50** A rigid body is supported by ball-and-socket joints at points $(4a; 3a; -3a)$ and $(3a; a; -a)$. The coordinates of the centre of mass are $(5a; 5a; -5a)$. A force $(3P, 5P, -2P)$ is applied at $(-2a; -3a; a)$. Another force $\boldsymbol{F}$ is applied at $(a; 2a; 0)$. What is the minimum value of $|\boldsymbol{F}|$ that can be permitted if the body is to be in equilibrium?

○ **2.51** A smooth sphere of mass m and radius r is suspended by a cord of length a. The cord is attached to a point A at the intersection between two perpendicular, vertical walls. Determine the tension in the cord and the reaction exerted on the sphere by each wall.

○ **2.52** A cube of weight Q is suspended by six cords attached at the vertices of the cube. The cords are stretched in the directions of various edges as shown. The top face is horizontal. Determine the tensions $S_1 \ldots S_6$.

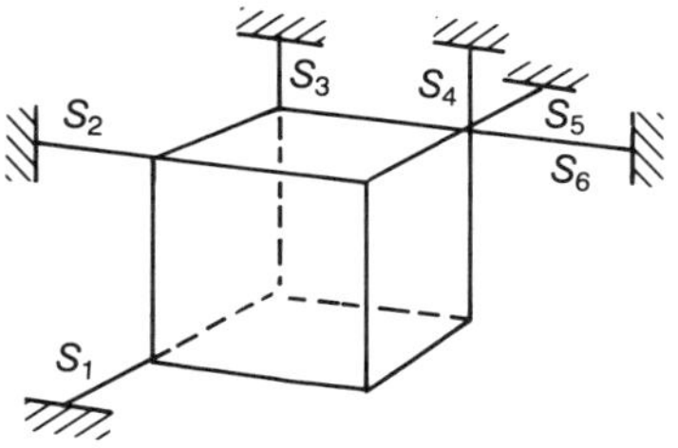

○ **2.53** The pipe OABC is welded at point O and loaded as shown. Determine the force and the couple exerted on the pipe by the support at O.

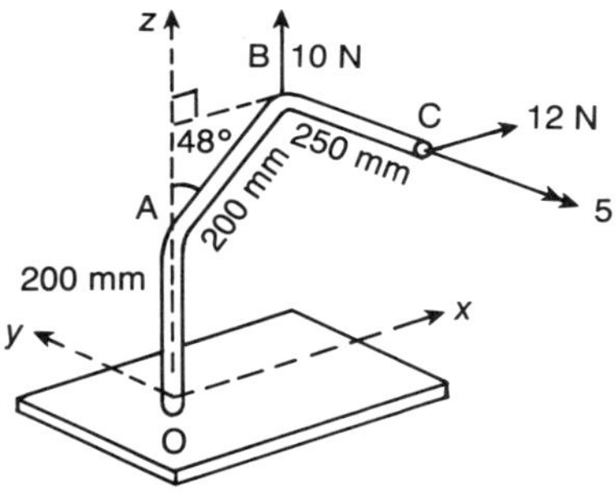

2.54 The mass of a pair of wheels is 270 kg, the axle included. The system is placed on a rough incline and is held in equilibrium in the position shown by a force P, which is parallel to the incline and perpendicular to the axle. Determine P and the forces exerted by the incline on the wheels at A and B. The components parallel to the axle are assumed to be zero.

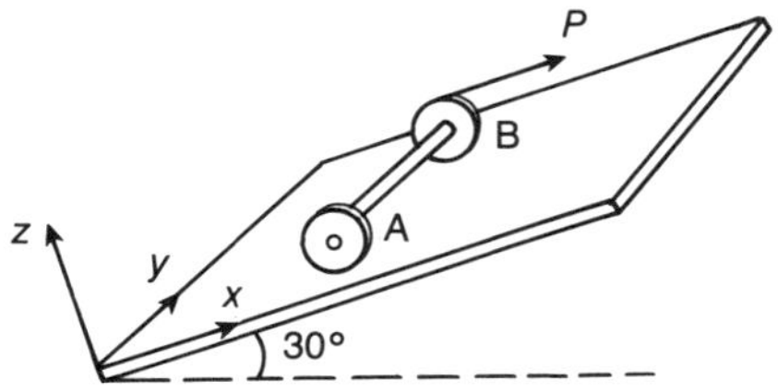

2.55 A flexible shaft is supported by three smooth loops A, B and C. A torsional moment $M = 5.0$ N m is applied at both ends of the shaft. Determine the x-, y- and z-components of the forces exerted on the loops by the shaft.

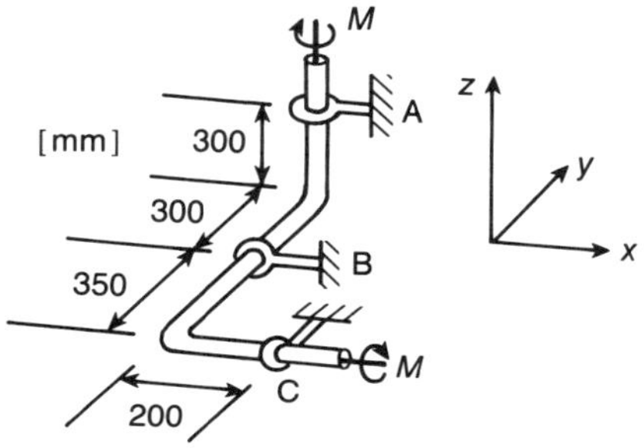

2.56 A bar ABCD of mass 42 kg and length 4.8 m is composed of three mutually perpendicular parts of equal length. The bar is supported by a ball-and-socket joint at point D. At point A the bar is in roller contact with a wall, meaning that the bar is free to move in the vertical direction. The bar is kept in the position shown by the cable CE. Determine the tension in the cable and the forces exerted by the bar on the supports at A and D.

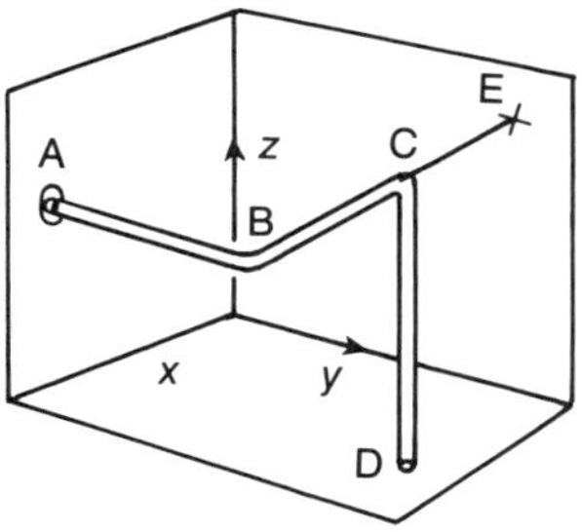

2.57 A uniform plate of mass m in the shape of an equilateral triangle is supported by three cords of equal length as shown. A small smooth ring of the same mass m is threaded on the cords. Determine the equilibrium position of the ring, i.e. the altitude above the plate.

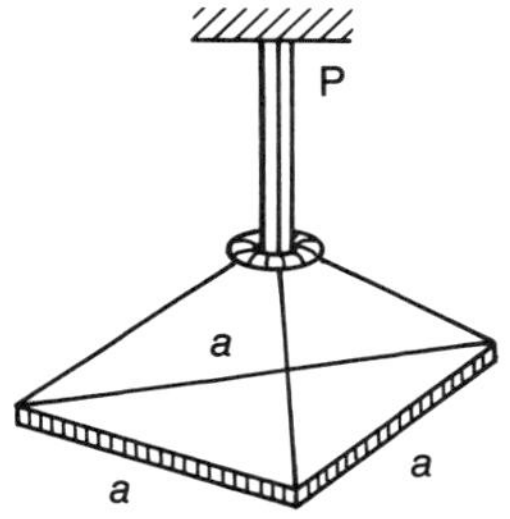

2.58 A body is subjected to forces such that all the lines of action intersect a certain axis. Show that it is impossible to obtain more than five linearly independent scalar equations from the conditions of equilibrium, $\Sigma \boldsymbol{F} = 0$ and $\Sigma \boldsymbol{M} = 0$.

2.59 A horizontal cable AB is attached to a pole as shown. The tension in AB is 0.70 kN. Two supporting wires, CD and CE, are attached to the ground at D and E, respectively (points D and E are both located in the xy-plane). The wires are stretched until no couple is exerted on the pole by the support at O.

(a) What are the corresponding tensions in the wires?

(b) What is the horizontal force component supported by the ground at O?

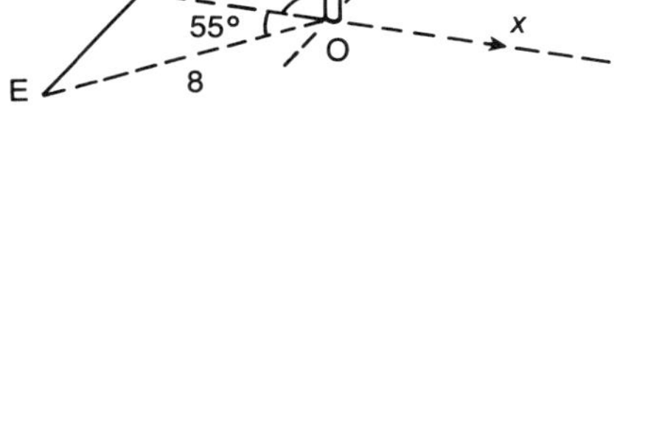

2.60 A cable of mass 1.2 kg/m is supported by poles. Between each pair of poles there is 55 m length of cable. The cable is fastened 13 m above the ground and makes a 12° angle with the horizontal at each pole as shown. At a certain pole the cable makes a 20° change of direction. To make sure that no couple is exerted on the corresponding pole at its support O a supporting wire CD of length 11 m is attached to the pole at a point C located 7 m above the ground and to a point D on the ground. Determine the location of D and the necessary tension in the wire.

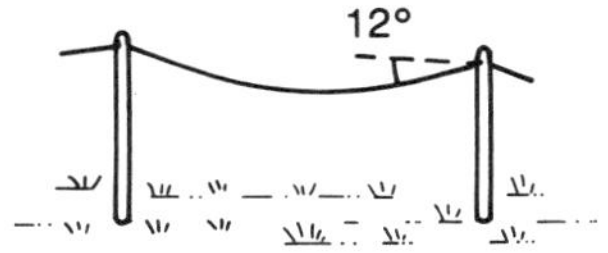

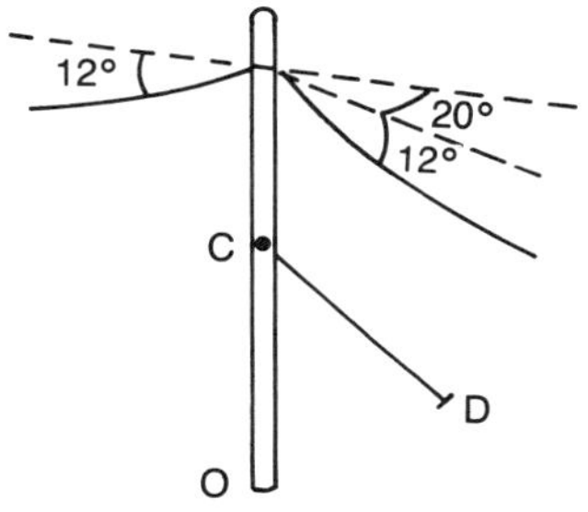

2.61 A uniform, rectangular sign of mass 120 kg is connected to a bar AB by two vertical wires. The bar is supported by a ball-and-socket joint at point A and by two wires CD and BE. Determine the tensions in wires CD and BE and the reaction at A.

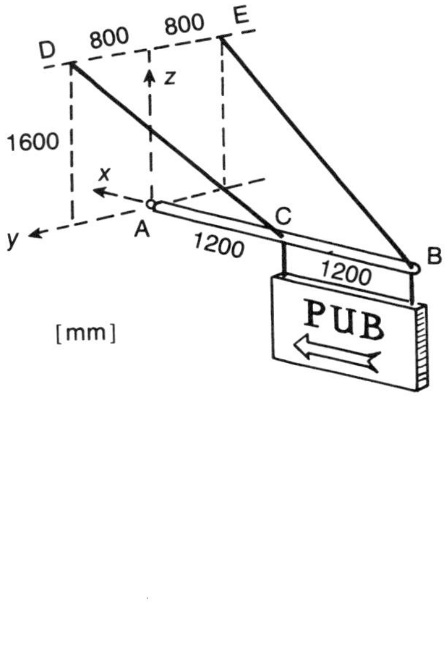

○ **2.62** A uniform cube of weight Q is supported by three cords as shown. One of the cords is connected to the midpoint of one of the edges; each one of the other two is connected to a vertex. The lengths of the cords are adjusted to make the top face horizontal. The shortest of the cords is of the same length as the edges of the cube. Determine the tensions in the cords.

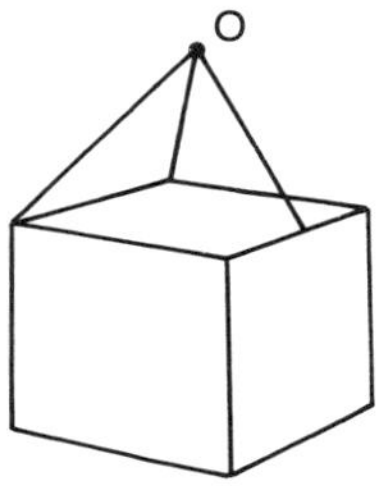

2.63 A circular plate of mass 17 kg and radius 0.38 m is held in a horizontal position by three cables of equal length. The cables are fastened at a point O located at an altitude of 0.95 m above the plate. Determine the tensions in the three cables as functions of α.

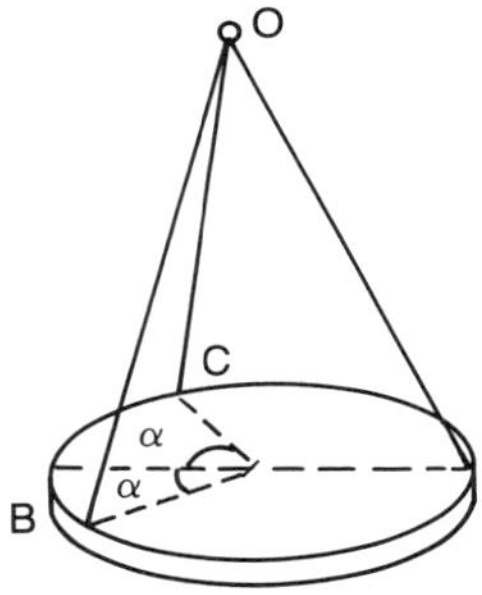

○ **2.64** A plate in the shape of an equilateral triangle is supported in the horizontal position by three vertical cords attached to the corners. A small body of mass 3.0 kg is placed on the plate at a point A, which is located at the midpoint of one of the altitudes as shown. How much do the tensions in the cords increase owing to the 3.0 kg loading?

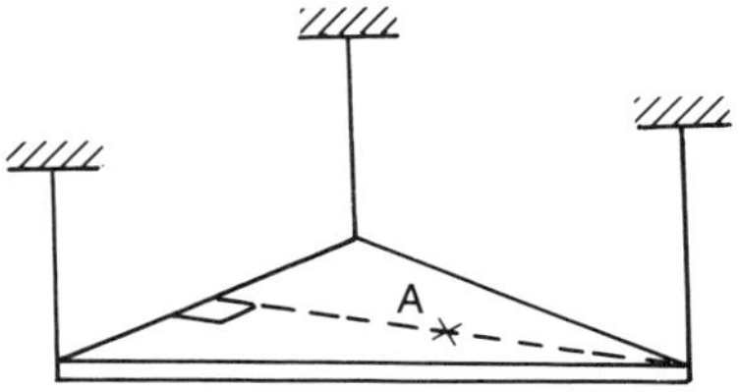

○ **2.65** A square table is loaded by three small bodies as shown. The sides of the top are of length $5a$. The table is only supported by the three legs shown. How are the forces supported by the legs changed owing to the loading?

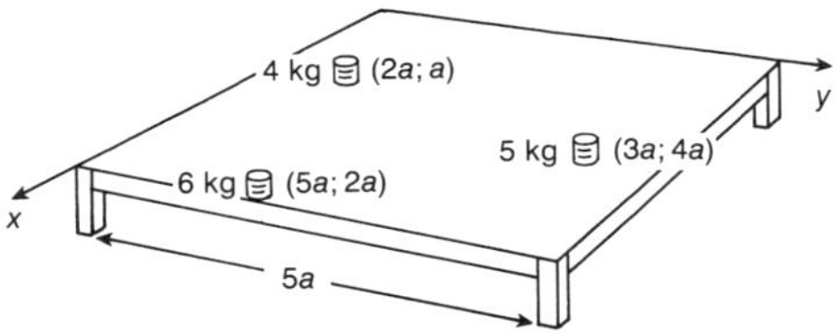

○ **2.66** The plate of Problem 2.63 is supported by *vertical* cables attached at points A, B and C. Determine the tensions in the cables as functions of α.

2.67 A sheet-metal section at a shipyard is resting on three supports in the same horizontal plane (the xy-plane). In order to calculate the location of the centre of gravity the vertical reactions at the supports are measured. The results of the measurements are given by the following table:

Coordinates of the support (m)	Force (kN)
(2.03; 0.61)	2.48
(2.55; −1.48)	3.78
(−1.51; −0.98)	3.05

Determine the x- and y-coordinates of the centre of gravity of the section.

2.68 A body is lifted by three vertical cables attached at points A, B and C. The tensions in the cables are $4P$, $5P$ and $3P$, respectively. The cable at C is replaced by a vertical cable attached at point D. What are the resulting tensions in the cables?

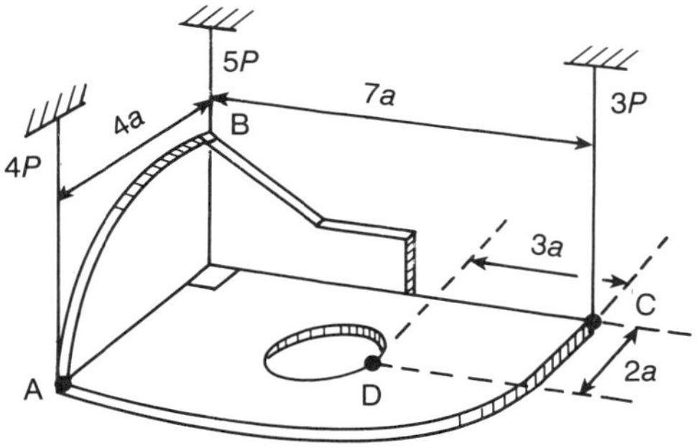

2.69 A door of mass 45 kg is supported by two hinges A and B that may be treated as ball-and-socket joints. Only hinge B can support vertical forces. Determine the magnitudes of the horizontal forces supported by the hinges.

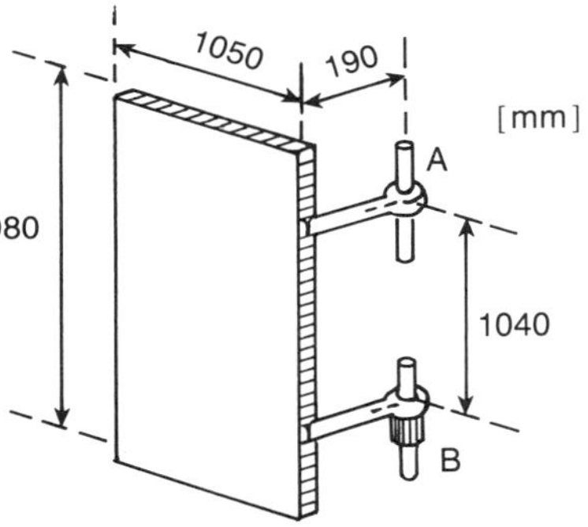

2.70 The force system acting on an opened car door, which is caught by the wind, is shown in the figure, where the rectangular frame ABCD represents the door. The door is supported by two hinges, E and F. Only the lower one can support vertical forces.

A resultant force of magnitude 320 N is exerted by the wind on the door at its centre. The weight of the door is 300 N. To keep the door in the position shown, a force P is applied to the handle. Determine P and the reactions at E and F assuming that the hinges may be treated as ball-and-socket joints.

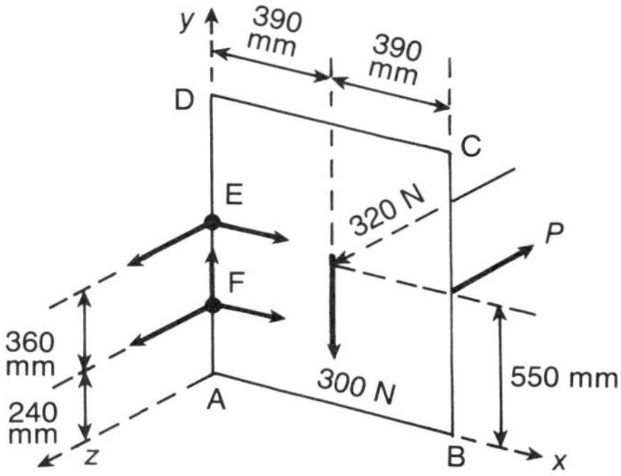

2.71 A square plate ABCD of weight Q is supported by ball-and-socket joints at A and B. The plate is kept in the horizontal position by a cord CED that runs through a smooth loop at E. Determine the vertical component of the reaction at A.

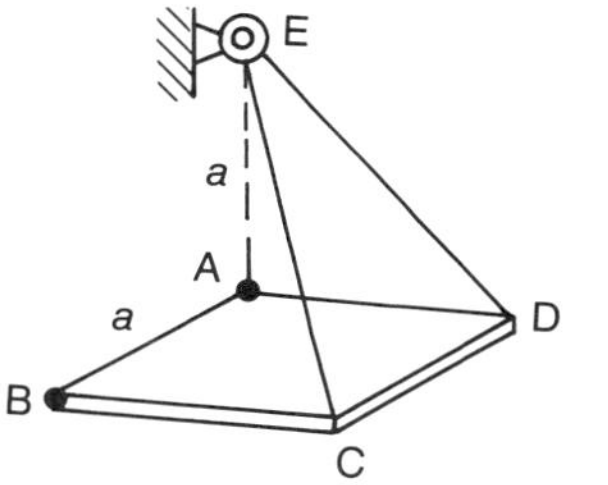

2.72 A square plate ABCD is kept in a horizontal position by a cord EF, which is attached to a point F exactly above vertex A. What is the minimum allowable value of the distance AF if the cord cannot withstand a tension that is larger than twice the weight of the plate?

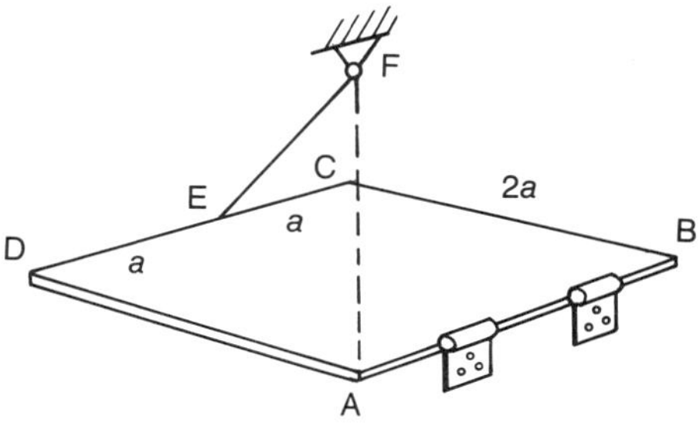

2.73 A cube of weight Q (see opposite) is supported by two hinges A and B, that may be treated as ball-and-socket joints. The cube is held in a position where the bottom face is horizontal by applying a force at vertex E in the plane BCEF. Determine the reactions at A and B assuming that the components along edge AB are zero.

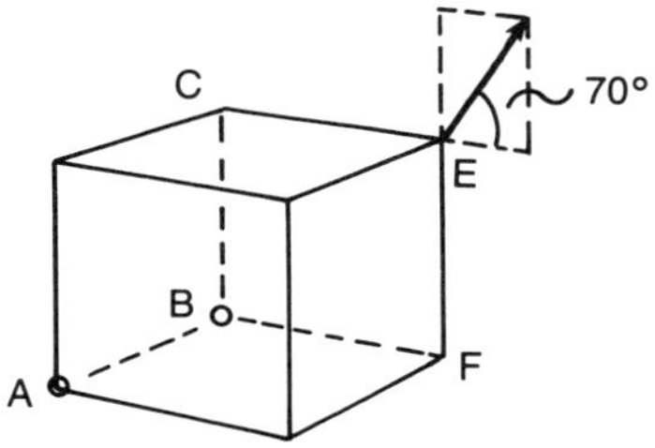

◎ **2.74** A uniform door of mass 28 kg may rotate freely about the horizontal axis AB. The position of the door is adjusted by a cable wrapped around a wheel that is welded to the axis. The door is kept in a position where it makes an angle α with the horizontal plane by applying a horizontal force F to the cable. Determine F and the reactions supported by the bearings at A and C assuming that these act as ball-and-socket-joints.

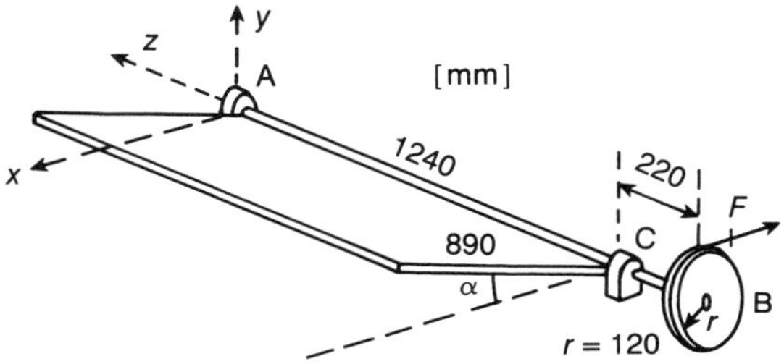

2.75 The base of a right prism (see opposite) is in the shape of a right-angled triangle. The prism, which is of weight Q, is supported by ball-and-socket joints at vertices A and B, which are located at the same altitude. It is desired to hold the body in a position where face ABCD is horizontal by applying a force at point E in a sense such that the magnitude of the force is minimal. Determine the force that must be applied and the x- and y-components of the corresponding reactions at A and B. (The centre of gravity is located at a distance of 0.4 m from face ECDF.)

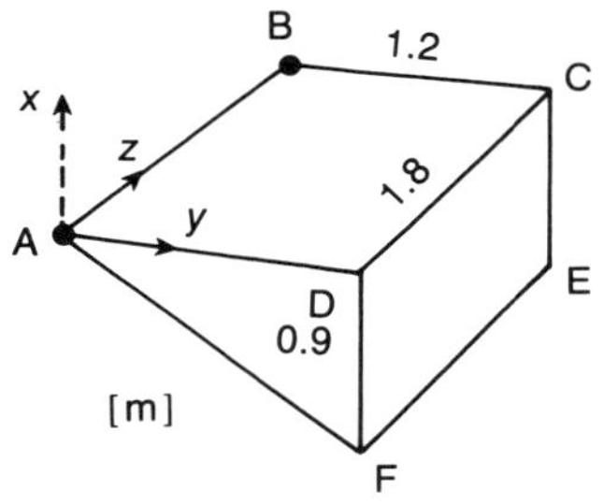

◎ **2.76** A door of weight Q is kept in a horizontal position by two hinges, A and B, which may be treated as ball-and-socket joints, and a support C at the midpoint of one of the edges. Two pins, each of which carries a small pulley, are mounted on the door as shown. A body of weight Q is supported by a cable which is attached to a point D located exactly above hinge A. Determine the x- and y-components of the reactions at A and B.

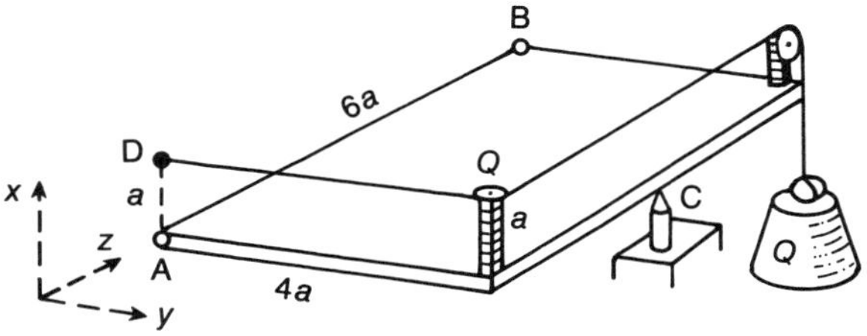

○ **2.77** Determine force F and the horizontal components of the reactions at A and C.

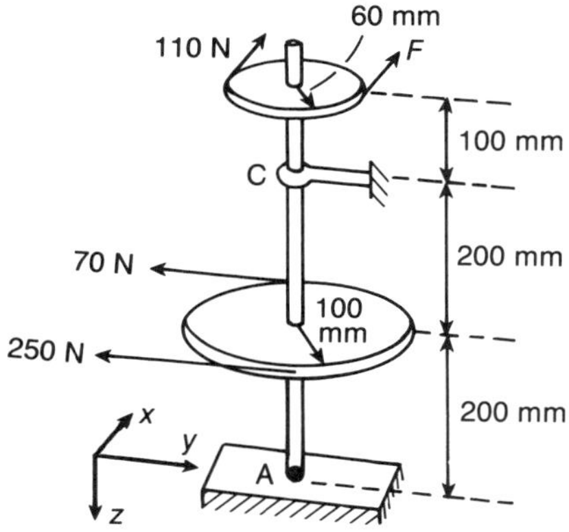

2.78 The drum of a manual winch is supported at A and B. A load L of mass 88 kg is supported by the wire. To keep the load in equilibrium a force of minimum magnitude is applied to the crank. What reactions are supported at A and B owing to the loading?

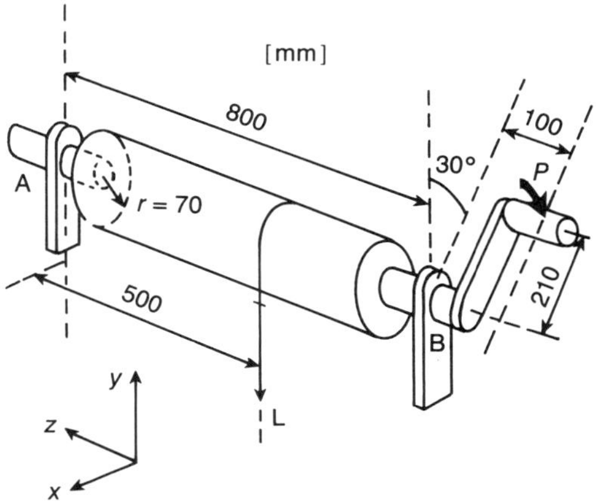

2.79 A woman who weighs 58 kg applies her entire weight to a bicycle pedal at the position where the crank is horizontal. Determine the tension in the chain and the magnitudes of the reactions supported by the bearings at A and B.

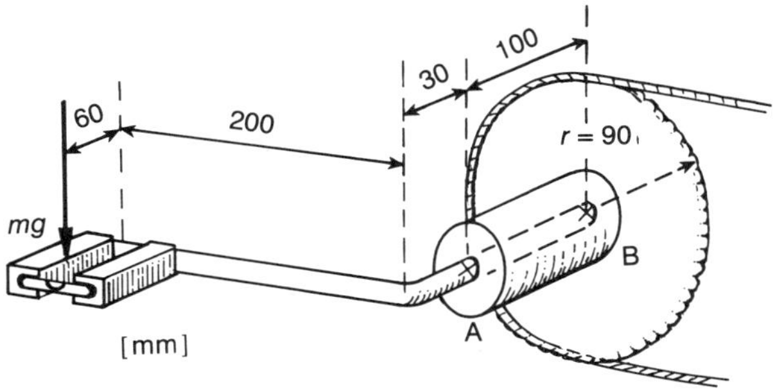

2.4 Section forces; beams, trusses

2.4.1 Section forces and couples

In Section 2.2.4 we saw how problems involving a more complex mechanical system could be solved by dividing the system into simpler portions. However, we restricted ourselves to systems of rigid bodies connected to each other in some way, cf. Worked Examples 2.11–13. The forces exerted by the rigid bodies on each other were then internal forces of the system.

However, there is nothing that prevents us from considering one single rigid body as being composed of two portions which are connected at a contact surface. It is then a matter of great practical interest to determine the forces exerted by the two portions of the body on each other. For this reason we study a completely general body B according to Figure 2.66. We assume that body B is subjected to a system of external forces, which is equivalent to zero when the body is in equilibrium.

We now consider the body to consist of two portions, B_1 and B_2, which are connected at a contact surface S. This surface is commonly called a *section*. The two bodies B_1 and B_2 exert forces on each other. These forces are in some way distributed over the surface S. In a general case this distribution may be fairly complex. However, we know from Section 1.3.5 that every force system, no matter how complicated it may be, can always be reduced to a simpler system which at most consists of *one* force $\boldsymbol{F}$ with an arbitrarily chosen point of application O and *one* couple $\boldsymbol{M}$.

If we isolate B_1 and B_2 we find from the free-body diagrams, Figure 2.67, that a force $\boldsymbol{F}$ and a couple $\boldsymbol{M}$ are exerted on body B_1 by body B_2. According to Newton's third law a force $-\boldsymbol{F}$ at the same point O and a couple $-\boldsymbol{M}$ are exerted on body B_2 by body B_1. It should be noted that the force $\boldsymbol{F}$ is uniquely determined for each section while the couple $\boldsymbol{M}$ depends on the choice of the point of application of $\boldsymbol{F}$. Force $\boldsymbol{F}$ is called the *section force* and $\boldsymbol{M}$ is the *section couple*. Sometimes the term 'section force' is used to denote the entire force system exerted by one body on the other, i.e. the force *and* the couple.

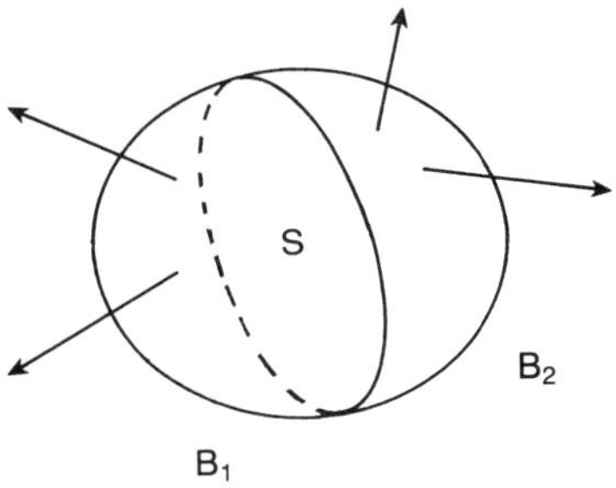

Figure 2.66

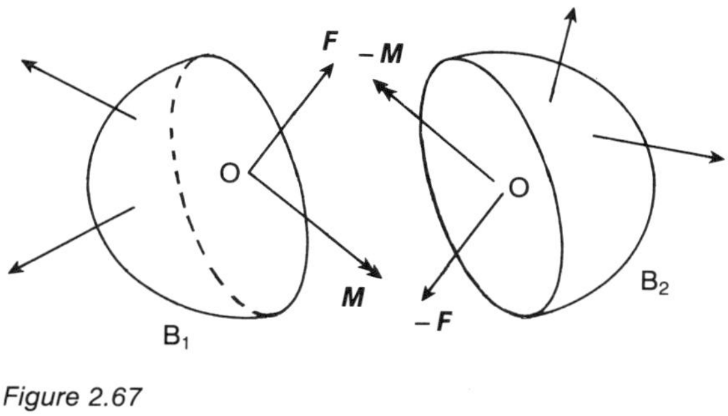

Figure 2.67

The system of external forces on body B_1 consists of those forces external to body B that are related to portion B_1 together with the section force $\boldsymbol{F}$ and the section couple $\boldsymbol{M}$. At equilibrium this force system must be equivalent to zero. An analogous reasoning may, of course, be applied to body B_2.

2.4.2 Beams

As an example of section forces and couples we shall now consider a very important technical application, namely *beams*. A beam is a long and narrow body, normally straight, whose length is much greater than the typical dimensions of the cross-section. A beam can then be considered to be one dimensional. Beams are used to carry loads, normally transverse loads, in various applications, especially within civil engineering. Here we shall assume that the loading consists of a two-dimensional force system according to Figure 2.68(a).

For dimensioning purposes it is necessary to determine section forces and couples at an arbitrary section as these quantities determine the ability of the beam to withstand the loading. From now on we assume that the normal of the section coincides with the tangential direction of the beam. In the case of a straight beam we indicate the location of the section by a coordinate x, see Figure 2.68(a).

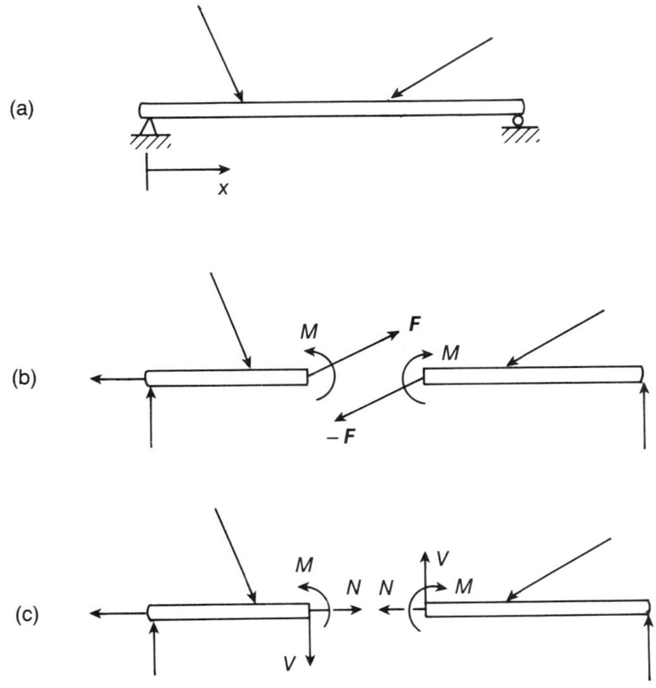

Figure 2.68

Our task is to determine the section force $\boldsymbol{F}$ and the section couple $\boldsymbol{M}$ as functions of x. As we have assumed that the beam is a one-dimensional body, the extension of the cross-section is negligible, i.e. the cross-section may be regarded as a point. This point, of course, has to be chosen as the point of application of force $\boldsymbol{F}$. If the applied force system is two dimensional it follows from the conditions of equilibrium that $\boldsymbol{F}$ must be located in the plane of the forces. Furthermore, couple $\boldsymbol{M}$ must be perpendicular to this plane and can then be considered as a scalar M, Figure 2.68(b).

It is customary to resolve $\boldsymbol{F}$ into two components, the *axial force* N directed along the beam and the *shear force* V normal to the beam, Figure 2.68(c). In many engineering applications the loading is normal to the beam, which means that the axial force N vanishes. The couple M is called the *bending moment*. In the following two worked examples we illustrate how these quantities are determined in some practical cases.

WORKED EXAMPLE 2.18

A cantilever beam of mass m and length L is clamped at the left end so that the beam makes an angle α with the vertical, Figure 2.69. The beam is only loaded by its own weight. Determine the axial force, the shear force and the bending moment as functions of the coordinate x.

SOLUTION

The simplest way to solve this problem is probably to isolate that portion of the beam that is located to the right of the section. In this way we do not have to determine the reaction exerted on the beam by the wall. The weight of the isolated portion is $mg(L - x)/L$, Figure 2.70.

There are various conventions when it comes to choosing the positive directions of N, V and M, but we will not concern ourselves with these here. Using the equilibrium conditions we obtain

$$\swarrow \quad N + \frac{L-x}{L} mg \cos\alpha = 0 \qquad \Rightarrow N = -\frac{L-x}{L} mg \cos\alpha$$

$$\nwarrow \quad V - \frac{L-x}{L} mg \sin\alpha = 0 \qquad \Rightarrow V = \frac{L-x}{L} mg \sin\alpha$$

$$\widehat{S} \quad M + \frac{L-x}{L} mg \sin\alpha \cdot \frac{L-x}{2} = 0 \Rightarrow M = -\frac{(L-x)^2}{2L} mg \sin\alpha$$

It is customary to present the results in the form of what is called axial force, shear force and moment diagrams as shown in Figure 2.71.

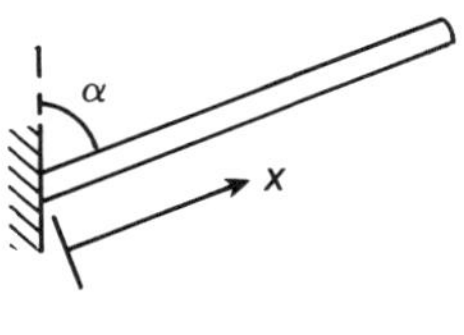

Figure 2.69

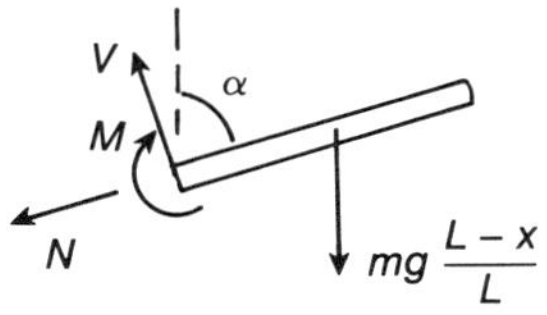

Figure 2.70

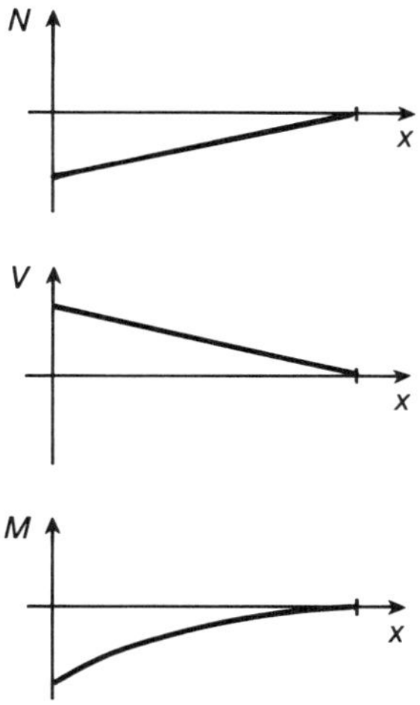

Figure 2.71

WORKED EXAMPLE 2.19

A beam of negligible weight is supporting a linearly distributed load as shown in Figure 2.72. The intensity q of the load, defined as force per unit length of the beam, increases from zero at the left end to a maximum value q_0 at the right end of the beam. Determine the shear force and the bending moment distributions.

SOLUTION

Our first step will be to determine the reactions denoted by R_A and R_B in the free-body diagram of Figure 2.73.

The intensity q depends on the coordinate x according to $q(x) = q_0 x/L$. We now consider a small element of the beam of length dx. The loading applied to this element is $q(x)$ dx. The total load is given by

$$Q = \int_0^L q(x)\,dx = \frac{q_0}{L}\int_0^L x\,dx = \frac{q_0 L}{2} \tag{2.40}$$

In an analogous way we determine the clockwise moment of the distributed load about A:

$$M_A = \int_0^L xq(x)\,dx = \frac{q_0}{L}\int_0^L x^2\,dx = \frac{q_0 L^2}{3} \tag{2.41}$$

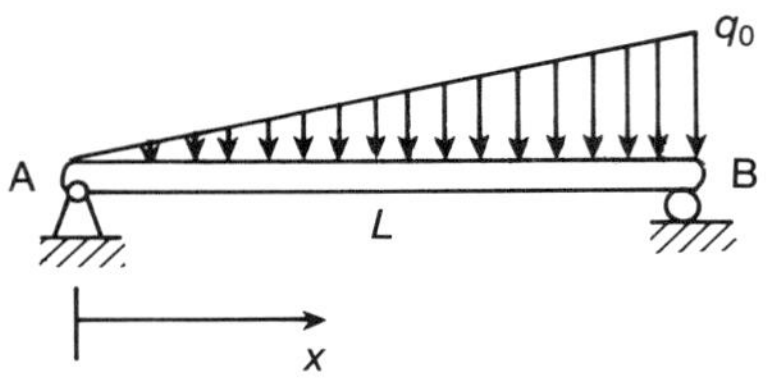

Figure 2.72

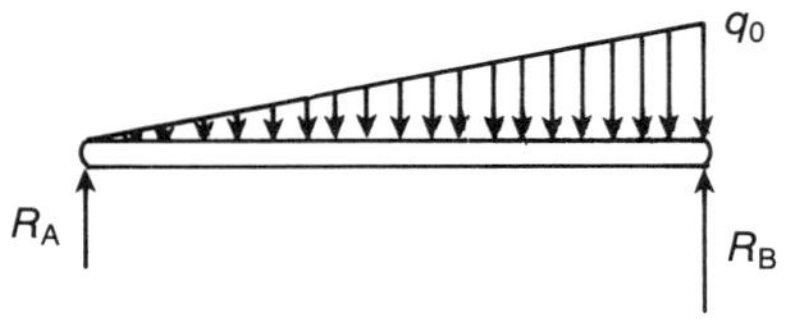

Figure 2.73

The equations of equilibrium for the beam yield

$$\uparrow \quad R_A + R_B - \frac{q_0 L}{2} = 0$$

$$\widehat{A} \quad -R_B L + \frac{q_0 L^2}{3} = 0$$

Solving for the reactions we obtain $R_A = q_0 L/6$ and $R_B = q_0 L/3$. The next step is to cut the beam at an arbitrary point C. Isolating portions AC and CB we obtain the free-body diagrams shown in Figure 2.74.

To obtain the equilibrium equations of AC we need expressions for the resultant force and the resultant moment of the distributed load applied to AC. Using the procedure leading to Equations (2.40) and (2.41), but integrating from 0 to x instead, we find that

$$\uparrow \quad R_A - V - \int_0^x q(\xi)\, d\xi = 0$$

$$\widehat{A} \quad -M + Vx + \int_0^x \xi q(\xi)\, d\xi = 0$$

The solution is

$$V = \frac{q_0(L^2 - 3x^2)}{6L}$$

$$M = \frac{q_0 x(L^2 - x^2)}{6L}$$

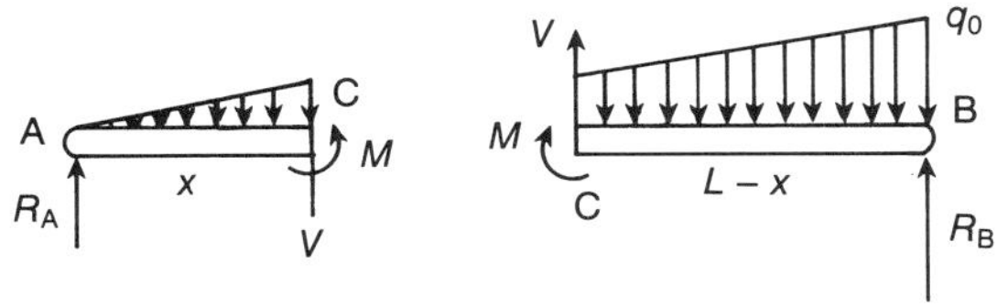

Figure 2.74

REMARK 1

We note that at $x = 0$ and $x = L$ the shear force is equal to R_A and $-R_B$, respectively, while the bending moments vanish. This can be checked by placing the section very close to each one of the supports and studying the equilibrium of the portion of the beam between the section and the support.

REMARK 2

All problems involving distributed loads may be treated using a similar procedure.

REMARK 3

We observe that M and V satisfy the equation $dM/dx = V$. It may be shown that this relation is generally valid provided that we have followed a certain convention when choosing the positive directions of M and V.

2.4.3 Trusses

Another important application of section forces is provided by *plane trusses*. A truss is a special type of construction that is used in civil engineering applications, e.g. bridges and roofs, to support various types of loads.

A truss consists of straight *members* that are connected by smooth *joints*. In most cases gravitational forces on the members may be neglected compared with the applied loading. We shall assume that the loading is applied at the joints only. An example is shown in Figure 2.75.

In the case of a plane truss all members are located in the same plane. In practical cases there are often two identical halves connected by transverse members. If the loading is applied symmetrically it will be equally distributed on both halves.

Our assumptions mean, of course, that certain approximations have been made. The weights of the members have been neglected, and the joints are in many practical cases far from smooth. Nevertheless, this crude model is considered to give useful results.

From the above assumptions it follows that each member of a truss is loaded at its end points only. Only two forces are applied to each member. From the discussion of two-force systems in Section 2.2.1 we conclude that these forces are of equal

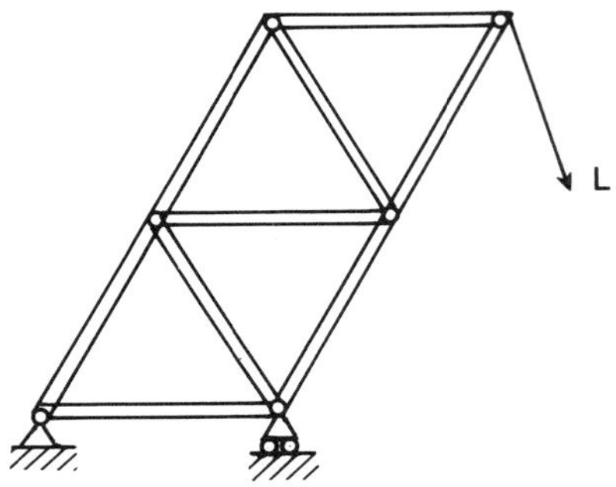

Figure 2.75

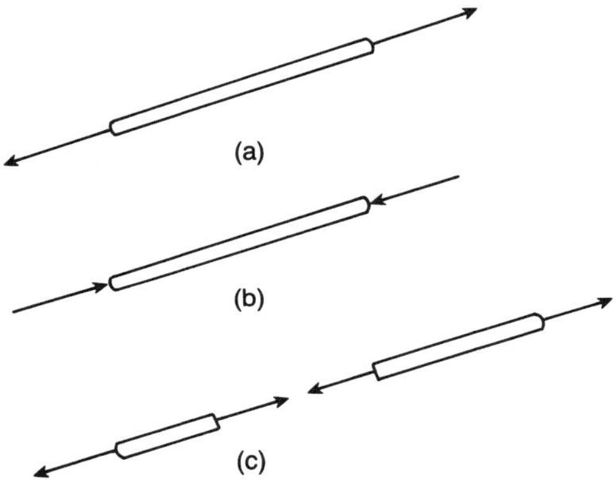

Figure 2.76

magnitude, collinear and of opposite direction. We have two possibilities according to Figure 2.76(a)–(b). The member is subjected to either tensile or compressive forces.

If we only consider a portion of a member we see that the section force is a purely axial force, either tensile or compressive, see Figure 2.76(c). From now on we shall draw all forces in free-body diagrams as tensile forces regardless of their true direction. If a force in a member turns out to be negative, it means that the force is compressive.

Each member of a truss must be dimensioned to make sure that it will resist the forces applied to it. To determine the necessary dimensions of the cross-sections and to choose a suitable material we need to know the forces in all members (magnitude and direction).

We shall utilize a systematic method, the *method of joints*, the idea of which is that each joint is isolated and the equilibrium of that joint is studied. The technique will be illustrated in the following worked example.

WORKED EXAMPLE 2.20

Determine the forces in all members of the truss shown in Figure 2.77.

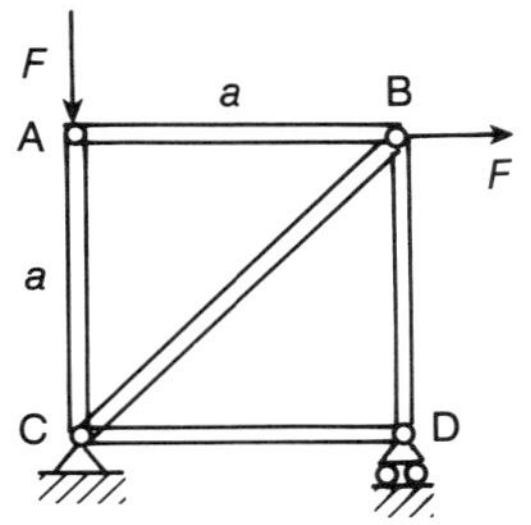

Figure 2.77

SOLUTION

As was mentioned above, all forces exerted on members of the truss by the joints will be treated as tensile forces in our free-body diagrams. According to Newton's third law the forces exerted by the members on the joints will then be directed outwards with respect to the joint.

Furthermore, we note that all forces applied to a joint intersect at one point, i.e. only two independent equations of equilibrium exist at each joint. Hence, it is appropriate to start by isolating a joint where only two members are connected, which in this case means joint A, Figure 2.78.

The equilibrium equations yield

$$\rightarrow \quad N_{AB} = 0$$

$$\downarrow \quad N_{AC} + F = 0 \Rightarrow N_{AC} = -F$$

Here, the forces have been denoted in an obvious manner. As the force N_{AC} is negative a compressive force is applied to member AC.

We continue by isolating joint B, Figure 2.79:

$$\leftarrow \quad \frac{N_{BC}}{\sqrt{2}} - F = 0$$

$$\downarrow \quad \frac{N_{BC}}{\sqrt{2}} + N_{BD} = 0$$

We then have $N_{BC} = F\sqrt{2}$ and $N_{BD} = -F$.

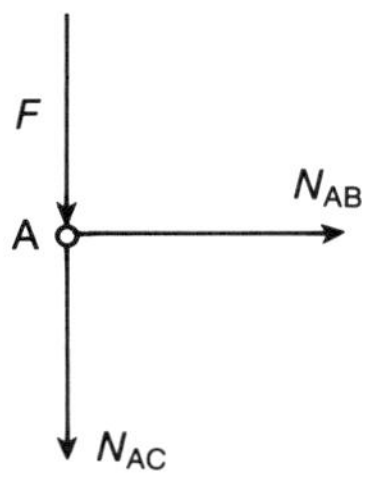

Figure 2.78

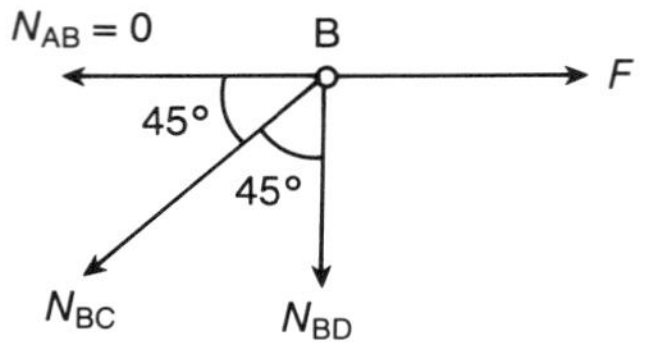

Figure 2.79

Moving to joint D, Figure 2.80, we find that

$$\leftarrow \quad N_{CD} = 0$$

$$\uparrow \quad V_D + N_{BD} = 0 \Rightarrow V_D = F$$

Here V_D is the reaction supported at D.

Finally, at joint C, Figure 2.81, we have

$$\uparrow \quad N_{AC} + \frac{N_{BC}}{\sqrt{2}} + V_C = 0$$

$$\leftarrow \quad H_C - \frac{N_{BC}}{\sqrt{2}} - N_{CD} = 0$$

which gives $V_C = 0$ and $H_C = F$.

To sum up we have

$$\begin{cases} N_{AB} = N_{CD} = 0 \\ N_{AC} = N_{BD} = -F \\ N_{BC} = F\sqrt{2} \end{cases}$$

Members AC and BD are loaded by compressive forces, each of magnitude F, BC by a tension $F\sqrt{2}$, while AB and CD are unloaded. This does not mean that these members are superfluous. Removing CD, for instance, and applying an arbitrarily small horizontal force at joint D will obviously cause the truss to collapse.

We also find that the reactions are $V_C = 0$ and $V_D = H_C = F$. It is left as an exercise to show that the equations of equilibrium for the truss, considered as one body, are satisfied.

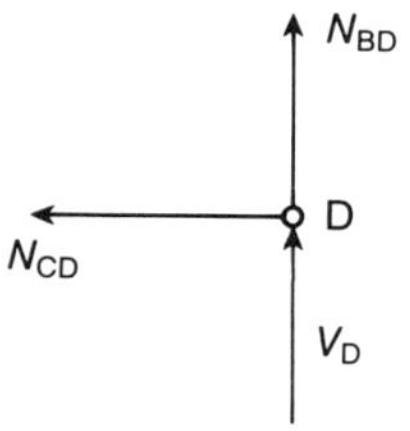

Figure 2.80

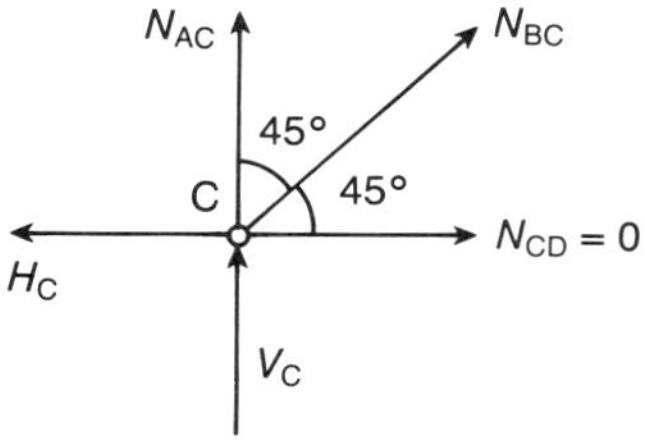

Figure 2.81

REMARK

We started to analyse the truss at joint A, because only two members are connected there. We then proceeded by taking the joints in such an order that the number of unknown forces was always two and thereby possible to determine from the equilibrium equations. An alternative procedure would have been to start by determining the reactions at the supports. Then we could have proceeded joint by joint starting at D.

PROBLEMS

○ **2.80** Draw axial force, shear force and moment diagrams for the two cases shown.

(a)

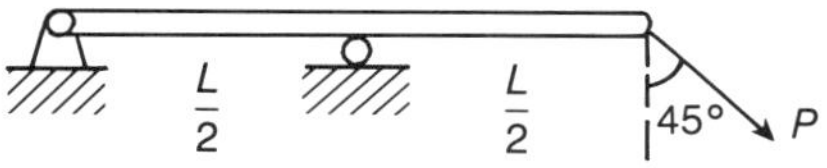

(b)

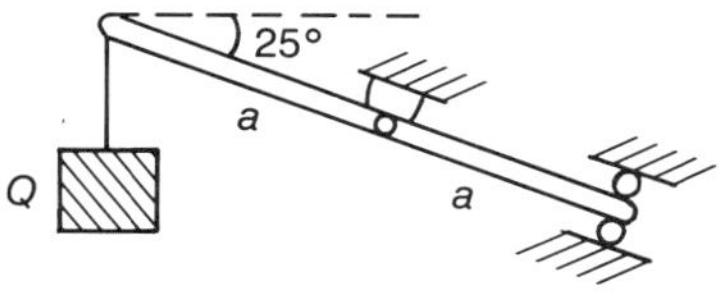

2.81 Draw shear force and moment diagrams for the beam AB.

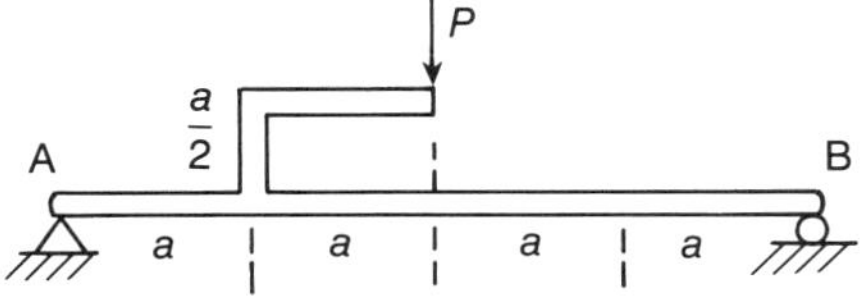

2.82 A load of weight Q is supported by a cable and two pulleys, each one of radius $a/4$, that are mounted on a beam of length $3a$. Draw axial force, shear force and moment diagrams for beam AB.

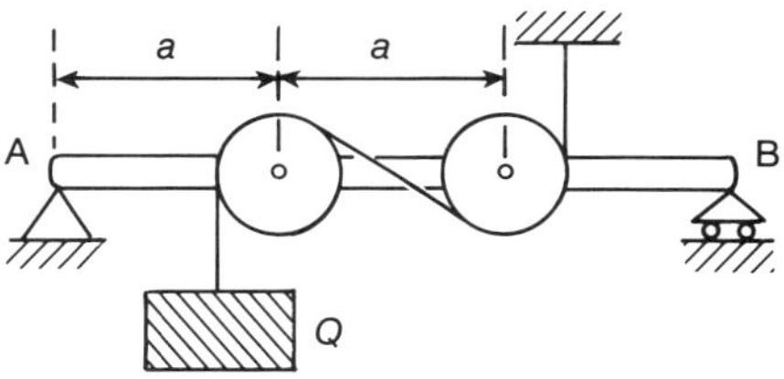

2.83 Draw axial force, shear force and moment diagrams for beam AB.

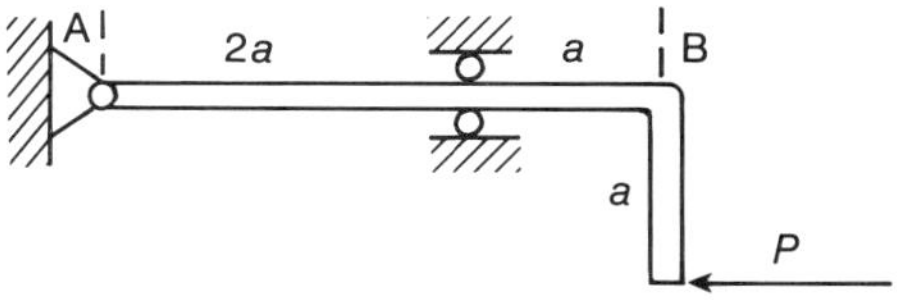

2.84 Two beams, each of mass m and length L, are welded together at right angles. The system is placed on a smooth table and is loaded by a vertical force P as shown overleaf. Determine the section forces and couples in one of the beams as functions of the distance from the vertex of the right angle.

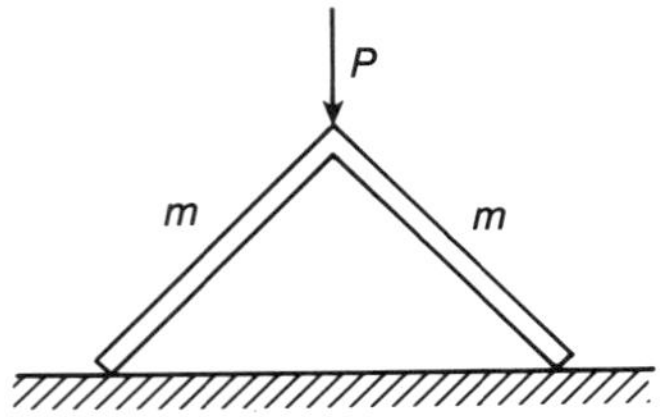

◎ **2.85** The system shown consists of two identical beams, each of mass m. Determine the section forces and couples for beam AB as functions of the distance from A.

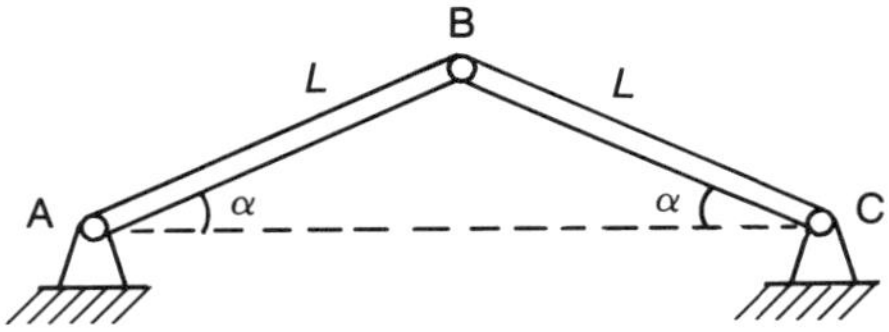

◎ **2.86** The frame shown is loaded by a force P. The weights of the bars may be neglected compared with P. Determine the axial force, shear force and bending moment distributions for bars BD and AC.

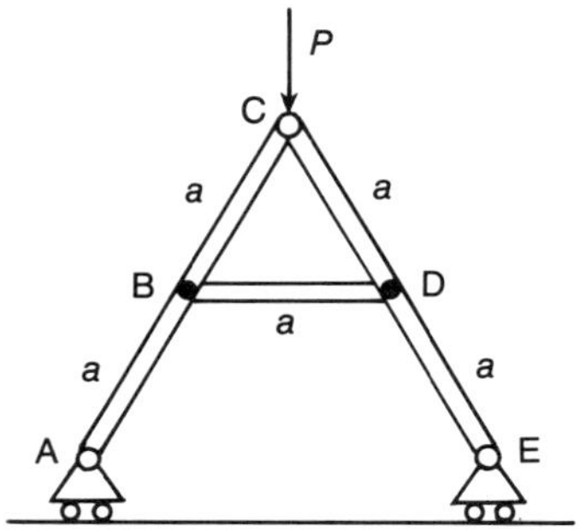

2.87 The ability of a beam to resist a bending moment is characterized by a quantity called the *section modulus*. For a section of rectangular shape the section modulus is proportional to the square of the height y of the rectangle. A horizontal cantilever beam of constant width and varying height is loaded by a force P as shown. Determine the height y as a function of x if there should be the same risk of fracture due to bending at every section of the beam.

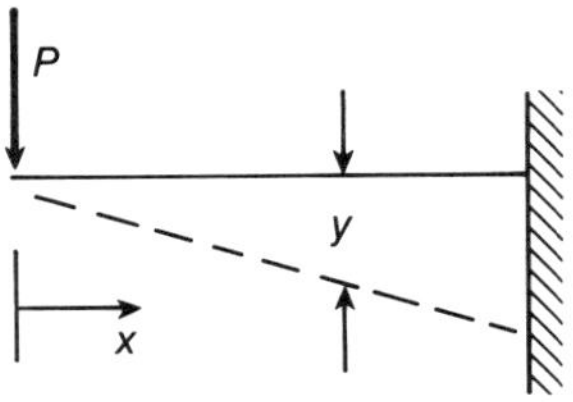

○ **2.88** Determine the forces in each member of the truss.

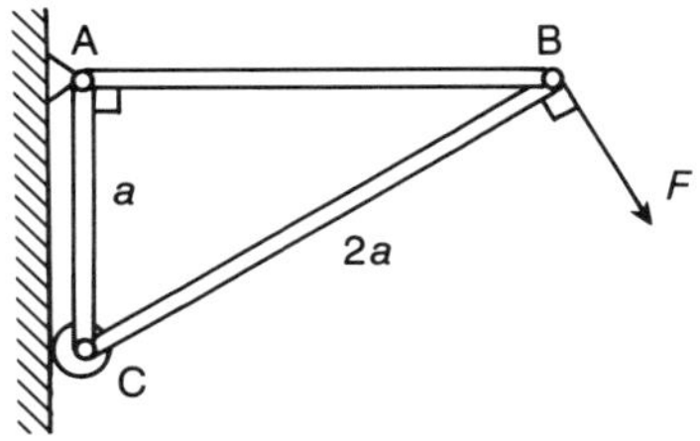

2.89 Determine the forces in each member of the truss. All members are of the same length.

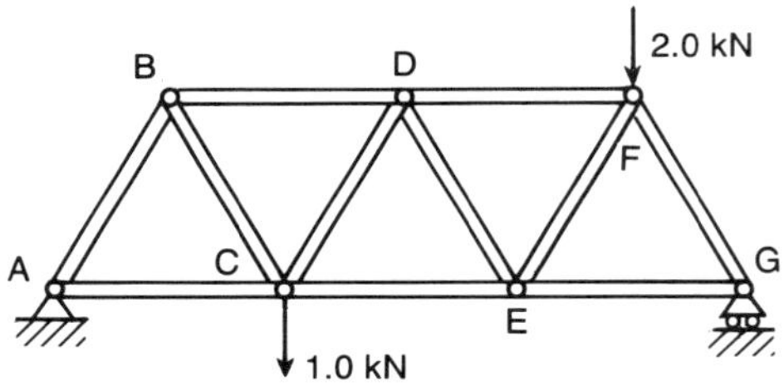

○ **2.90** Calculate the forces in members AB and AC.

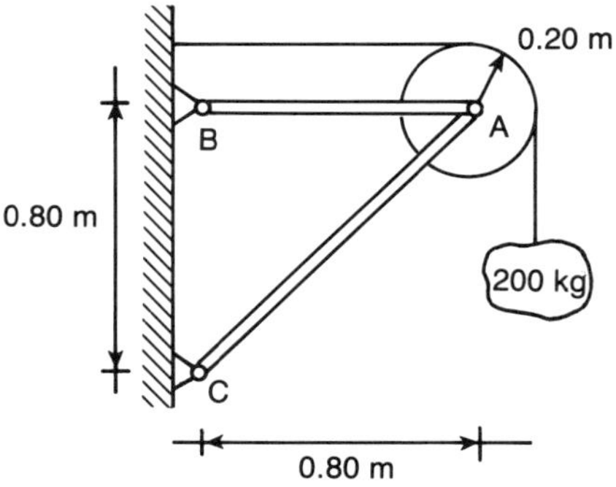

◎ **2.91** Determine the force in member BD (force F is parallel to AD).

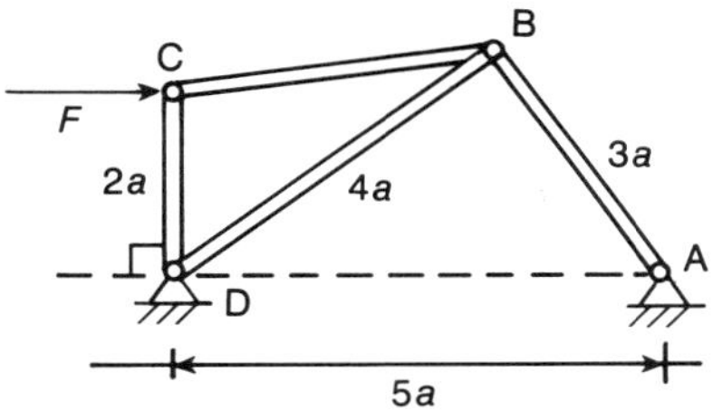

2.92 Determine the forces in members AB, AC and AD if
(a) $\alpha = 30°$
(b) $\alpha = 45°$.

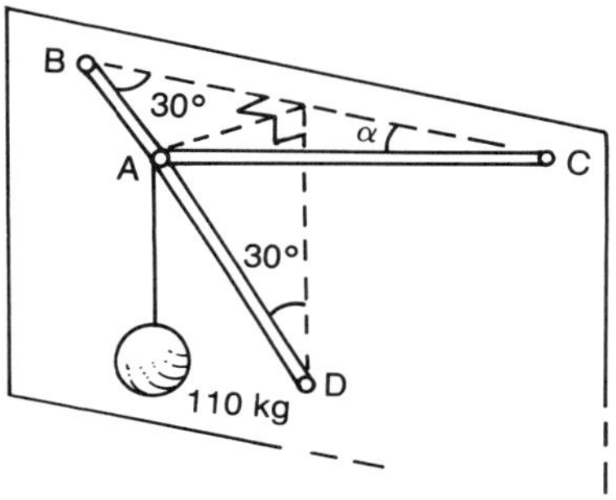

2.93 The legs of a light camera tripod OABC are assumed to be connected by ball-and-socket joints at A, B, C and O. What are the forces in the legs caused by the following loads?
(a) A 12 kg camera (Q)
(b) A horizontal force (P) in the vertical plane OCD.

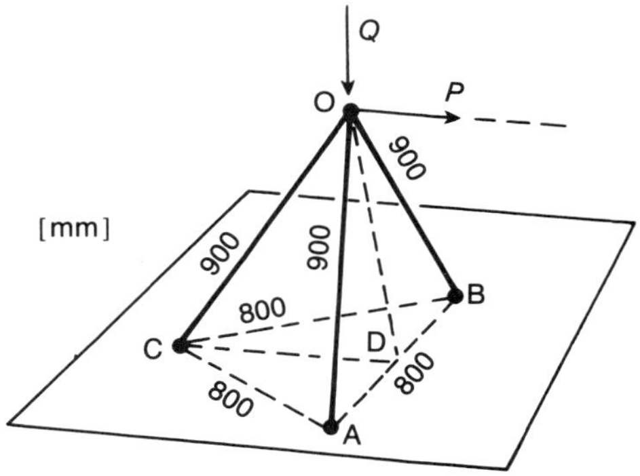

◎ **2.94** ABC is an equilateral triangle composed of three bars in a horizontal plane. The vertical members AD, BE and CF are all of the same length as the sides of triangle ABC. The loads applied at B and C are horizontal. Determine the forces in all members.

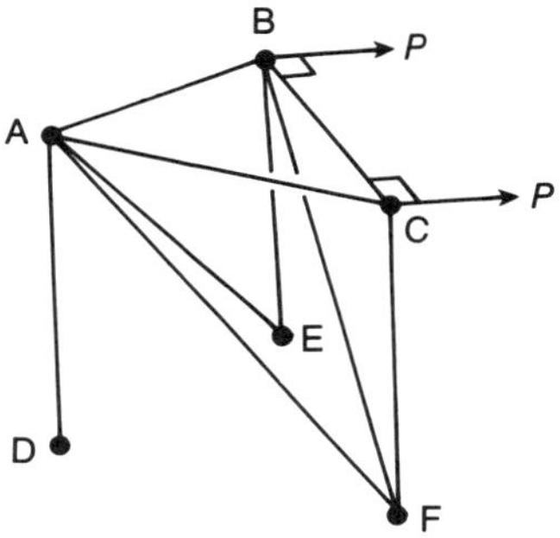

2.5 Friction

2.5.1 Static and kinetic friction

In this section we shall be concerned with a phenomenon that we briefly touched upon in Section 2.2.2, namely friction. Let us start by considering a body placed on a horizontal surface and subjected to a horizontal force P, Figure 2.82(a). From experiments we may conclude that the body is in equilibrium as long as the magnitude of P does not exceed a certain value.

From the free-body diagram of Figure 2.82(b) we see that, as well as force P and the weight Q, a reaction exerted by the horizontal surface is acting on the body. As the conditions of equilibrium are fulfilled, this reaction must have a component $F = P$

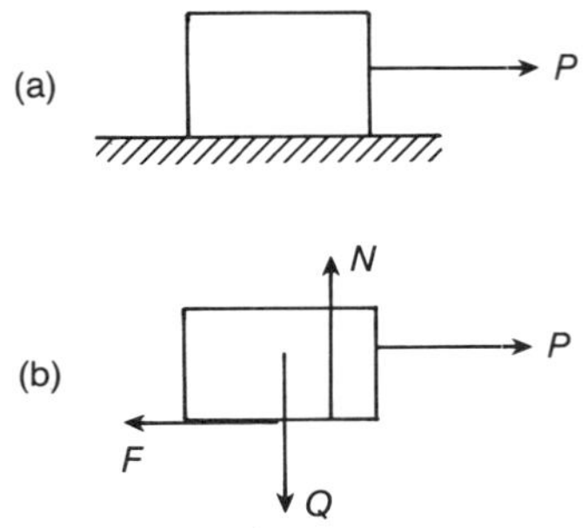

Figure 2.82

parallel to the contact surface and a component $N = Q$ normal to this surface. The components F and N are called the *friction force* and the *normal force*, respectively. The origin of the friction force is rather obvious if we consider an enlarged view of the contact surfaces, Figure 2.83. Even if the surfaces look perfectly smooth to the human eye, their structure is considerably rough in reality. All these irregularities become enmeshed in one another and thus prevent the displacement that force P is trying to initiate.

In this situation *friction* occurs at the contact. As the body is not free to move horizontally but is subjected to a constraint, a corresponding constraining force, namely the friction force F, must be exerted on the body. This force is always directed in the opposite sense with respect to the tendency to slip. From Figure 2.83 it is evident that there is contact between the body and the supporting surface at a large number of points. Consequently, a correspondingly large number of reactions are exerted on the body by the surface. The above-mentioned reaction with components F and N is, in fact, the resultant to this system of forces. Contact surfaces of this type are called *rough* surfaces in contrast to *smooth* surfaces, where friction may be neglected.

We now assume that we perform experiments where a vertical force T is applied to the body in addition to the above-mentioned forces, see Figure 2.84. As long as the body is in equilibrium, we have

$$F = P \qquad N = T + Q$$

If T is held constant and P is increased gradually, we will find that the ratio $F/N = P/(T + Q)$ will also increase up to a certain maximum value where the body is starting to move.

Figure 2.83

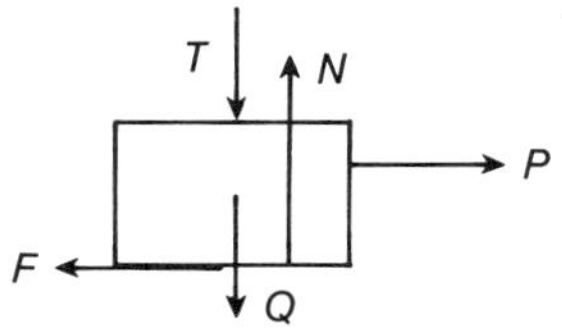

Figure 2.84

If the experiment is repeated for other values of T we will find that the critical value of the ratio F/N corresponding to impending motion is practically unchanged, i.e. independent of the normal force N. From experience it is known that this result applies fairly generally to cases of two bodies in contact with each other, even though we have chosen a rather simple experiment to describe the result.

The reason for the occurring motion is, of course, that the material of the two bodies in contact cannot withstand arbitrarily large forces without being deformed. When the irregularities of the contact surfaces, see Figure 2.83, become sufficiently deformed, motion can no longer be prevented by the enmeshing effect described above.

The bodies of Figure 2.85 have a common tangential plane at the point of contact. The forces exerted by the bodies on each other can always be resolved into components: $\mathbf{F}$ and $-\mathbf{F}$ in the tangential plane, $\mathbf{N}$ and $-\mathbf{N}$ normal to the plane. When the contact occurs at a plane surface rather than at a point, $\mathbf{F}$ and $\mathbf{N}$ must be interpreted as components of the resultant of a complex system of reactions that are distributed over the contact surface.

Equilibrium can only exist if:

- the equilibrium equations are satisfied for both bodies,
- the ratio $|\mathbf{F}|/|\mathbf{N}|$ does not exceed a certain value, the *coefficient of static friction*, which is denoted by μ_S.

This *law of static friction* is commonly expressed as

$$\frac{|\mathbf{F}|}{|\mathbf{N}|} < \mu_S \text{ at equilibrium}$$

In two-dimensional problems it is normally superfluous to use vector formalism. Instead we may write

$$\frac{|F|}{|N|} < \mu_S \text{ at equilibrium} \tag{2.42}$$

Thus the maximum value attainable by the ratio $|F|/|N|$ at equilibrium is μ_S, i.e.

$$|F|_{max} = \mu_S |N| \tag{2.43}$$

This situation *only* occurs when the body is on the verge of slipping. In a general case the magnitude of the friction force is *less than* this value at equilibrium.

The value of the coefficient of friction μ_S depends primarily on the nature of the contact surfaces, i.e. material, roughness, etc. However, the area of the contact surface

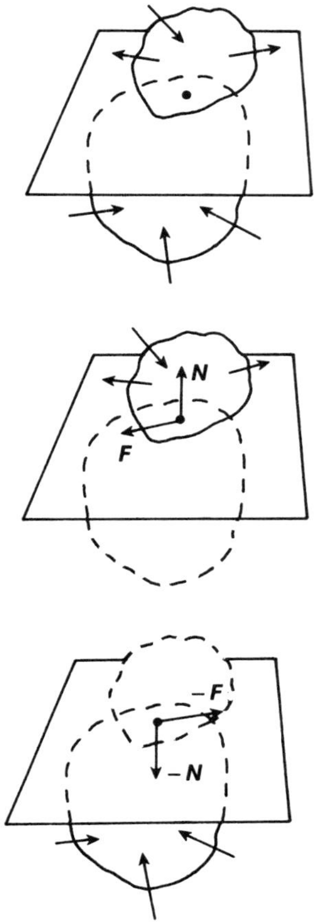

Figure 2.85

is of less importance. Some typical numerical values are given in Table 2.1. It should be noted that such values are *very* approximate. It should also be observed that there is no theoretical reason why the coefficient of friction should be less than unity. We stated above that the coefficient of friction is independent of the normal force N. In reality, however, it is observed that μ_s increases slightly for large values of N.

If friction occurs at more than one surface of contact the law of static friction, Equation (2.42), must be satisfied at each of these. See Worked Example 2.22 below.

When solving problems we often face a situation where forces F and N can be expressed as functions of some variable, say φ. This variable can, for instance, be an angle defining the position of the system studied. In such cases the ratio F/N can be expressed as a function $f(\varphi)$. The condition $|F|/|N| < \mu_s$ then yields $|f(\varphi)| < \mu_s$ which, in its turn, yields

$$-\mu_s < f(\varphi) < \mu_s$$

Table 2.1 **Coefficients of friction**

	μ_S	μ_K
Steel/steel	0.1–0.3	0.03–0.25
Metal/wood	0.5–0.7	0.2–0.5
Leather/metal	0.3–0.6	0.18–0.25
Brake lining/cast iron	0.4	0.3
Hemp rope/metal	0.3	0.2
Teflon/Teflon	0.04	0.04
Rubber/metal	0.4	0.3
Rubber/asphalt	0.7–1.0	0.5–0.8

These conditions, one of which is often trivially satisfied, give an interval of φ within which equilibrium is possible – an *interval of equilibrium* (see Worked Example 2.24 below).

Kinetic friction

We now return to the introductory example of Figure 2.82. As long as the body is in equilibrium, we have $F = P$, $N = Q$ and $|F|/|N| < \mu_S$. Consequently, $P < \mu_S Q$, see Figure 2.86.

If P, and thus F, is gradually increased, then the ratio F/N will sooner or later reach the value μ_S. This happens when $P = \mu_S Q$. The body is then starting to move. It is then observed that the force P required to keep the body in motion with constant speed is less than $\mu_S Q$. At the same time the equation $F = P$ must still be satisfied, since the body will be at rest relative to some inertial frame as long as it is moving with constant speed, cf. Section 2.1.1. Evidently, the value of the friction force F is less than the equilibrium value $\mu_S Q$.

From experience it is known that

$$F = \mu_K N \tag{2.44}$$

when slipping occurs at a contact surface. Obviously, $\mu_K < \mu_S$. The quantity μ_K is called the *coefficient of kinetic friction*. Some numerical values are given in Table 2.1.

There are several simple applications of the theory, e.g. problem solving in elementary courses, where the difference between μ_K and μ_S is disregarded and the same numerical value is used for both of them, normally denoted by μ.

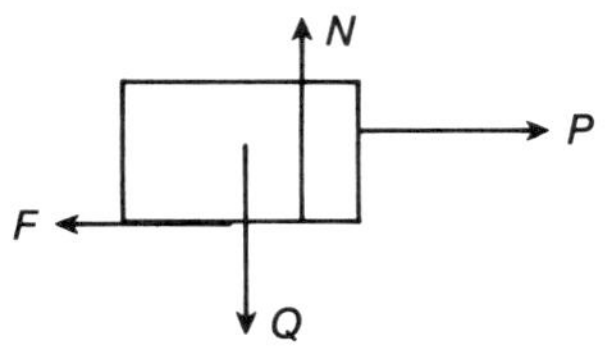

Figure 2.86

The coefficient of kinetic friction μ_k is not entirely independent of the speed of the body. If force P is further increased after slipping has occurred, the body will accelerate. It is then observed that μ_k decreases as the speed increases.

All that has been said about friction so far is applicable to *dry contact surfaces*. This type of friction is known as *dry friction* or *Coulomb friction*. In many technical applications, however, lubricants of various types are used. The contact surfaces are then separated by a thin liquid film. The purpose is to reduce the effects of friction and to prevent the surfaces from wearing. This type of friction is called *fluid friction*, and the theory presented here is not applicable.

In this chapter we shall only be concerned with statics problems. Whenever we refer to the coefficient of friction from now on, we will have the coefficient of static friction μ_s in mind. The subscript 's' can then be excluded, and the coefficient is simply denoted by μ.

Applications

WORKED EXAMPLE 2.21

Two identical bars, AB and BC, are connected as shown in Figure 2.87. A and B are smooth joints. What is the minimum value of the coefficient of friction at C to maintain equilibrium?

SOLUTION

To determine the coefficient of friction required at C, we first have to determine the friction and normal forces exerted on bar BC by the wall. To do this we isolate the system AB + BC.

From the free-body diagram of Figure 2.88 we see that the normal force N can be determined immediately:

$$\overset{\frown}{A} \quad N \cdot 2l \sin\varphi - 2Q\frac{l}{2}\cos\varphi = 0 \Rightarrow N = \frac{1}{2}Q\cot\varphi \qquad (2.45)$$

To determine the friction force F, however, we have to isolate bar BC separately, Figure 2.89.

To avoid including the reactions at B, we take the sum of moments about B:

$$\overset{\frown}{B} \quad Fl\cos\varphi - Q\frac{l}{2}\cos\varphi - Nl\sin\varphi = 0$$

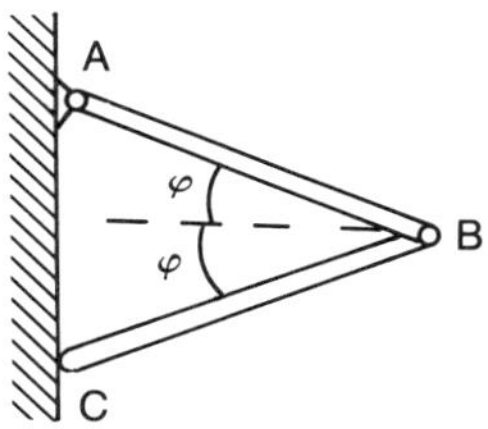

Figure 2.87

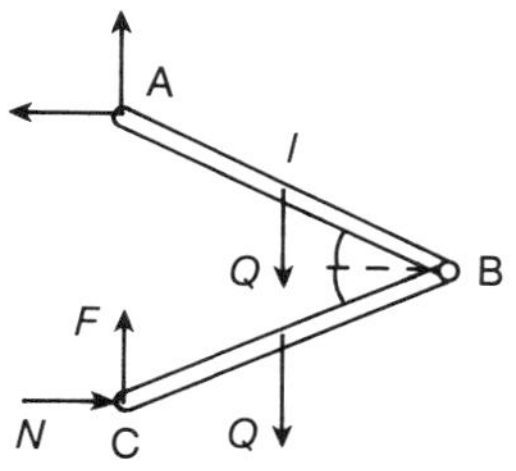

Figure 2.88

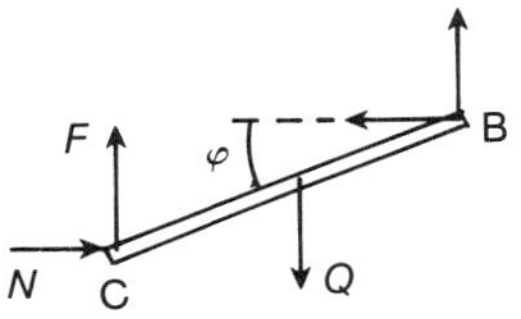

Figure 2.89

Using (2.45) we obtain

$$F = \frac{1}{2}Q + N \tan \varphi = Q \tag{2.46}$$

From Equations (2.45) and (2.46) we have $F/N = 2 \tan \varphi$.

From the law of static friction, Equation (2.42), we finally obtain

$$\mu > \frac{|F|}{|N|} = 2 \tan \varphi \Rightarrow \mu_{\min} = 2 \tan \varphi$$

WORKED EXAMPLE 2.22

Two bodies A and B of masses 10 kg and 25 kg, respectively, are placed on a rough horizontal surface, Figure 2.90. At the contact between A and B the coefficient of friction is 0.30, while it is 0.10 between B and the supporting surface. What is the maximum allowable value of force P if the system is to remain in equilibrium?

SOLUTION

In this problem we do not know in advance if slipping is most likely to occur between A and B or between B and the supporting surface. Consequently, the ratio between the friction and normal forces must be compared with the corresponding coefficient of friction at both contact surfaces. For this reason, we isolate the two bodies and obtain the free-body diagrams shown in Figure 2.91.

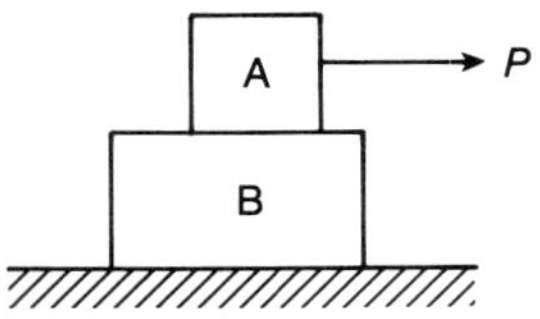

Figure 2.90

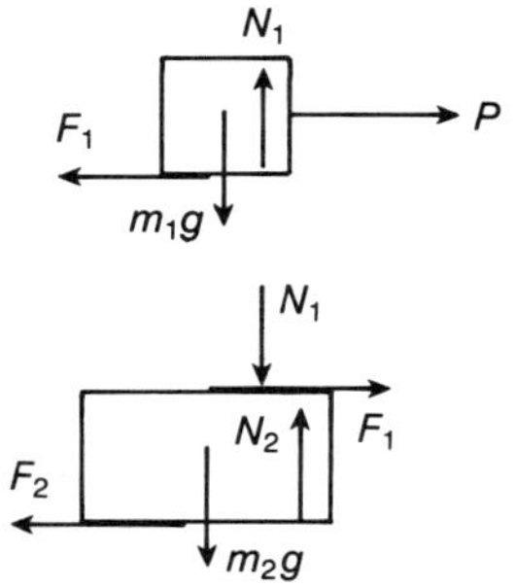

Figure 2.91

The following equilibrium equations follow immediately:

Body A:

$$\rightarrow \quad P - F_1 = 0 \tag{2.47}$$

$$\uparrow \quad N_1 - m_1 g = 0 \tag{2.48}$$

Body B:

$$\rightarrow \quad F_1 - F_2 = 0 \tag{2.49}$$

$$\uparrow \quad N_2 - N_1 - m_2 g = 0 \tag{2.50}$$

Equations (2.47) and (2.48) give $F_1 = P$ and $N_1 = m_1 g$. Hence,

$$\frac{F_1}{N_1} = \frac{P}{m_1 g}$$

Equations (2.47) and (2.49) give $F_2 = P$, while Equations (2.48) and (2.50) give $N_2 = (m_1 + m_2)g$. Hence,

$$\frac{F_2}{N_2} = \frac{P}{(m_1 + m_2)g}$$

To maintain equilibrium both of the following conditions must be satisfied:

$$\frac{|F_1|}{|N_1|} < 0.30 \Rightarrow P < 0.30\, m_1 g \approx 29.4 \text{ N}$$

$$\frac{|F_2|}{|N_2|} < 0.10 \Rightarrow P < 0.10(m_1 + m_2)g \approx 34.3 \text{ N}$$

As the law of static friction must be satisfied at both contacts simultaneously the maximum allowable value of force P is:

$$P_{max} \approx 29 \text{ N}$$

The maximum risk of slipping thus occurs at the contact between the bodies.

REMARK

Note that $|F_1|/|N_1| = 0.30$ while $|F_2|/|N_2| < 0.10$ when slipping is impending. This shows the importance of using the inequalities when the law of friction is utilized. In many cases it is common practice, especially among students, to replace the inequality $|F|/|N| < \mu$ by the equation $|F|/|N| = \mu$ in order to study the limiting case of impending motion. Obviously, this procedure may lead to an erroneous result in this case.

WORKED EXAMPLE 2.23

A small wedge of negligible weight has been pressed under a horizontal beam of mass 20 kg, Figure 2.92. The coefficient of friction μ is 0.20 at both contact surfaces of the wedge.

(a) Determine the force P required to keep the wedge in its place if $\alpha = 24°$.
(b) Determine the maximum value of α if the wedge is self-locking, i.e. if equilibrium can be maintained even if $P = 0$.

SOLUTION

Our first step is to isolate the beam and the wedge separately.

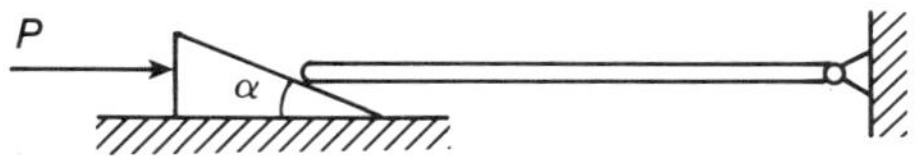

Figure 2.92

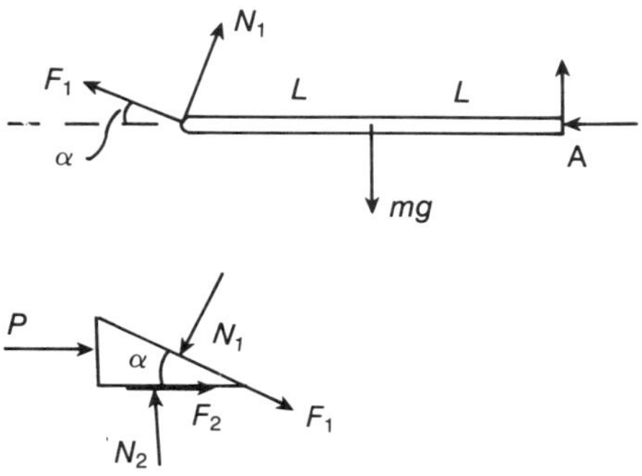

Figure 2.93

From the free-body diagrams of Figure 2.93 we obtain the following set of equilibrium equations:

The beam:

$$\overset{\frown}{A} \quad F_1 \sin \alpha \cdot 2L + N_1 \cos \alpha \cdot 2L - mgL = 0 \tag{2.51}$$

The wedge:

$$\uparrow \quad N_2 - F_1 \sin \alpha - N_1 \cos \alpha = 0 \tag{2.52}$$

$$\rightarrow \quad P + F_2 + F_1 \cos \alpha - N_1 \sin \alpha = 0 \tag{2.53}$$

This situation differs from the one treated in the previous worked example in that slipping must occur at *both* contact surfaces simultaneously in this case; that is, if the wedge is on the verge of slipping at one contact surface, then the same must be true at the other one.

We now consider the limiting case where force P has the minimum allowable value to maintain equilibrium. Then $F_1 = \mu N_1$ and $F_2 = \mu N_2$. These two equations together with Equations (2.51)–(2.53) yield the minimum value of P at equilibrium:

$$P_{\min} = mg \frac{(1-\mu^2) \sin \alpha - 2\mu \cos \alpha}{2(\mu \sin \alpha + \cos \alpha)} \tag{2.54}$$

(a) Inserting $\mu = 0.20$ and $\alpha = 24^\circ$ into Equation (2.54) we obtain $P_{\min} \approx 2.47$ N.
(b) In this case $P_{\min} = 0$ which, when inserted into Equation (2.54), yields

$$\tan \alpha = 2\mu/(1-\mu^2).$$

For $\mu = 0.20$ we have $\alpha \approx 22.6^\circ$.
This is the maximum value of α that can be allowed if the wedge is to be self-locking.

WORKED EXAMPLE 2.24

A uniform hemisphere of radius r is suspended by a cord of length $2r$. The body is supported by a rough wall, Figure 2.94. The coefficient of friction is 0.20. For what values of the angle φ, such that $0 < \varphi < \pi/2$, is it possible to maintain equilibrium?

SOLUTION

Isolating the hemisphere we obtain the free-body diagram shown in Figure 2.95. Note that the direction of the friction force is not obvious in this problem. The force is always in the opposite sense with respect to the tendency of slipping. Contrary to the previous worked examples we cannot tell in advance in what direction the body is likely to move, as this depends on the value of φ. Hence, we have to assume that the force is directed in a certain sense, e.g. upwards.

From the conditions of equilibrium we obtain

$$\overset{\frown}{A} \quad N \cdot 4r \cos \varphi - Q\left(r \sin \varphi + \frac{3}{8} r \cos \varphi\right) = 0 \tag{2.55}$$

$$\rightarrow \quad N - S \sin \varphi = 0 \tag{2.56}$$

$$\uparrow \quad F - Q + S \cos \varphi = 0 \tag{2.57}$$

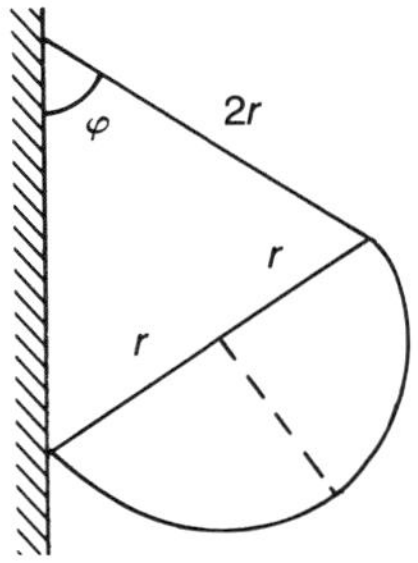

Figure 2.94

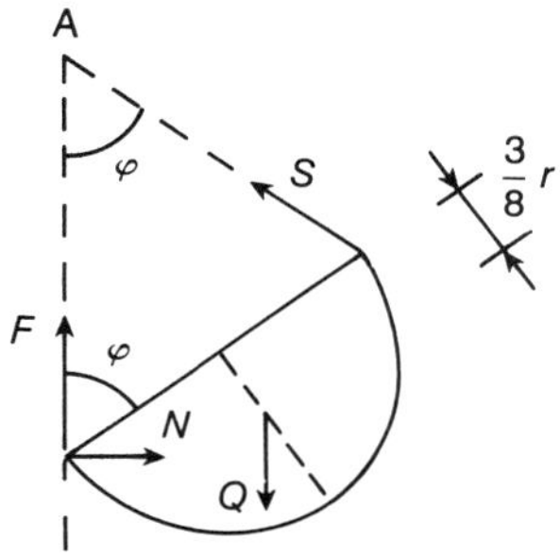

Figure 2.95

Equation (2.55) yields:

$$N = \frac{Q}{4}\left(\tan\varphi + \frac{3}{8}\right)$$

which together with Equations (2.56) and (2.57) gives

$$F = \frac{3Q}{4}\,\frac{\tan\varphi - \frac{1}{8}}{\tan\varphi}$$

Now it is obvious that the sign of F depends on the value of φ as was stated above. This means that

$$\frac{F}{N} = \frac{3\left(\tan\varphi - \frac{1}{8}\right)}{\tan\varphi\left(\tan\varphi + \frac{3}{8}\right)} = \frac{3\left(x - \frac{1}{8}\right)}{x\left(x + \frac{3}{8}\right)} = f(x)$$

where $x = \tan\varphi$.

From the law of static friction it follows that $|F|/|N| < 0.20$, i.e. $-0.20 < f(x) < 0.20$. All values of x ($= \tan\varphi$) that satisfy these inequalities correspond to possible positions of equilibrium. The graph of the function $f(x)$ is shown in Figure 2.96. As $0 < \varphi < \pi/2$, it follows that $\tan\varphi = x > 0$.

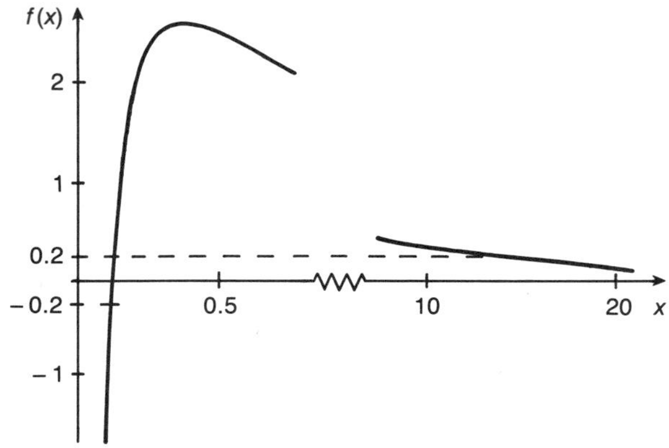

Figure 2.96

Equilibrium positions can only exist for those values of x corresponding to values of $f(x)$ between -0.20 and $+0.20$. From the graph we can only obtain a rough estimate. An exact computation, however, shows that the solution is given by

$$\frac{-123+\sqrt{15\,609}}{16} < x < \frac{117-\sqrt{13\,209}}{16} \quad \text{and} \quad x > \frac{117+\sqrt{13\,209}}{16}$$

This corresponds to

$$\begin{cases} 6.9^\circ < \varphi < 7.4^\circ \\ \varphi > 86.1^\circ \end{cases}$$

Note that one of the inequalities and thus one part of the solution is lost if we do not use the absolute values of F and N in the law of static friction.

REMARK

We saw in Worked Example 2.9 that there is a position of equilibrium at $\varphi \approx 7.13^\circ$ in the case of a smooth wall. Owing to friction equilibrium is now possible in an interval in the neighbourhood of this value. It is not clear intuitively that equilibrium may exist for large values of φ. This has to do with the fact that the normal force N tends to infinity as φ approaches 90°, while F tends to a finite value. This means that the ratio $|F|/|N|$ becomes very small for large values of φ, and thus equilibrium is possible.

2.5.2 Belt friction

There are a number of technical applications, e.g. belt drives, braking devices, etc., where a flexible belt or a cable is wrapped around a drum. In such cases friction plays an important part at the contact surface between the belt and the drum.

We consider a belt of negligible weight that is wrapped around a portion of a rough circular cylinder so that the belt is located in a plane normal to the axis of the cylinder, see Figure 2.97. If the coefficient of friction between the cylinder and the belt is sufficiently large, the belt may be subjected to tensions S_1 and S_2 of different magnitudes without slipping.

We introduce an angle θ and consider a small element of the belt of length $R\,d\theta$ as shown in Figure 2.98. The tension in the belt is denoted by $S(\theta)$. We consider the case $S_2 > S_1$ to start with. The belt then tends to slide clockwise with respect to the cylinder. The forces acting on the element are the tensions S and $S + dS$, the normal force dN and the friction force dF.

The equations of equilibrium are

$$\rightarrow \quad (S + dS)\cos\frac{d\theta}{2} - S\cos\frac{d\theta}{2} - dF = 0$$

$$\downarrow \quad (S + dS)\sin\frac{d\theta}{2} + S\sin\frac{d\theta}{2} - dN = 0$$

Since $d\theta$ is very small we have

$$\cos\frac{d\theta}{2} \approx 1 \qquad \sin\frac{d\theta}{2} \approx \frac{d\theta}{2}$$

Furthermore, dS may be neglected compared with S. We then have

$$\begin{cases} dS - dF = 0 \\ S\,d\theta - dN = 0 \end{cases}$$

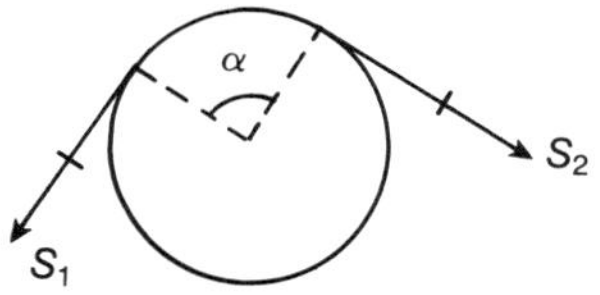

Figure 2.97

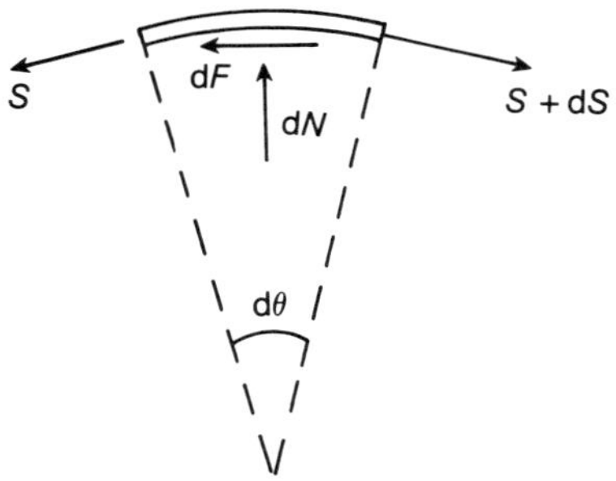

Figure 2.98

In the limiting case when S_2 is so large that the belt is on the verge of slipping, we know that $dF = \mu\ dN$. Hence,

$$dS = \mu S\ d\theta \Rightarrow \int_{S_1}^{S_2} \frac{dS}{S} = \mu \int_0^{\alpha} d\theta \Rightarrow \ln \frac{S_2}{S_1} = \mu\alpha \Rightarrow S_2 = S_1\ e^{\mu\alpha}$$

If the belt is to slip clockwise the minimum value of S_2 required is $S_1 e^{\mu\alpha}$. Consequently, no slipping occurs if

$$S_2 < S_1\ e^{\mu\alpha} \tag{2.58}$$

If the derivation is repeated for the case $S_2 < S_1$ we find that slipping counterclockwise does not occur if

$$S_1 < S_2\ e^{\mu\alpha} \tag{2.59}$$

The inequalities (2.58) and (2.59) can be summarized as follows:

$$S_2\ e^{-\mu\alpha} < S_1 < S_2\ e^{\mu\alpha} \tag{2.60}$$

where subscripts 1 and 2 are interchangeable. The tension at one end of the belt thus has to assume a value between two limiting values to prevent the belt slipping either way.

When deriving the above conditions of equilibrium we considered a circular cylinder. The results apply, however, equally well to a non-circular cylinder. To see this, we replace the radius R by the radius of curvature of the cylinder everywhere in the derivation. The meaning of the contact angle α in this case is shown in Figure 2.99.

If a belt is wrapped around two or more cylinders according to Figure 2.100, we have in the case $S_1 > S_2$ that no slipping occurs if

$$\begin{cases} S_1 < S'\ e^{\mu_1\alpha_1} \\ S' < S_2\ e^{\mu_2\alpha_2} \end{cases}$$

Here μ_1 and μ_2 are the coefficients of friction. Eliminating S' we obtain

$$S_1 < S_2\ e^{\mu_1\alpha_1 + \mu_2\alpha_2}$$

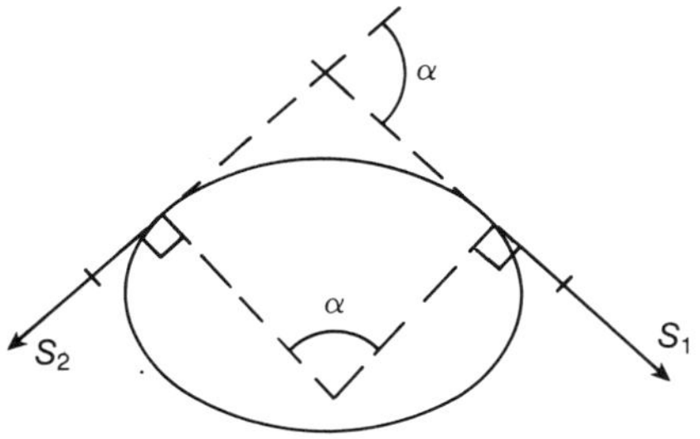

Figure 2.99

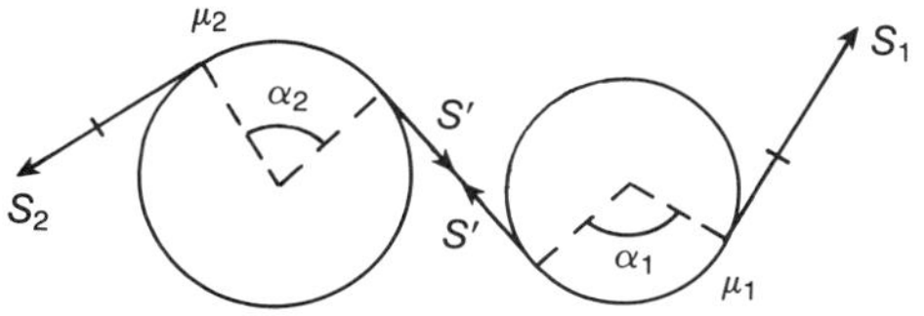

Figure 2.100

In the special case where the coefficients μ_1 and μ_2 are equal ($\mu_1 = \mu_2 = \mu$) we have

$$S_1 < S_2\, e^{\mu\alpha}$$

where $\alpha = \alpha_1 + \alpha_2$ is the total contact angle.

WORKED EXAMPLE 2.25

A cable is wrapped one and a half turns around a giant branch of a tree. A car of mass 1 tonne is suspended by the cable, Figure 2.101. The coefficient of friction between the branch and the cable is assumed to be 0.5. What is the minimum value of the force S_1 required to prevent slipping of the cable?

SOLUTION

The tension at the other end of the cable is equal to the weight of the car. Hence, $S_2 = 1000 \cdot 9.81$ N. The contact angle $\alpha = 3\pi$. From Equation (2.60) we obtain:

$$S_{1,\min} = 9810 \cdot e^{-0.5 \cdot 3\pi} = 88 \text{ N}$$

which is approximately equal to the weight of a 9 kg body.

WORKED EXAMPLE 2.26

A flywheel of radius $3a/2$ is driven by an electric motor. A torque M is exerted by the motor on the wheel as shown in Figure 2.102. In order to brake the wheel a belt is wrapped around the rim of the wheel as shown. The coefficient of friction between the belt and the wheel is 0.25.

What is the minimum value of force P applied to the lever which is required to prevent the wheel from rotating?

SOLUTION

Isolating the wheel and the lever we obtain the free-body diagrams shown in Figure 2.103. The wheel:

$$\widehat{O} \quad S_1 \frac{3a}{2} - S_2 \frac{3a}{2} - M = 0 \tag{2.61}$$

The lever:

$$\widehat{A} \quad P \cdot 8a - S_1 \sin 60^\circ \cdot 2a - S_2 \sin 60^\circ \cdot a = 0 \tag{2.62}$$

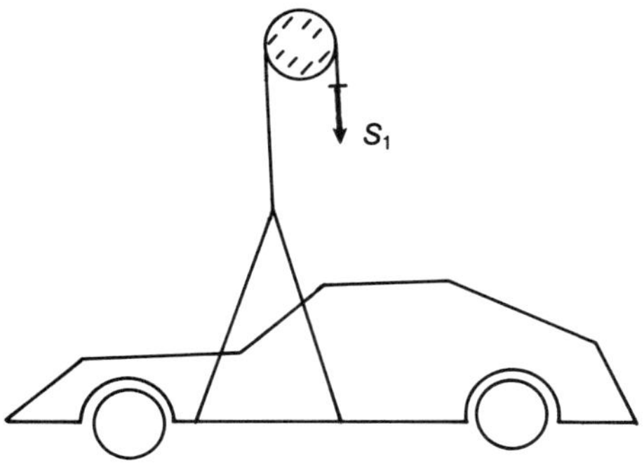

Figure 2.101

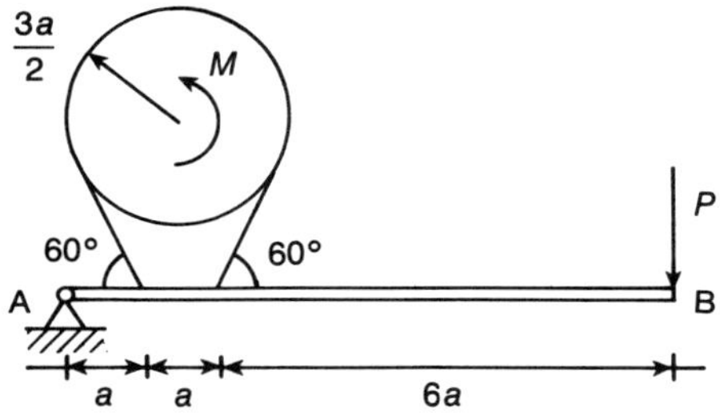

Figure 2.102

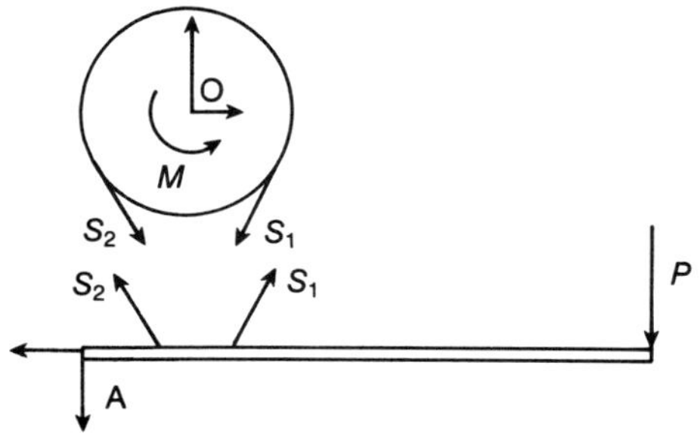

Figure 2.103

Solving for S_1 and S_2 we have:

$$\begin{cases} S_1 = \dfrac{16P}{3\sqrt{3}} + \dfrac{2M}{9a} \\ S_2 = \dfrac{16P}{3\sqrt{3}} - \dfrac{4M}{9a} \end{cases} \tag{2.63}$$

Obviously, S_1 is greater than S_2. To prevent the belt from slipping with respect to the wheel it is necessary that

$$S_1 < S_2\, e^{0.25 \cdot 4\pi/3}$$

where $4\pi/3$ is the contact angle. Inserting the results from Equations (2.63) and simplifying we have

$$P > \frac{M}{8\sqrt{3}a} \frac{2e^{\pi/3} + 1}{e^{\pi/3} - 1} \approx 0.26 \frac{M}{a}$$

PROBLEMS

○ **2.95** A body of mass 2.5 kg is placed on a rough 15° incline. Determine the minimum value of the coefficient of friction required to prevent the body from slipping in cases (a)–(d) shown here.

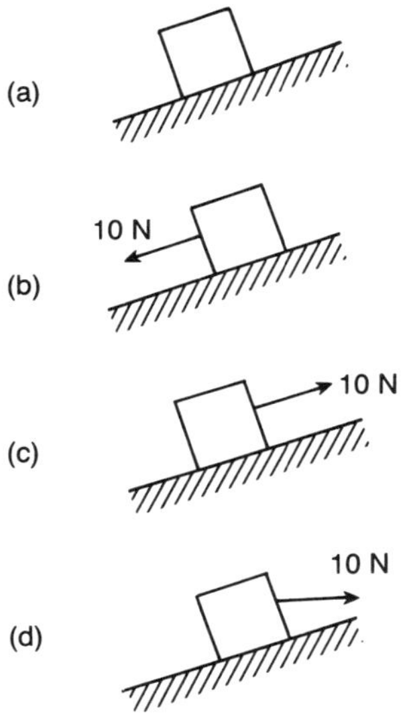

○ **2.96** A small ring of mass 0.12 kg is threaded on a rough horizontal bar (see next figure). A force of magnitude $P = 0.90$ N is applied to the ring in the vertical plane including the bar. Determine the minimum value of the coefficient of friction to prevent slipping if

(a) P is directed obliquely downwards as shown,

(b) P is directed obliquely upwards but still making a 20° angle with the bar.

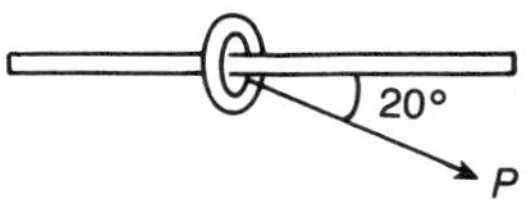

○ **2.97** A rectangular plate is supported by a rough horizontal surface. The plate is kept in the position shown by a cord. What is the minimum value of the coefficient of friction to maintain equilibrium?

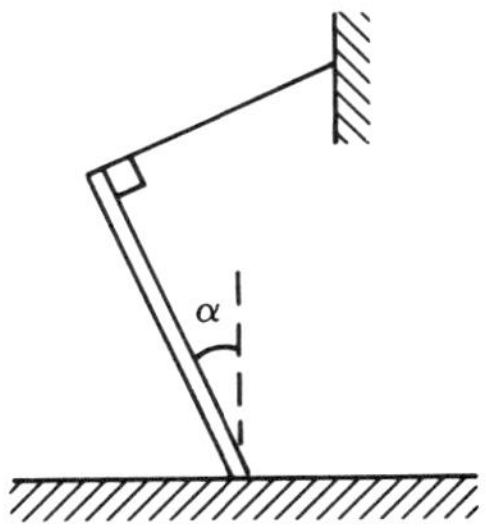

○ **2.98** Two small bodies, A and B, of weights Q and λQ, respectively, are connected by a cord. A is resting at the bottom of a hemispherical bowl while B is suspended freely. The coefficient of friction between A and the bowl is denoted by μ. The effects of friction between the cord and the rim of the bowl may be neglected. Determine whether equilibrium is possible or not in the following cases:

(a) $\lambda = 0.25$, $\mu = 0.30$

(b) $\lambda = 0.50$, $\mu = 0.30$

(c) $\lambda = 100$, $\mu = 1.2$

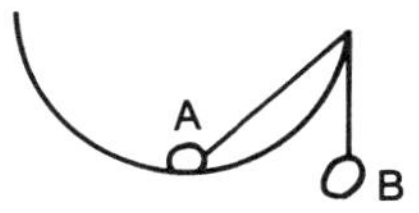

2.99 A uniform hemisphere is supported by a smooth wall and a rough floor. For what values of α can equilibrium be maintained if the coefficient of friction is
(a) 0.30
(b) 0.40?

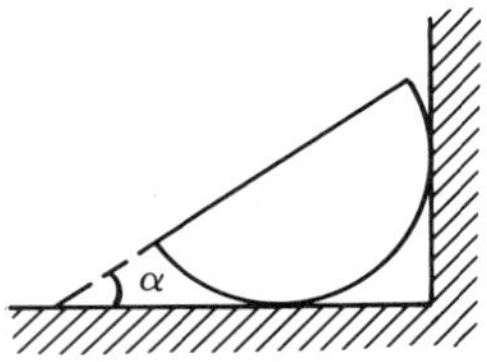

2.100 What is the minimum value of the coefficient of friction at C to prevent slipping?

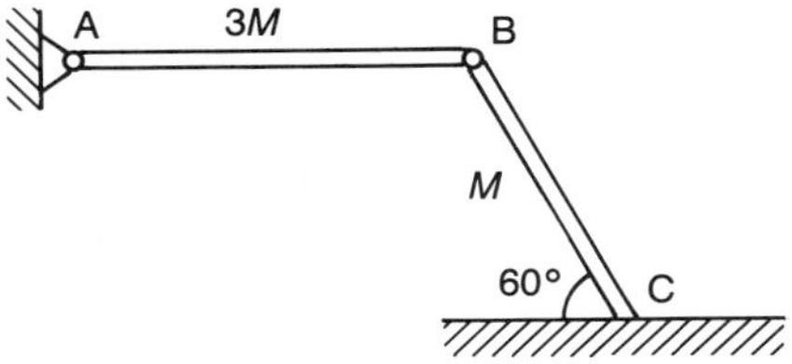

○ **2.101** A cylinder is kept in equilibrium on a rough incline by a cord that is wrapped around it. Determine the minimum value of the coefficient of friction required if the cord is attached at
(a) A, meaning that the cord is horizontal,
(b) B, meaning that the cord is parallel to the incline.

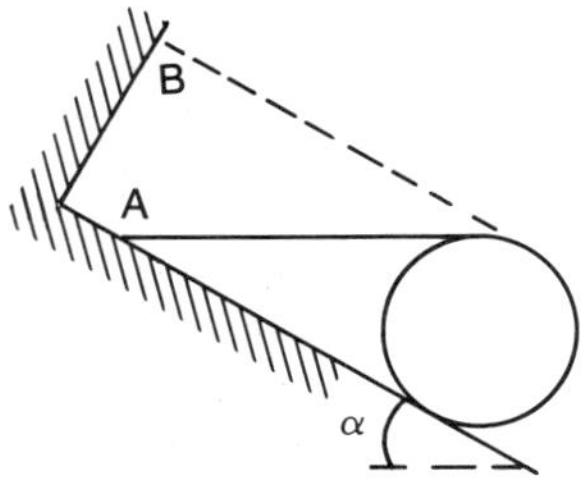

○ **2.102** A cylinder is fitted into a smooth vertical pipe. To keep it in the position shown, it is supported by a bar which is hinged to the cylinder. What is the minimum value of the coefficient of friction required at the lower end of the bar if the weight of the bar
(a) may be neglected,
(b) is 1/10 of the weight of the cylinder?

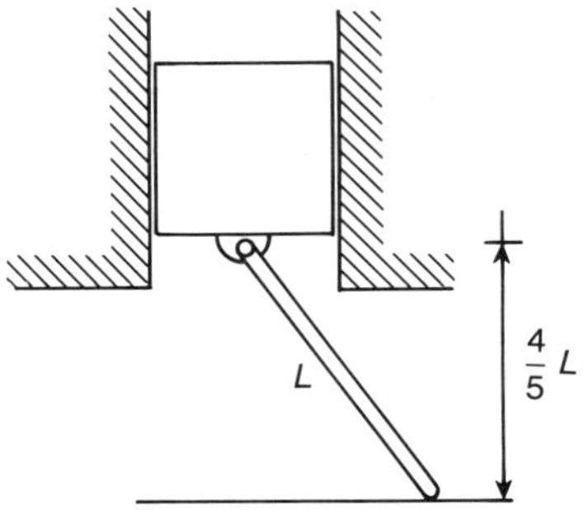

2.103 A uniform hemisphere is resting on a rough incline in the position where the plane face is vertical. What minimum value of the coefficient of friction is required?

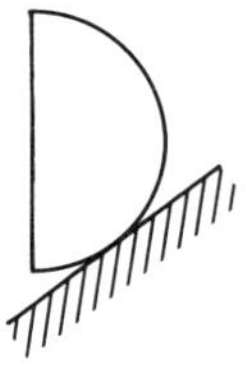

2.104 A device for lifting the lids of gully-holes is shown here. What is the maximum allowable value of the distance b (expressed in terms of a, d and the coefficient of friction μ) to make the device function?

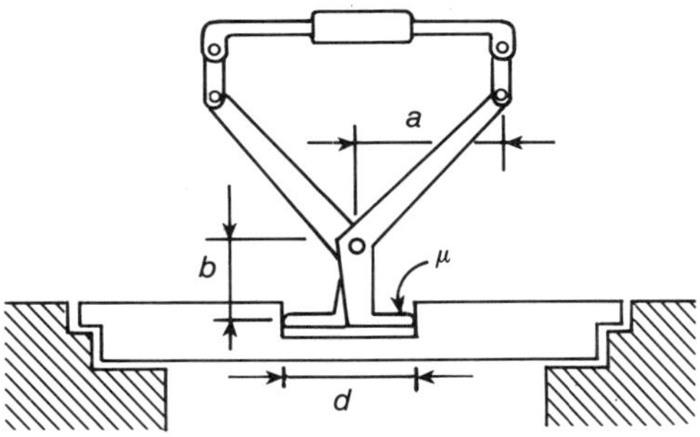

2.105 A paper-clip consists of a fixed frame and a rough roll of negligible mass. The coefficients of friction have the same value μ at both contacts, i.e. between the roll and the frame and between the roll and the paper. What is the minimum value of μ if the paper is to remain in its place when a force P is applied to it?

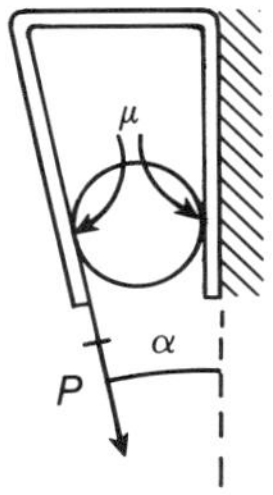

○ **2.106** A wedge is used to cleave a log. If the coefficient of friction is sufficiently large the wedge will be self-locking, i.e. the friction forces acting on the wedge are large enough to prevent it from slipping out. What minimum value of the coefficient of friction is required?

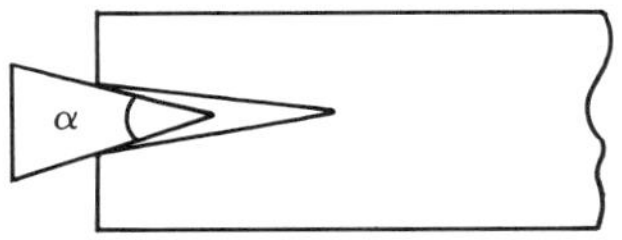

○ **2.107** A ladder of mass 10 kg and length 4.0 m is leaning against a wall. What is the maximum allowable value of the angle φ to maintain equilibrium if the coefficient of friction is 0.35 at the wall as well as at the floor?

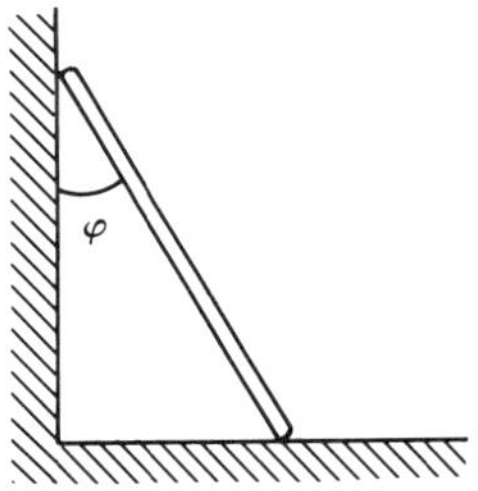

2.108 The ladder of Problem 2.107 makes a 30° angle with the wall. A 60 kg man is slowly climbing up the ladder. At what height h does the ladder start to slip?

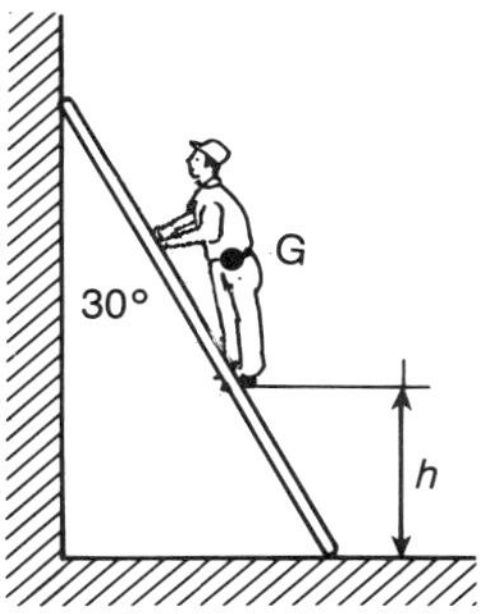

○ **2.109** Consider the wedge and the beam of Worked Example 2.23 in the case where angle $\alpha = 24°$. What force is required to press the wedge inwards?

2.110 Two knobs A and B are mounted on a drawer as shown. It is desired to pull out the drawer by applying a force to one of the knobs only. What is the minimum distance x that can be allowed? The coefficient of friction is 0.50, and it is assumed that there is a small space between the drawer and the walls of the chest. The weight of the drawer is neglected.

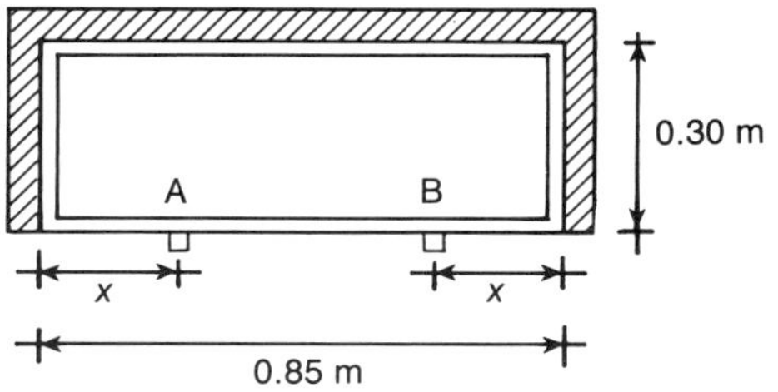

2.111 A cylinder of mass m is in contact with a rough floor and a rough wall as shown. The coefficient of friction is μ at both contacts. What is the minimum value of the torque M required to rotate the cylinder?

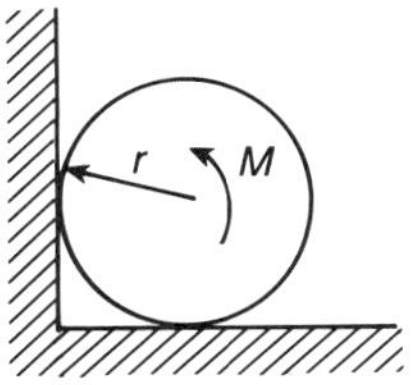

2.112 Two bodies of weights Q_1 and Q_2 are connected as shown. What are the maximum and minimum values of the ratio Q_1/Q_2 that can be permitted at equilibrium if the coefficient of friction is 0.15 at both contact surfaces?

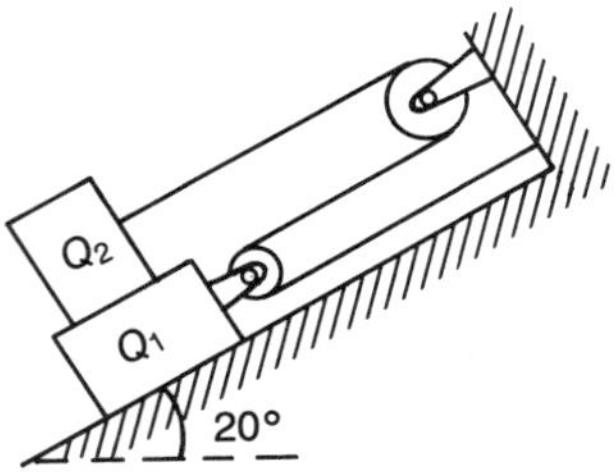

2.113 What is the force P required
(a) to pull the wedge outwards
(b) to press the wedge inwards?
The coefficient of friction is 0.25 at both contact surfaces. The gravitational forces may be neglected.

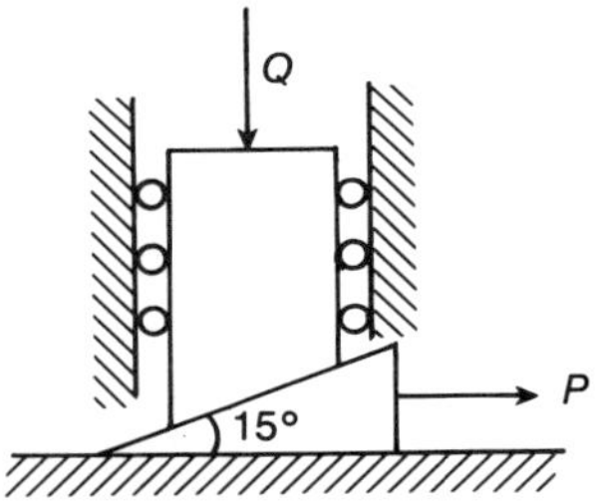

○ **2.114** A body of mass 2.0 kg is placed on a rough incline ($\mu = 0.25$). It is connected to a body of mass m by a cord and a pulley as shown. For what values of m can equilibrium be maintained?

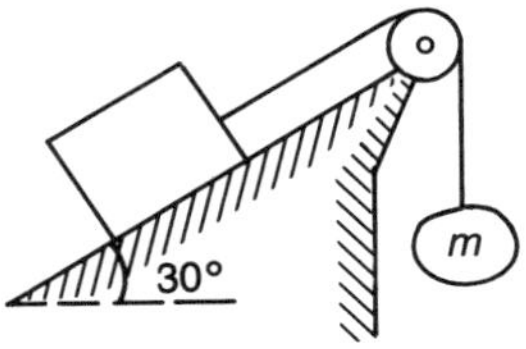

2.115 A bar is subjected to a horizontal force P which is equal to one-third of the weight of the bar. Force P is applied at the centre of gravity of the bar. For what values of φ can equilibrium be maintained in the two cases shown, if the coefficient of friction μ is 0.50?

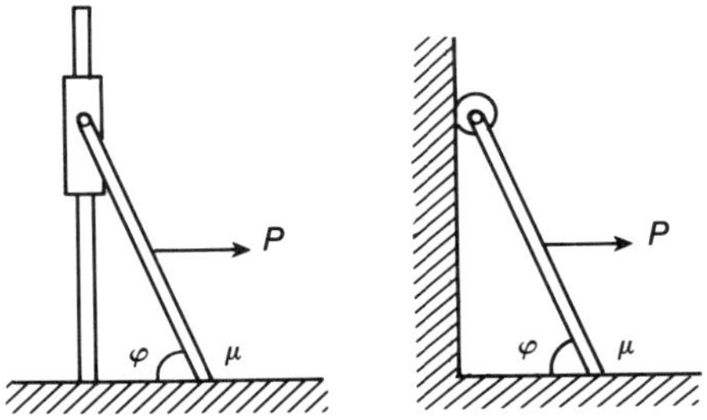

(a) (b)

◎ **2.116** A bar is supported by a wall and a rough incline ($\mu = 0.25$). What values of φ are allowable if the bar is in equilibrium?

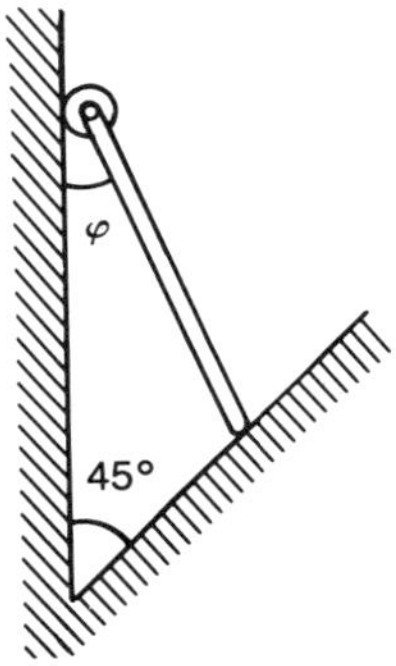

2.117 A beam of weight Q is resting on two horizontal rails as shown.
(a) Determine the force P corresponding to impending motion of the beam.
(b) Determine the corresponding values of the friction forces at A and B. The coefficient of friction is μ at both A and B.

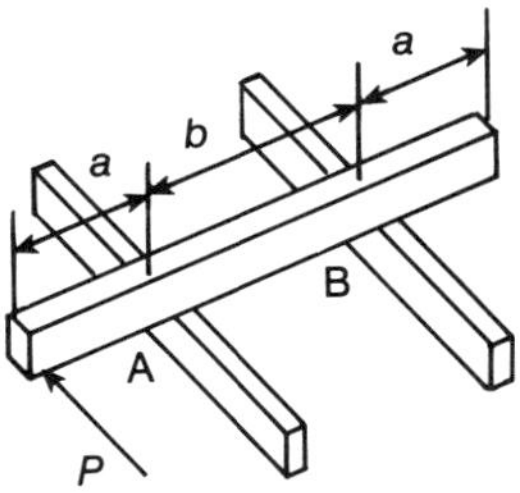

◎ **2.118** A bar is supported by a cylinder of the same weight. The coefficients of friction are equal ($=\mu$) at both contacts A and B. What are the minimum allowable values of μ to maintain equilibrium in the two cases shown?

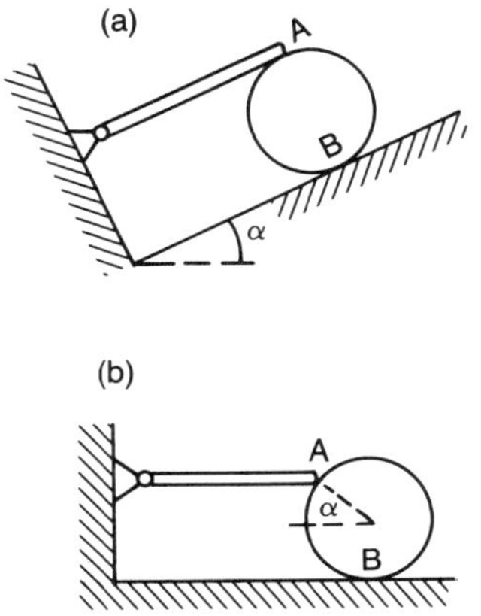

◎ **2.119** A ladder consists of two identical halves AB and AC, each of mass 10 kg, connected by a smooth hinge at A. The ladder is placed on a floor, so that the coefficients of friction are 0.40 at B and 0.35 at C.

To what height h can a man who weighs 75 kg climb before the ladder begins to slip? The centre of gravity is directly above the feet.

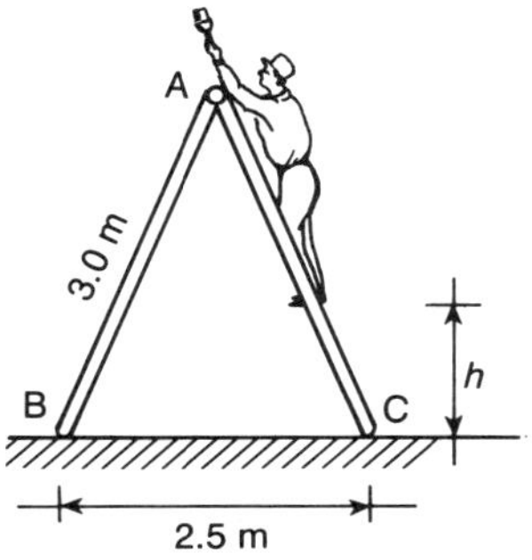

2.120 A pipe wrench may be considered to consist of two rigid portions connected by a smooth joint B as shown. If the coefficient of friction μ between the wrench and the pipe is large enough it turns out that no slipping occurs regardless of the magnitude of force P. What is the minimum allowable value of μ?

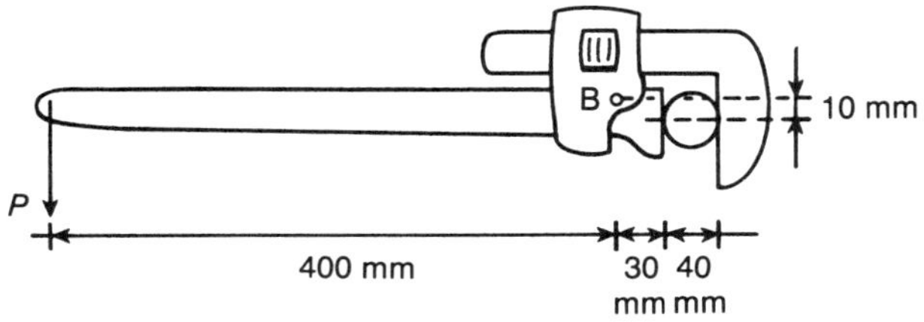

◎ **2.121** A uniform cube is divided into two parts in the ratio 2:1. The parts are connected by smooth hinges along an edge as shown. The system is placed on a rough horizontal surface ($\mu = 1/3$). For what values of α is equilibrium possible?

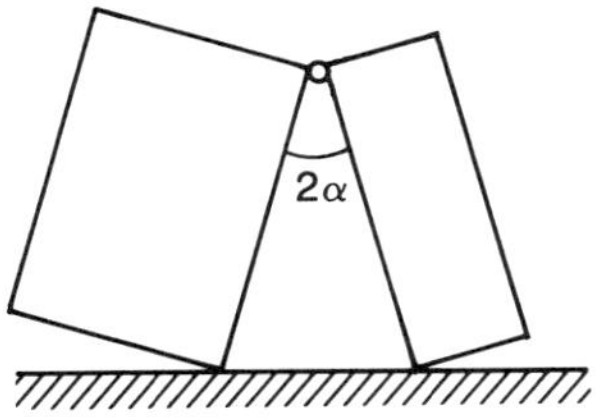

○ **2.122** A bale of paper (uniform rectangular parallelepiped) is placed on a floor. The coefficient of friction is μ. To move the bale along the floor a horizontal force P is applied at a distance h above the floor. What is the maximum value of h that can be permitted if the bale is not to tip?

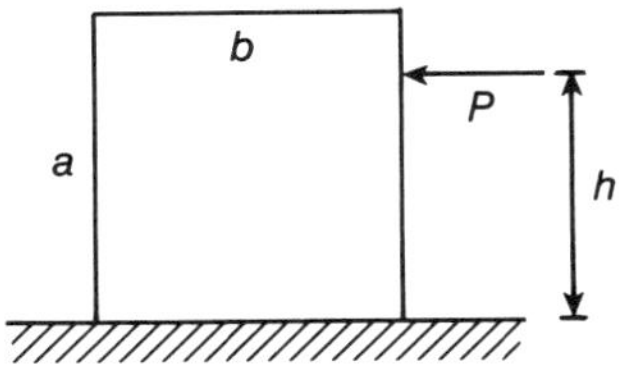

2.123 A uniform rectangular parallelepiped is placed on an inclined plate, the slope of which may be varied. What is the maximum value of angle β which can be permitted at equilibrium if the coefficient of friction is
(a) 0.60
(b) 0.80?

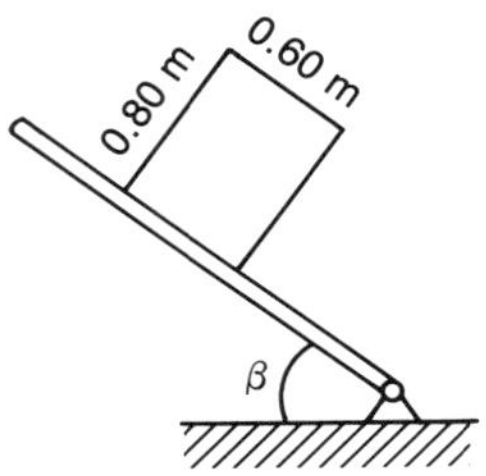

2.124 We now return to Problem 2.96. Assume that force P is horizontal but still makes a 20° angle with the bar. What is the minimum allowable value of the coefficient of friction to maintain equilibrium?

2.125 A small body of mass m is resting on a rough incline. A horizontal force P is applied to the body parallel to the incline as shown. What is the minimum allowable value of the coefficient of friction to maintain equilibrium?

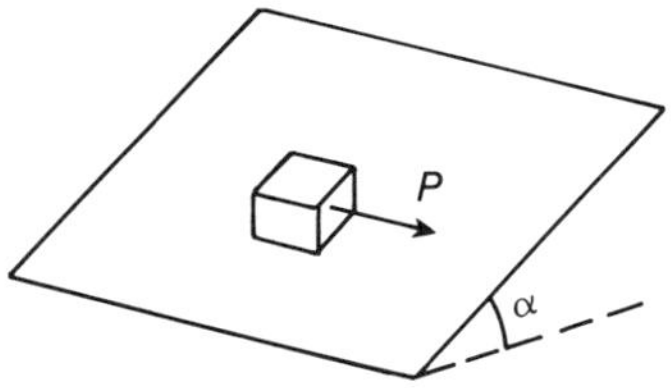

◎ **2.126** A small body of mass 2.0 kg is placed on a rough incline. The coefficient of friction is 0.45. A cord is attached to the body and runs through a smooth loop A as shown. What is the maximum force P that can be applied to the cord without moving the body?

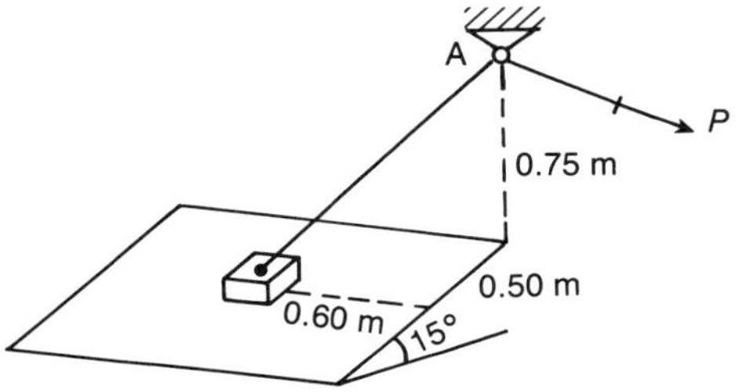

◎ **2.127** Four identical balls, each of radius r, are placed on a table so that the centres of the balls form a square of sides $2r$. A fifth ball, identical to the other ones, is placed on top of these. The coefficient of friction has the same value μ at all contact points. For what values of μ can equilibrium be maintained?

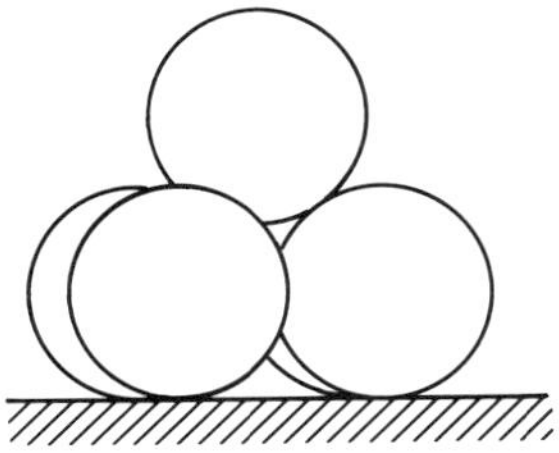

◎ **2.128** Solve Problem 2.127 assuming that the 'pyramid' consists of four balls, all in contact with one another, instead of five.

○ **2.129** A cable of negligible weight is wrapped around a fixed circular cylinder with a horizontal axis. A load L is suspended at one end of the cable. At the other end a person applies a force of 200 N. What is the maximum weight of L that can be supported by the person, if the cable is wrapped
(a) a half turn
(b) one and a half turns
(c) two and a half turns
around the cylinder? $\mu = 0.47$.

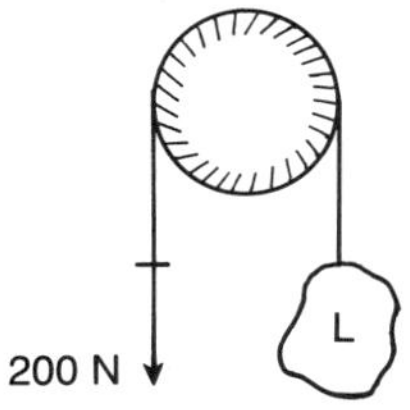

○ **2.130** What is the maximum load that the person in Problem 2.129 (a)–(c) can hoist?

2.131 In the next figure determine the upper and lower limits of the mass of B to ensure equilibrium. The coefficient of friction between the rope and the cylinders is 0.30 everywhere.

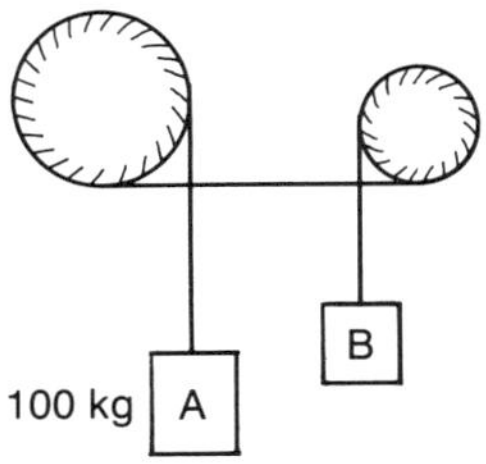

○ **2.132** Determine all possible values of force P to maintain equilibrium if the coefficient of friction is 0.20.

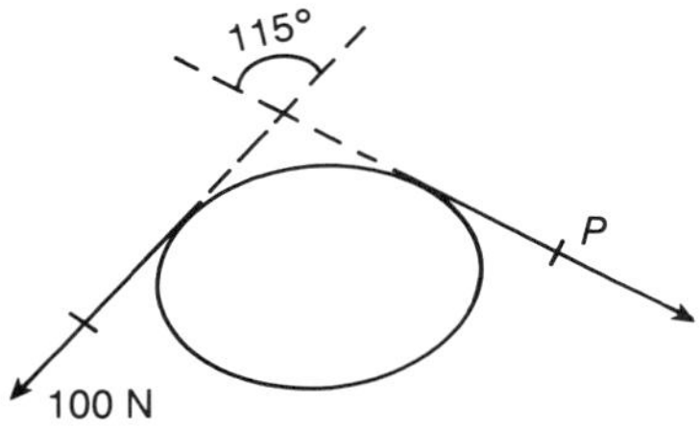

◎ **2.133** A woman is pulling a crate of mass 60 kg along a floor with the aid of a rope and a rotating drum, driven by a motor. The rope is wrapped a good turn around the drum. The coefficient of friction between the crate and the floor is 0.40; between the rope and the drum it is 0.45. Determine the force that must be applied by the woman. Compare with the case where the woman pulls in the same direction (25° angle to the floor) without the aid of the rotating drum.

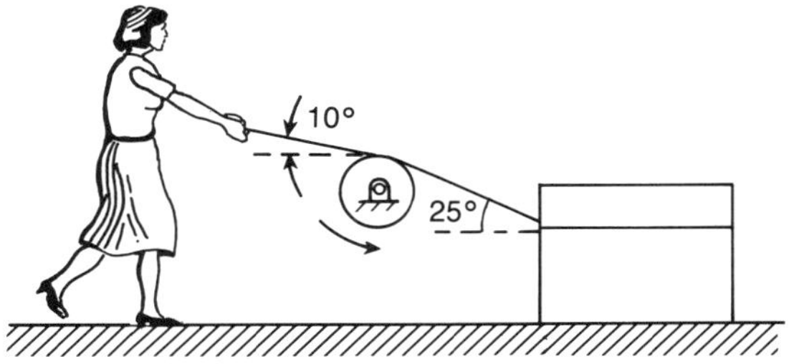

2.134 Determine the force F required to brake the wheel. The coefficient of friction is μ.

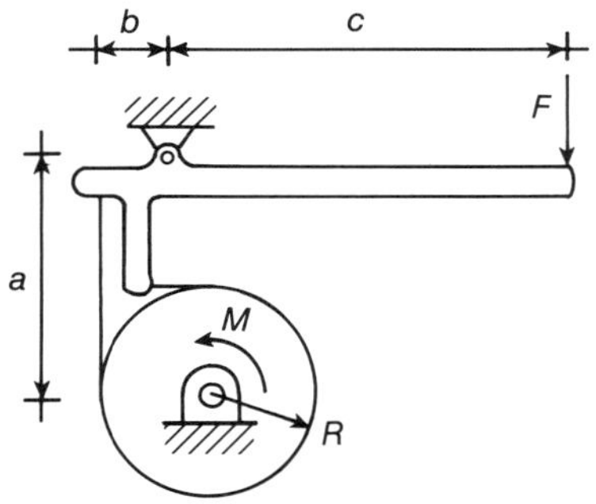

◎ **2.135** A cylinder of mass 20 kg is kept in the position shown by a cord and a body of mass m suspended by the cord. The coefficient of friction between the cylinder and the wall is 0.50, while it is 0.60 between the cord and the cylinder. What is the minimum value of m to maintain equilibrium?

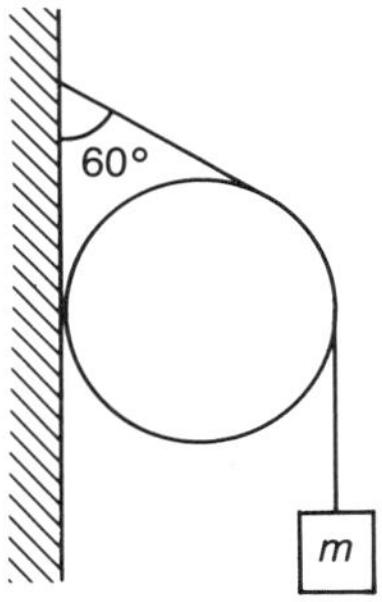

3

Centre of mass – centre of gravity

3.1 Fundamental concepts and relations

In the subject of mechanics we study bodies, systems of bodies or, more generally, what are called material systems. In *statics* we study systems at rest, while systems in motion are treated in *dynamics*. However, these two types of situation have a lot in common.

When studying a body at rest or in motion we do not necessarily have to be concerned with the position, the velocity or the mass of every individual part of the body. Sometimes it may be sufficient to consider quantities that characterize the body as a whole. Such a quantity is the total mass (m) of the body. The mass will be needed if we are interested, for example, in the total gravitational force on the body. This is, as is well known, given by the product mg.

There are various relations in statics and dynamics that are most easily expressed using the total mass m. However, it is not always sufficient to use the total mass to represent the distribution of mass. When it is desired to compute reactions at supports or bearings, for example, it is necessary to know more about the distribution of mass in space.

We shall now introduce certain quantities that will prove to be very useful later on. They will first be presented using a mathematical language, and they may be regarded as some kind of averages. To begin with, the procedure may seem somewhat abstract but, as we shall see in the concrete applications that follow, the quantities to be defined here will correspond to physical concepts of which we already have a certain, intuitive, comprehension. The reader is reminded that there is always a relation between a physical concept and the mathematical model used to represent the concept in question.

3.1.1 First moments of mass

When describing a distribution of matter we use two different concepts in classical physics: the *particle* and the *continuous distribution of mass*. In discussions of a more fundamental nature it may be advantageous to consider the distributed mass as composed of a large number of small parts that may be treated like particles. If we adopt this view we can consider these small parts as enumerative and numbered.

The mass of particle number i is denoted by m_i and its coordinates by x_i, y_i and z_i. The position vector of the particle is $\mathbf{r}_i = (x_i, y_i, z_i)$. See Figure 3.1.

The total mass of the body is given by

$$m = \sum_i m_i \tag{3.1}$$

Equation (3.1) may be considered as a definition of the total mass of a body assuming that the mass of a particle is a familiar concept.

The next quantity is introduced merely by its definition.

DEFINITION

The *first moment of mass* or *static moment of mass* with respect to the *yz*-plane is denoted by S_{yz} and defined by

$$S_{yz} = \sum_i m_i x_i \tag{3.2a}$$

In an analogous manner we may define the first moment of mass with respect to any plane. Thus the first moments of mass with respect to the *xz*- and *xy*-planes are given by

$$S_{xz} = \sum_i m_i y_i \qquad S_{xy} = \sum_i m_i z_i \tag{3.2b}$$

It should be noted that the first moments of mass are additive quantities just like masses. This means that if a body is composed of various parts, then the first moment of mass of the entire body is equal to the sum of the first moments of mass of the individual parts. The same is, of course, true for the mass of the body.

The character and meaning of the first moments of mass will be obvious in due course. At this stage we only point out that their values in some way tell us something about the distribution of mass with respect to the plane in question.

As an illustration we consider Figure 3.2 where a uniform body and its approximate location with respect to a coordinate system are shown.

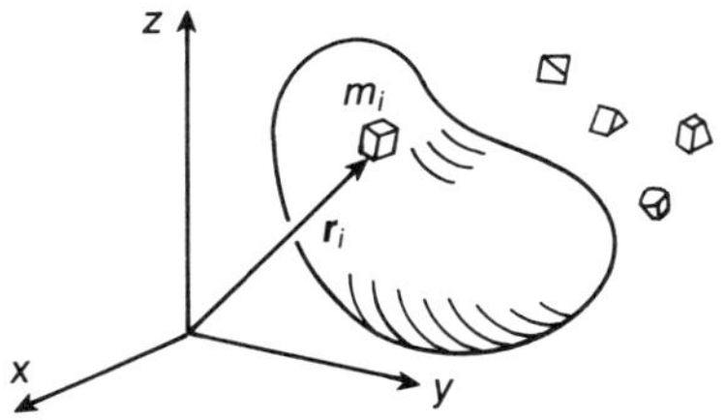

Figure 3.1

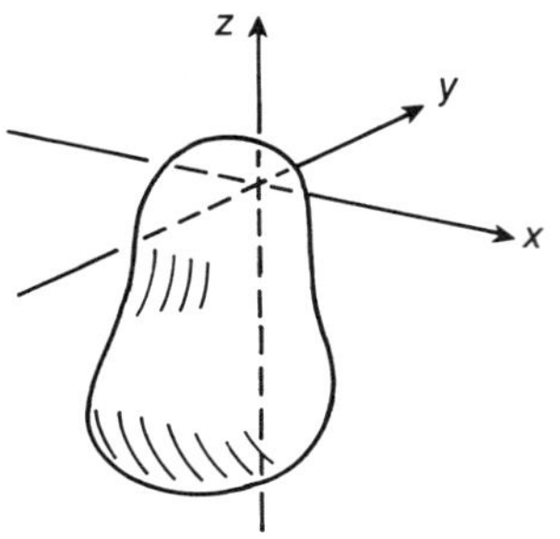

Figure 3.2

Let us assume that various parts of the body have mainly positive x-coordinates, negative y-coordinates and negative z-coordinates. If this is correct, we then have $S_{yz} > 0$, $S_{xz} < 0$ and $S_{xy} < 0$. From this example we deduce that a first moment of mass may be greater than zero, less than zero or, in a limiting case, equal to zero.

3.1.2 Centre of mass

The centre of mass of a system of particles is a point whose coordinates are weighted averages of the coordinates of the particles. The concept is defined as follows.

DEFINITION

The centre of mass is a point whose coordinates $(\bar{x}; \bar{y}; \bar{z})$ are given by

$$\bar{x} = \frac{S_{yz}}{m} \qquad \bar{y} = \frac{S_{xz}}{m} \qquad \bar{z} = \frac{S_{xy}}{m} \tag{3.3}$$

This means that the coordinates of the centre of mass are

$$\bar{x} = \frac{\sum m_i x_i}{\sum m_i} \qquad \bar{y} = \frac{\sum m_i y_i}{\sum m_i} \qquad \bar{z} = \frac{\sum m_i z_i}{\sum m_i} \tag{3.4}$$

where the coordinates of each individual point $(x_i \ldots)$ have been multiplied by a weighting factor m_i when taking the average.

As the position vector of the centre of mass is $\bar{\boldsymbol{r}} = (\bar{x}, \bar{y}, \bar{z})$ we may summarize Equations (3.4) as follows:

$$\bar{\boldsymbol{r}} = \frac{\sum m_i \boldsymbol{r}_i}{m} \tag{3.5}$$

This vector expression may be interpreted as an alternative, more compact, definition of the centre of mass.

WORKED EXAMPLE 3.1

Three identical particles, each of mass μ, are placed at the vertices of an equilateral triangle of side a, Figure 3.3. Determine the coordinates of the centre of mass using

(a) the $Oxyz$ system
(b) the $O'x'y'z'$ system.

SOLUTION

(a) The coordinates of the particles are:

$$x_1 = -\frac{a}{2} \qquad x_2 = 0 \qquad x_3 = \frac{a}{2}$$

$$y_1 = a \qquad y_2 = a + \frac{a\sqrt{3}}{2} \qquad y_3 = a$$

$$z_1 = 0 \qquad z_2 = 0 \qquad z_3 = 0$$

Using Equation (3.4) we obtain

$$\bar{x} = \frac{\mu(-a/2) + \mu \cdot 0 + \mu(a/2)}{3\mu} = 0$$

$$\bar{y} = \frac{\mu a + \mu(a + a\sqrt{3}/2) + \mu a}{3\mu} = a + \frac{a\sqrt{3}}{6}$$

$$\bar{z} = 0$$

The position of the centre of mass is most easily described as that point on the altitude located at a distance $h/3$ from the base, Figure 3.4. Here $h = a\sqrt{3}/2$ is the altitude of the triangle.

(b) The coordinates of the particles are

$$x_1' = a \qquad x_2' = \frac{3a}{2} \qquad x_3' = 2a$$

$$y_1' = 0 \qquad y_2' = \frac{a\sqrt{3}}{2} \qquad y_3' = 0$$

$$z_1' = 0 \qquad z_2' = 0 \qquad z_3' = 0$$

From Equations (3.4) it follows that

$$\bar{x}' = \frac{3a}{2} \qquad \bar{y}' = \frac{a\sqrt{3}}{6} \qquad \bar{z}' = 0$$

In other words, the centre of mass has *the same location with respect to the particles*, on the altitude at a distance $h/3$ from the base of the triangle, regardless of our choice of coordinates. This is a general property of the centre of mass which is not proven here, however.

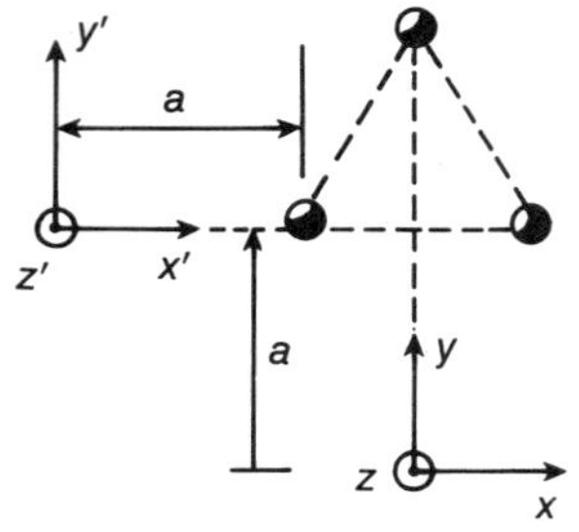

Figure 3.3

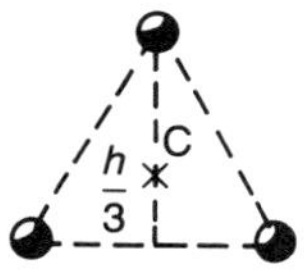

Figure 3.4

3.1.3 Centre of gravity

The centre of mass of a material system was introduced without any reference to forces. We shall now see that this point plays a very important part in every situation where a *uniform gravitational field* exists.

- If a gravitational field exists at a certain point, then a particle at this point is subjected to a force that is proportional to the mass (m) of the particle.
- If, in addition, the gravitational field is uniform within a region, then the constant of proportionality (g) is the same everywhere in that region.

These properties can be summarized in the equation

$$\boldsymbol{F} = m\boldsymbol{g} \tag{3.6}$$

As long as we are concerned with a region of relatively small extension in the immediate vicinity of the surface of the earth, the magnitude of the vector $\boldsymbol{g}$ is close to 9.8 N/kg and the direction is downwards. The reader is referred to Section I.4 for more details about gravitational forces.

We now consider a system of particles of masses m_i. The individual gravitational forces form a parallel force system. As was shown in Chapter 1 these may be replaced by one single force, the resultant $\boldsymbol{R}$. The direction of $\boldsymbol{R}$ coincides with the direction of the individual forces and the magnitude is equal to the algebraic sum of these ($\boldsymbol{R} = \Sigma \boldsymbol{F}_i$).

Hence, the resultant of the system of gravitational forces is given by

$$\boldsymbol{R} = \sum_i m_i \boldsymbol{g} = m\boldsymbol{g} \tag{3.7}$$

Consequently, the magnitude and the direction of $\boldsymbol{R}$ are determined. What remains is to find the correct point of application. This must be chosen such that the moment of $\boldsymbol{R}$ is equal to the sum of moments of the original forces.

We shall now show that this is the case if the resultant $\boldsymbol{R}$ is applied at the centre of mass. To prove this we choose an arbitrary moment point O, see Figure 3.5. The sum of moments of the system of gravitational forces is given by

$$\sum \boldsymbol{M}_O = \sum_i [\boldsymbol{r}_i \times m_i \boldsymbol{g}]$$

which can be rearranged in the following way:

$$\sum \boldsymbol{M}_O = \sum_i [m_i \boldsymbol{r}_i \times \boldsymbol{g}] = \left[\sum_i m_i \boldsymbol{r}_i\right] \times \boldsymbol{g}$$

Using Equation (3.5) this can be rewritten as

$$\sum \boldsymbol{M}_O = m\bar{\boldsymbol{r}} \times \boldsymbol{g} = \bar{\boldsymbol{r}} \times m\boldsymbol{g} = \bar{\boldsymbol{r}} \times \boldsymbol{R}$$

In other words, the sum of moments is equal to the moment of the resultant $\boldsymbol{R}$ if this is applied at the centre of mass according to Figure 3.5. For this reason the centre of mass is often called the *centre of gravity*.

In statics this point is of interest as the point of application of the resultant gravitational force. It is then natural to use the term *centre of gravity*.

In dynamics, as we shall see later on in Section 6.1, we are interested in the same point as the *centre of mass*. In that case this term is the more appropriate.

The centre of mass is one concept – the centre of gravity is another one. In a uniform gravitational field they coincide; in a non-uniform field they do not. However, it should be admitted that in current usage the term *centre of gravity* has come to dominate in such a way that it is often used in cases where the *centre of mass* is the more correct term.

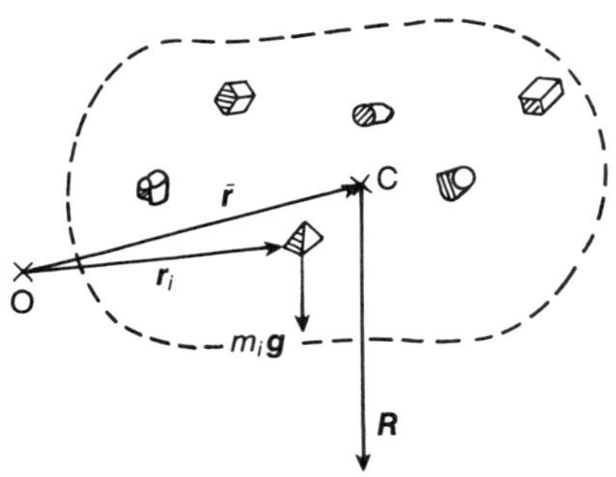

Figure 3.5

PROBLEMS

○ **3.1** Locate the centre of mass of the following molecules:
(a) NaCl, bond length 2.78 Å (angstroms), where $1\ \text{Å} = 1 \cdot 10^{-10}$ m
(b) H_2O; the length of the OH-bond is 0.958 Å and the angle between the bonds is 104.45°.
Use the following atomic masses: Na 23, Cl 35, H 1, O 16.

3.2 A system of particles consists of three identical particles which are placed at the vertices of a triangle. Another system of three identical particles is placed at the midpoints of the sides of the same triangle.

Show by use of Equation (3.5) that the two centres of mass coincide regardless of the shape of the triangle.

○ **3.3** Five lead balls are located at various vertices or midpoints of the edges of a right parallelepiped as shown. The origin of the coordinate system is at the centre of the parallelepiped. Calculate the coordinates of the centre of mass.

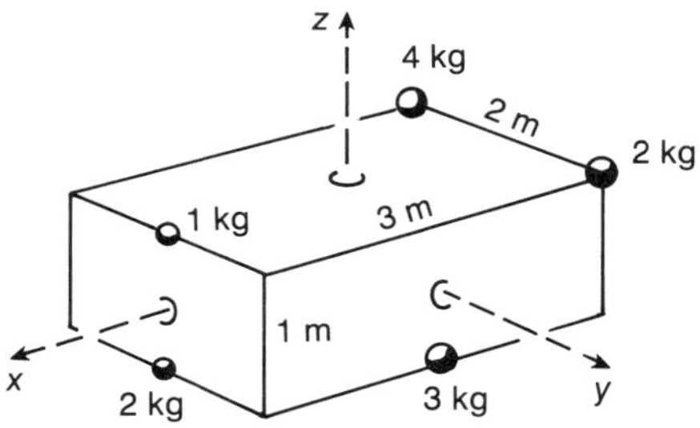

○ **3.4** A material system consists of a system of particles according to the table below:

Mass	Location
m	$(a; 0; a)$
$2m$	$(2a; 0; -2a)$
m	$(a; 0; -a)$
$3m$	$(-a; -a; -a)$
m	$(2a; 0; 0)$

Determine the coordinates of the centre of mass.

3.2 Computation methods

In this section we shall develop methods that can be used to locate the centre of gravity (centre of mass) for a certain body.

3.2.1 Symmetry considerations

In many cases it is possible to draw certain conclusions by studying the symmetry properties of a body. In a typical situation the computations may be simplified or even avoided.

A body possesses a *plane of symmetry* with respect to the distribution of mass if each particle at one side of the plane corresponds to an image, i.e. a particle of the same mass, at the other side of the plane. As an example we consider the body of Figure 3.6. The particle of mass m_j and coordinates $(x_j; y_j; z_j)$ corresponds to an image of mass $m_k = m_j$ and coordinates $(x_k; y_k; z_k) = (-x_j; y_j; z_j)$. This is true for every particle belonging to the body. Hence, the yz-plane is a plane of symmetry of the body.

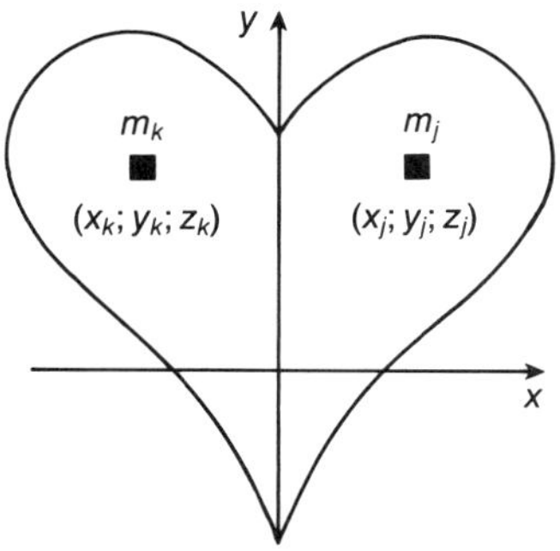

Figure 3.6

The first moment of mass with respect to the *yz*-plane is

$$S_{yz} = \sum_i m_i x_i = 0$$

as every term of the sum with a positive value of x_i is cancelled by a corresponding term with a negative value of x_i. The *x*-coordinate of the centre of mass is then given by

$$\bar{x} = \frac{S_{yz}}{m} = 0$$

We have thus shown the following: *If a body possesses a plane of symmetry with respect to the distribution of mass, then the centre of mass will be located in that plane.*

It should be noted that it is not sufficient in a general case that the plane is a plane of symmetry from a geometrical point of view. The distribution of mass must be the same at both sides of the plane. In the case of a body of constant density, however, it is sufficient that the plane is a geometric plane of symmetry.

In the case of a body in the shape of a thin disk, where the entire mass is located in a plane, it follows readily that the centre of mass will be located in the plane of the disk.

Sometimes we may conclude that the centre of mass is located in a certain plane even though there is no plane of symmetry. Consider the body of Figure 3.7. If every particle of mass m_j and coordinates $(x_j; y_j; z_j)$ corresponds to a particle of mass $m_k = m_j$ and coordinates $(x_k; y_k; z_k) = (-x_j; y_k; z_k)$, then the first moment of mass becomes

$$S_{yz} = \sum_i m_i x_i = 0 \Rightarrow \bar{x} = \frac{S_{yz}}{m} = 0$$

meaning that the centre of mass is located in the *yz*-plane, which is *not* a plane of symmetry, however.

As an example of a body with *one* plane of symmetry a half, right circular cone is shown in Figure 3.8(a).

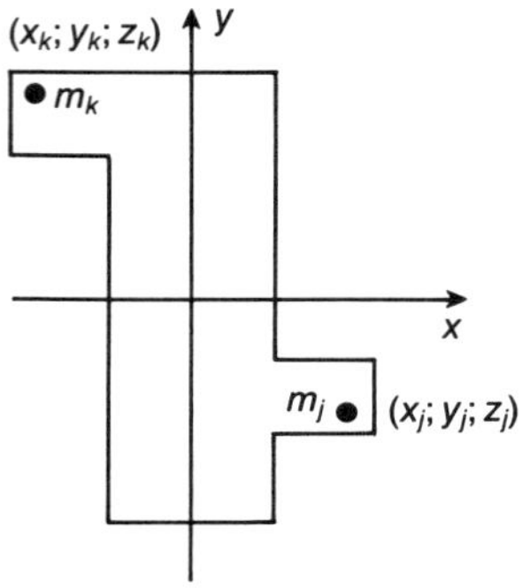

Figure 3.7

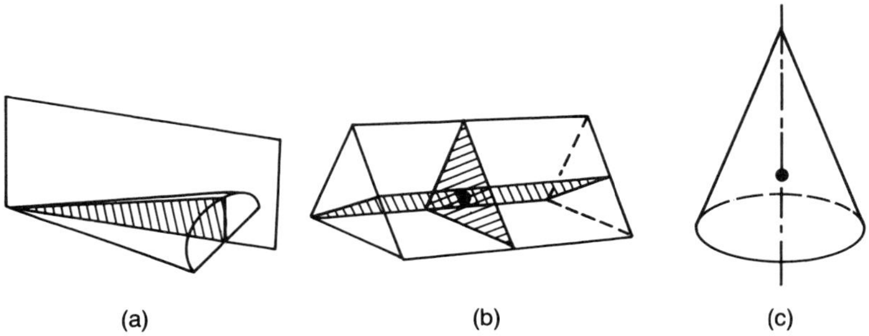

Figure 3.8

If a body possesses two planes of symmetry the centre of mass is located at the intersection of these planes. An example is given in Figure 3.8(b) where a right prism with a cross-section in the shape of an isosceles triangle is shown.

If there are three or more planes of symmetry intersecting at one point, then this point is the centre of mass. This is true, for example, for a right parallelepiped or a sphere.

In the special case of a body of revolution, every plane that contains the axis of revolution is a plane of symmetry. Hence, the centre of mass is located on this axis, the *axis of symmetry*. As an example a right circular cone is shown in Figure 3.8(c).

3.2.2 Continuous distribution of mass

The symmetry properties of a body can seldom give us the exact location of the centre of mass. In most cases we can conclude that the centre of mass is located in a certain plane or at a certain axis but nothing else. To continue we have to leave our original model where we consider the body to be composed of a very large number of discrete particles.

Instead we shall assume that matter, and thus mass, is continuously distributed over the entire volume of the body, i.e. every small volume of the body will be completely filled with matter. This assumption, also called the continuum hypothesis, does not take the atomic structure of the material into consideration, but as long as we do not consider extremely small volumes this assumption is perfectly acceptable. It is also widely used in rigid-body mechanics as well as in the mechanics of deformable solids and fluids.

To proceed from our previous system of particles to a continuous distribution of mass we let the number of particles tend to infinity at the same time as the mass of each particle tends to zero in such a way that the total mass of the body remains constant. Instead of considering individual particles of mass m_i we consider small elements according to Figure 3.9 of 'infinitely small' or infinitesimal mass $\mathrm{d}m = \rho \, \mathrm{d}V$. Here ρ is the density and $\mathrm{d}V$ is the volume of the element. The position vector of the mass point $\boldsymbol{r}_i = (x_i; y_i; z_i)$ is replaced by $\boldsymbol{r} = (x; y; z)$.

The expression for the mass of the body

$$m = \sum_i m_i$$

is replaced by an integral

$$m = \int \mathrm{d}m = \int \rho \, \mathrm{d}V \tag{3.8}$$

where the integral should be considered as the limiting value of the sum of a large number of very small quantities, where the number of terms tends to infinity while each term tends to zero. The expression $\int \rho \, \mathrm{d}V$ can be considered as a volume integral.

The expressions for the first moments of mass

$$S_{yz} = \sum_i m_i x_i \qquad S_{xz} = \sum_i m_i y_i \qquad S_{xy} = \sum_i m_i z_i$$

are replaced in a corresponding way by

$$\begin{aligned} S_{yz} &= \int x \, \mathrm{d}m = \int x\rho \, \mathrm{d}V \\ S_{xz} &= \int y \, \mathrm{d}m = \int y\rho \, \mathrm{d}V \\ S_{xy} &= \int z \, \mathrm{d}m = \int z\rho \, \mathrm{d}V \end{aligned} \tag{3.9}$$

The coordinates of the centre of mass

$$\bar{x} = \frac{S_{yz}}{m} = \sum_i m_i x_i$$

$$\bar{y} = \frac{S_{xz}}{m} = \sum_i m_i y_i$$

$$\bar{z} = \frac{S_{xy}}{m} = \sum_i m_i z_i$$

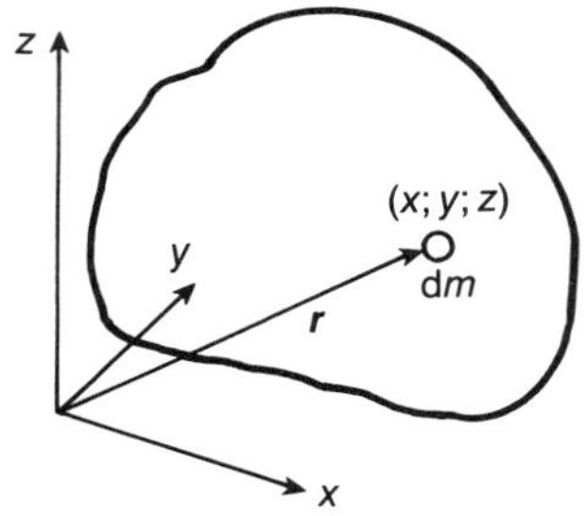

Figure 3.9

are replaced by

$$\bar{x} = \frac{S_{yz}}{m} = \frac{1}{m} \int x \, \mathrm{d}m = \frac{1}{m} \int x\rho \, \mathrm{d}V \text{ etc.} \tag{3.10}$$

Equations (3.10) can be summarized as

$$\bar{\boldsymbol{r}} = \frac{1}{m} \int \boldsymbol{r} \, \mathrm{d}m = \frac{1}{m} \int \boldsymbol{r}\rho \, \mathrm{d}V \tag{3.11}$$

Centroid of a volume

In a special case the body considered may be homogeneous, i.e. the density ρ is constant. In this case the mass of the body is given by $m = \rho V$ where V is the volume of the body. Simplifying Equation (3.11) we obtain

$$\bar{\boldsymbol{r}} = \frac{1}{V} \int \boldsymbol{r} \, \mathrm{d}V \tag{3.12}$$

or, in terms of coordinates,

$$\bar{x} = \frac{1}{V} \int x \, \mathrm{d}V$$

$$\bar{y} = \frac{1}{V} \int y \, \mathrm{d}V \tag{3.13}$$

$$\bar{z} = \frac{1}{V} \int z \, \mathrm{d}V$$

Obviously, the centre of mass of a homogeneous body is a purely geometrical property which is also known as the *centroid* of the volume. Even if the body is not homogeneous, it is still possible to determine the location of the centroid from Equations (3.13). In this case, however, the centre of mass and the centroid in general do not coincide.

We now define the *first moment of volume* with respect to the *yz*-plane as

$$S^{V}_{yz} = \int x \, dV \tag{3.14}$$

In an analogous way we define S^{V}_{xz} and S^{V}_{xy}. From Equations (3.13) we then have

$$\bar{x} = \frac{S^{V}_{yz}}{V} \qquad \bar{y} = \frac{S^{V}_{xz}}{V} \qquad \bar{z} = \frac{S^{V}_{xy}}{V} \tag{3.15}$$

Centroid of an area

Bodies in the shape of thin shells make up another important special case. A shell may be divided into small elements of mass $dm = \sigma \, dA$ according to Figure 3.10. Here dA is the area of the element and σ is the mass per unit area, i.e. $\sigma = \rho t$, where t is the thickness of the shell.

We then have

$$m = \int dm = \int \sigma \, dA \tag{3.16}$$

The coordinates of the centre of mass are given by

$$\bar{x} = \frac{1}{m} \int x\sigma \, dA, \text{ etc.} \tag{3.17}$$

In a special case σ may be a constant. Then $m = \sigma A$, where A is the area of the shell. Simplifying we obtain

$$\begin{aligned} \bar{x} &= \frac{1}{A} \int x \, dA \\ \bar{y} &= \frac{1}{A} \int y \, dA \\ \bar{z} &= \frac{1}{A} \int z \, dA \end{aligned} \tag{3.18}$$

The point determined by Equations (3.18) is called the *centroid of the area*.

The integral $\int x \, dA$ is known as the *first moment of area* with respect to the *yz*-plane. It is denoted by S^{A}_{yz}. Together with the corresponding expressions for S^{A}_{xz} and S^{A}_{xy} we now have

$$\bar{x} = \frac{S^{A}_{yz}}{A} \qquad \bar{y} = \frac{S^{A}_{xz}}{A} \qquad \bar{z} = \frac{S^{A}_{xy}}{A} \tag{3.19}$$

Centroid of a line

A thin wire according to Figure 3.11 may be treated in a similar way to the thin shell above. The mass of a small element is given by $dm = \lambda \, dL$, where λ is the mass per unit

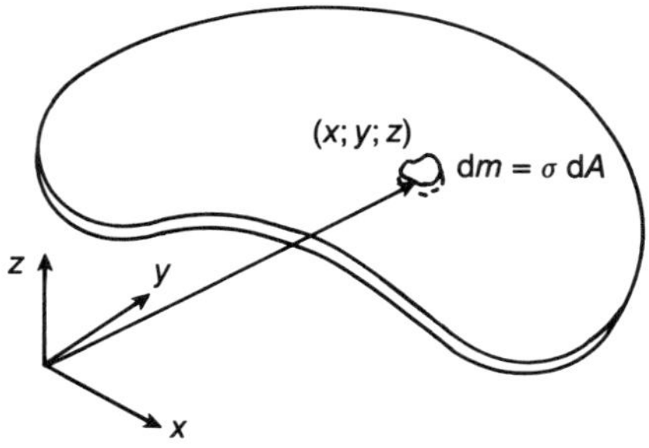

Figure 3.10

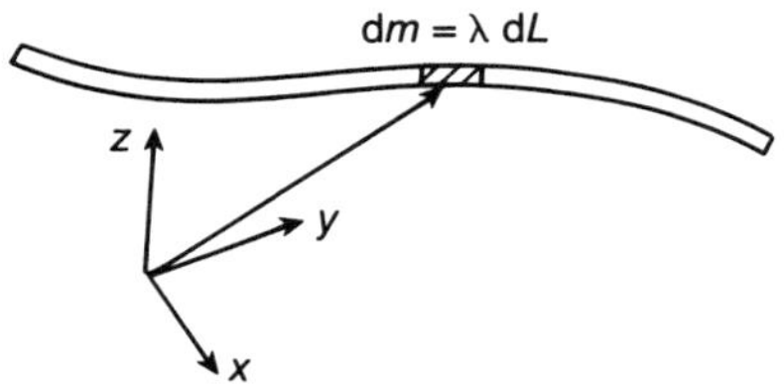

Figure 3.11

length and dL is the length of the element. Here $\lambda = \rho A$, where A is the area of the cross-section of the wire. Hence,

$$m = \int dm = \int \lambda \, dL \tag{3.20}$$

$$\bar{x} = \frac{1}{m} \int x\lambda \, dL \text{ etc.} \tag{3.21}$$

If λ is a constant, then $m = \lambda L$, where L is the length of the wire. Consequently,

$$\bar{x} = \frac{1}{L} \int x \, dL \text{ etc.} \tag{3.22}$$

In this case the centre of mass of the wire coincides with the *centroid of the line* as defined by Equations (3.22). The integral $\int x \, dL$ is called the *first moment of line* with respect to the *yz*-plane. Using denotations analogous to those used above for the first moments of volume and area we have

$$S_{yz}^{L} = \int x \, dL \qquad S_{xz}^{L} = \int y \, dL \qquad S_{xy}^{L} = \int z \, dL \tag{3.23}$$

The coordinates of the centroid of the line are then given by

$$\bar{x} = \frac{S_{yz}^{L}}{L} \qquad \bar{y} = \frac{S_{xz}^{L}}{L} \qquad \bar{z} = \frac{S_{xy}^{L}}{L} \tag{3.24}$$

WORKED EXAMPLE 3.2

Locate the centroid of a circular arc of radius r and central angle 2α, Figure 3.12.

SOLUTION

From the symmetry we conclude that the centroid is located on the x-axis. The arc is divided into elements as shown in Figure 3.13. The length of an arbitrary element is $dL = r\,d\varphi$ and the x-coordinate is $r \cos \varphi$. The length of the arc is $L = 2r\alpha$.

The x-coordinate of the centroid is determined from Equations (3.22):

$$\bar{x} = \frac{1}{L}\int x\,dL = \frac{1}{2r\alpha}\int_{-\alpha}^{\alpha} r\cos\varphi \cdot r\,d\varphi = \frac{r}{2\alpha}\,[\sin\varphi]_{-\alpha}^{\alpha} = r\,\frac{\sin\alpha}{\alpha}$$

Note that the angle α must be measured in radians.

REMARK

We see that in the special case where $\alpha = \pi$ (complete circle), we obtain $\bar{x} = 0$ as expected. For small values of α we find that $\bar{x} \approx r$, which we also could have anticipated. In the case of a semicircular arc ($\alpha = \pi/2$) we see that $\bar{x} = 2r/\pi$.

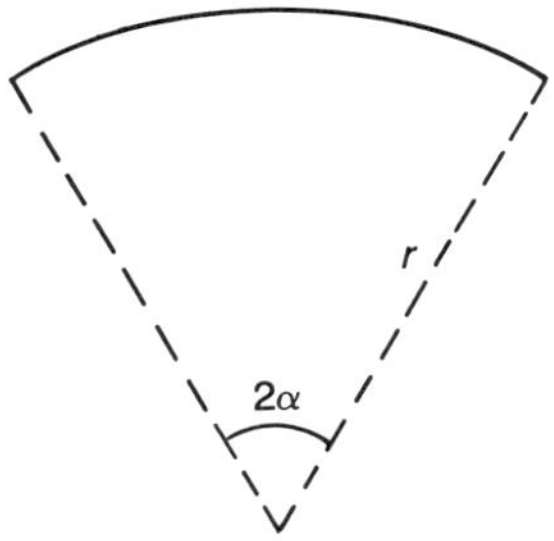

Figure 3.12

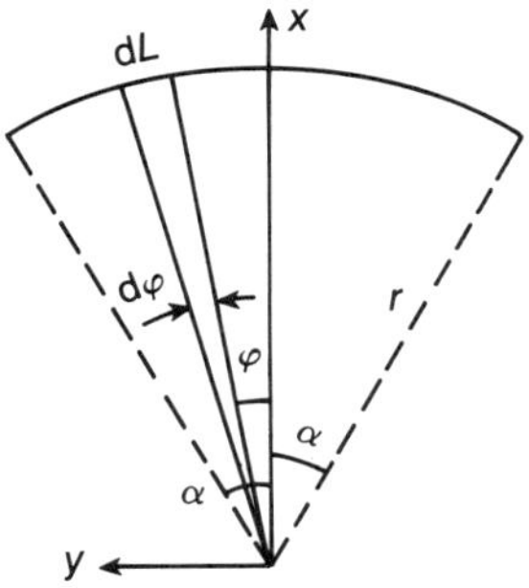

Figure 3.13

3.2.3 Composite bodies

A common problem in the computation of centres of mass occurs when a body is composed of a number of parts or elements with known masses and centres of mass.

We assume that the body shown in Figure 3.14 is composed of N elements of masses $m_1, m_2, \ldots, m_N$. The position vectors of the centres of mass are $\bar{r}_1 = (\bar{x}_1, \bar{y}_1, \bar{z}_1)$, $\bar{r}_2 = (\bar{x}_2, \bar{y}_2, \bar{z}_2), \ldots, \bar{r}_N = (\bar{x}_N, \bar{y}_N, \bar{z}_N)$, respectively.

We know that the mass of the composite body is equal to the sum of the masses of the elementary bodies, i.e.

$$m = \sum_{i=1}^{N} m_i \tag{3.25}$$

Furthermore, the first moments of mass of the composite body are equal to the sum of the first moments of mass of the elements. Thus,

$$S_{yz} = \sum_{i=1}^{N} S_{iyz} \qquad S_{xz} = \sum_{i=1}^{N} S_{ixz} \qquad S_{xy} = \sum_{i=1}^{N} S_{ixy} \tag{3.26}$$

where

$$S_{iyz} = m_i\,\bar{x}_i \qquad S_{ixz} = m_i\,\bar{y}_i \qquad S_{ixy} = m_i\,\bar{z}_i$$

according to Equations (3.3).

For the composite body we have

$$\bar{x} = \frac{S_{yz}}{m} = \frac{\sum S_{iyz}}{\sum m_i} = \frac{m_1\,\bar{x}_1 + m_2\,\bar{x}_2 + \ldots + m_N\,\bar{x}_N}{m_1 + m_2 + \ldots + m_N} \tag{3.27}$$

and similar expressions for $\bar{y}$ and $\bar{z}$, of course. This means that when determining the position of the centre of mass we may replace each element by a particle of the same mass located at the centre of mass of the corresponding element.

For homogeneous bodies we have from Equations (3.15) that $\bar{x} = S_{yz}^{V}/V$ etc. If the volumes of the elements are denoted by $V_1, V_2, \ldots, V_N$ and the first moments of volume by S_{1yz}^{V} etc., we have

$$S_{yz}^{V} = \sum_{i=1}^{N} S_{iyz}^{V} = \sum_{i=1}^{N} V_i\,\bar{x}_i \qquad V = \sum_{i=1}^{N} V_i$$

$$\therefore \quad \bar{x} = \frac{S_{yz}^{V}}{V} = \frac{V_1\,\bar{x}_1 + V_2\,\bar{x}_2 + \ldots + V_N\,\bar{x}_N}{V_1 + V_2 + \ldots + V_N} \tag{3.28}$$

Analogous expressions apply to $\bar{y}$ and $\bar{z}$, of course.

In the same way we can derive expressions for thin shells of constant mass per unit area and thin wires of constant mass per unit length, i.e. for centroids of areas and lines. We find that

$$\bar{x} = \frac{S_{yz}^{A}}{A} = \frac{A_1\,\bar{x}_1 + A_2\,\bar{x}_2 + \ldots + A_N\,\bar{x}_N}{A_1 + A_2 + \ldots + A_N} \tag{3.29}$$

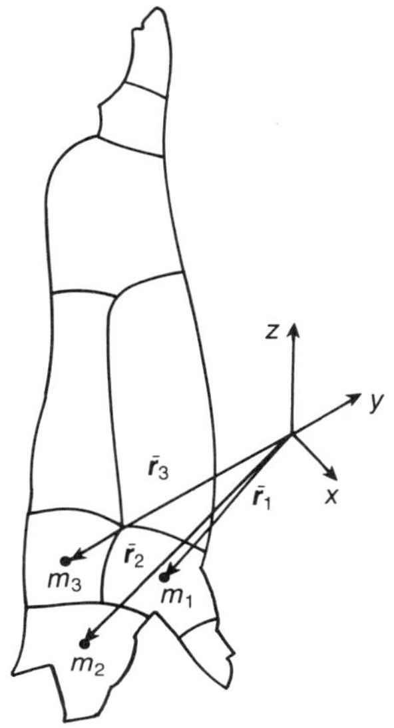

Figure 3.14

and

$$\bar{x} = \frac{S^{L}_{yz}}{L} = \frac{L_1\bar{x}_1 + L_2\bar{x}_2 + \ldots + L_N\bar{x}_N}{L_1 + L_2 + \ldots + L_N} \tag{3.30}$$

respectively, with corresponding expressions for the y- and z-coordinates.

In many cases a body may be considered as created by removing parts with known centres of mass from an original body whose centre of mass is also known. When computing the centre of mass of the remaining body we may utilize the fact that its first moment of mass is equal to the first moment of mass of the original body reduced by the first moments of mass of the parts that have been cut off. This is, of course, also true for the masses or, alternatively, the volumes, the areas or the lengths. An example will be given in the following worked example.

WORKED EXAMPLE 3.3

Determine the coordinates of the centroid of the area shown in Figure 3.15.

SOLUTION 1

The area may be considered to be composed of two rectangles as shown in Figure 3.16. The centre of mass of a rectangle is, of course, known from the symmetry properties. From Equation (3.29) we then obtain

$$\bar{x} = \frac{A_1\bar{x}_1 + A_2\bar{x}_2}{A_1 + A_2} = \frac{2a^2 \cdot a/2 + 4a^2 \cdot 3a}{2a^2 + 4a^2} = \frac{13}{6}a$$

$$\bar{y} = \frac{A_1\bar{y}_1 + A_2\bar{y}_2}{A_1 + A_2} = \frac{2a^2 \cdot a + 4a^2 \cdot a/2}{2a^2 + 4a^2} = \frac{2}{3}a$$

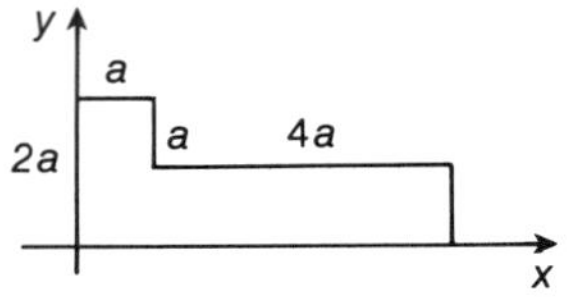

Figure 3.15

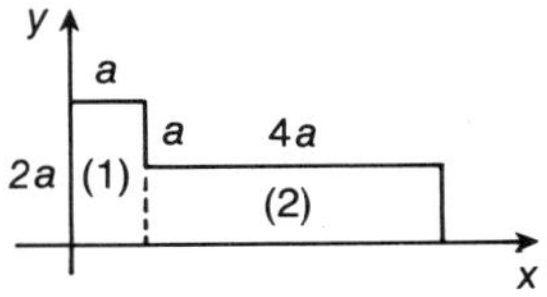

Figure 3.16

SOLUTION 2

Alternatively, the area may be regarded as the difference between a large rectangle (1) and a smaller rectangle (2) as shown in Figure 3.17.

We then have

$$\bar{x} = \frac{S^A_{yz}}{A} = \frac{S^A_{1yz} - S^A_{2yz}}{A_1 - A_2} = \frac{A_1\bar{x}_1 - A_2\bar{x}_2}{A_1 - A_2} = \frac{10a^2 \cdot 5a/2 - 4a^2 \cdot 3a}{10a^2 - 4a^2} = \frac{13}{6}a$$

$$\bar{y} = \frac{S^A_{xz}}{A} = \frac{S^A_{1xz} - S^A_{2xz}}{A_1 - A_2} = \frac{A_1\bar{y}_1 - A_2\bar{y}_2}{A_1 - A_2} = \frac{10a^2 \cdot a - 4a^2 \cdot 3a/2}{10a^2 - 4a^2} = \frac{2}{3}a$$

The above computations can be systematized by using a table as shown in Figure 3.18 (the table applies to Solution 1).

We then have

$$\bar{x} = \frac{S^A_{yz}}{A} = \frac{A_1\bar{x}_1 + A_2\bar{x}_2}{A_1 + A_2} \text{ etc.}$$

Division into differential elements

Equations (3.27) for the centre of mass of a body which is divided into elements may also be used in the limiting case where the number of elements tends to infinity. As an example a body which is divided into an infinite number of 'infinitely thin' shells

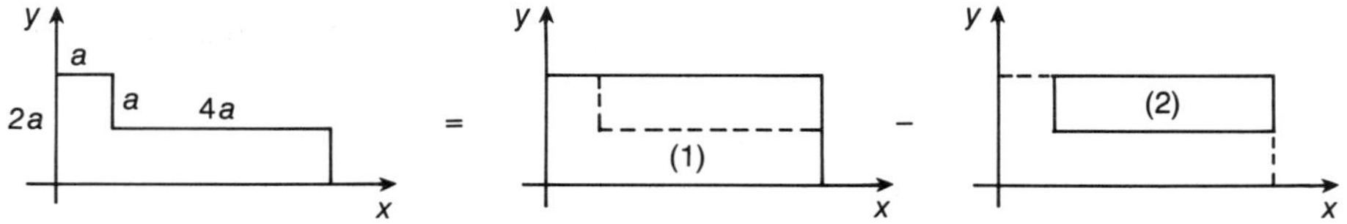

Figure 3.17

Element	Area	$\bar{x}$	$\bar{y}$	S_{yz}^{A}	S_{xz}^{A}
(1) (2)	A_1 A_2	$\bar{x}_1$ $\bar{x}_2$	$\bar{y}_1$ $\bar{y}_2$	$A_1\bar{x}_1$ $A_2\bar{x}_2$	$A_1\bar{y}_1$ $A_2\bar{y}_2$
(1) + (2)	$A_1 + A_2$	–	–	$A_1\bar{x}_1 + A_2\bar{x}_2$	$A_1\bar{y}_1 + A_2\bar{y}_2$

Figure 3.18

is shown in Figure 3.19. Only one shell is shown, however. In a special case the body may be divided into thin, plane disks.

Equations (3.27) are then replaced by

$$\bar{x} = \frac{1}{m} \int x_c \, dm \qquad \bar{y} = \frac{1}{m} \int y_c \, dm \qquad \bar{z} = \frac{1}{m} \int z_c \, dm \tag{3.31}$$

Here $r_c = (x_c, \ y_c, \ z_c)$ is the position vector of the *centre of mass* of the differential element and $dm = \rho \ dV$ is the mass of the element. We assume that the density ρ is constant within each shell. Different shells may have different densities, however. Equations (3.31) are very similar to the definition of the centre of mass as expressed in Equations (3.10). The difference is that the extension of the elements used in the definition is small in every direction, which is not necessary for the shells used here. In the definition of the centre of mass each element can be considered as a particle and its location is given by a position vector $\boldsymbol{r} = (x, \ y, \ z)$. In Equations (3.31), however, it is important that the denotation r_c is referred to the centre of mass of the element.

In the case of a homogeneous body Equations (3.28) are replaced by

$$\bar{x} = \frac{1}{V} \int x_c \, dV \qquad \bar{y} = \frac{1}{V} \int y_c \, dV \qquad \bar{z} = \frac{1}{V} \int z_c \, dV \tag{3.32}$$

For a thin shell with constant mass per unit area Equations (3.29) are replaced by

$$\bar{x} = \frac{1}{A} \int x_c \, dA \qquad \bar{y} = \frac{1}{A} \int y_c \, dA \qquad \bar{z} = \frac{1}{A} \int z_c \, dA \tag{3.33}$$

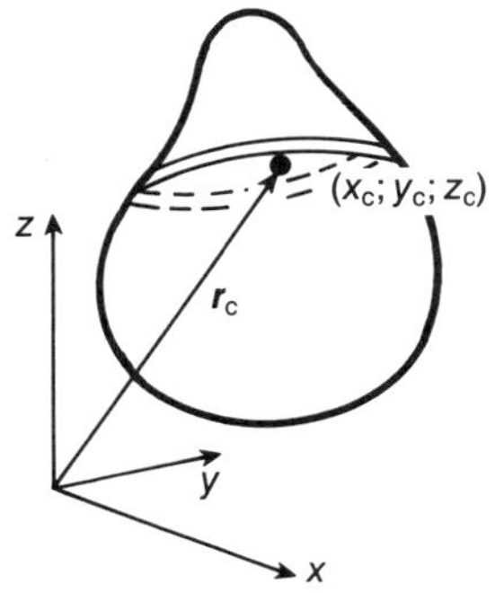

Figure 3.19

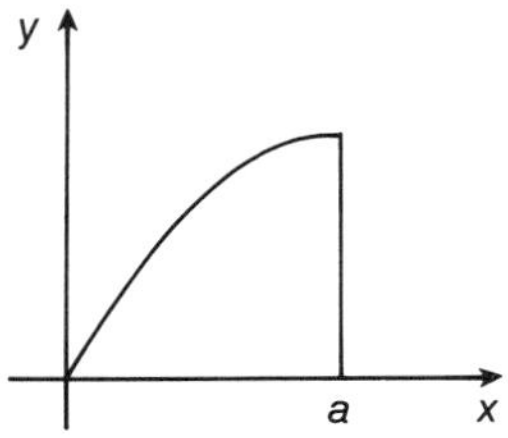

Figure 3.20

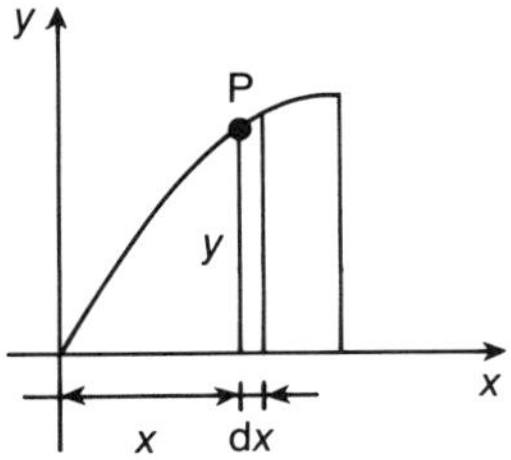

Figure 3.21

The above equations are, of course, also an expression of the fact that the first moment of mass (volume, area) of the composite body is equal to the sum of the first moments of mass (volume, area) of the elements.

WORKED EXAMPLE 3.4

Locate the centroid of the area under the curve $y = b\sin(\pi x/2a)$ from $x = 0$ to $x = a$, Figure 3.20.

SOLUTION

The area is divided into elements as shown in Figure 3.21. These are approximately rectangles of height y and infinitesimal base dx.

The area of the element is $dA = y\,dx = b\sin(\pi x/2a)\,dx$. It should be noted that x and y are the coordinates of point P shown in the figure. The coordinates of the centre of mass, however, are x_c and y_c. In this case we have

$$x_c = x$$

$$y_c = \frac{y}{2} = \frac{1}{2} b \sin\frac{\pi x}{2a}$$

The total area is

$$A = \int dA = \int_0^a b\sin\frac{\pi x}{2a}\,dx = \left[-\frac{2a}{\pi} b\cos\frac{\pi x}{2a}\right]_0^a = \frac{2ab}{\pi}$$

We now use Equations (3.33) to determine the coordinates of the mass centre of the area:

$$\bar{x} = \frac{1}{A}\int x_c \, dA = \frac{\pi}{2ab}\int_0^a xb \sin\frac{\pi x}{2a}\, dx$$

Integrating by parts we have

$$\bar{x} = \frac{\pi}{2a}\left[-x\frac{2a}{\pi}\cos\frac{\pi x}{2a}\right]_0^a + \frac{\pi}{2a}\int_0^a \frac{2a}{\pi}\cos\frac{\pi x}{2a}\, dx = \left[\frac{2a}{\pi}\sin\frac{\pi x}{2a}\right]_0^a = \frac{2a}{\pi}$$

In a similar way we obtain

$$\bar{y} = \frac{1}{A}\int y_c \, dA = \frac{\pi}{2ab}\int_0^a \frac{1}{2} b^2 \sin^2\frac{\pi x}{2a}\, dx$$

$$= \frac{\pi b}{4a}\int_0^a \frac{1}{2}\left(1 - \cos\frac{\pi x}{a}\right) dx = \frac{\pi b}{8a}\left[x - \frac{a}{\pi}\sin\frac{\pi x}{a}\right]_0^a = \frac{\pi b}{8}$$

WORKED EXAMPLE 3.5

Determine the distance from the base of a cone of height h to the centroid.

SOLUTION

We divide the cone into thin slices or disks parallel to the base, Figure 3.22.

We denote by $A(z)$ the area of an arbitrary slice of thickness dz at a distance z from the vertex of the cone. The volume of the element is given by $dV = A(z)\, dz$.

As the top cone of height z is similar to the whole cone we have

$$\frac{A(z)}{A_0} = \left(\frac{z}{h}\right)^2 \Rightarrow dV = A_0\left(\frac{z}{h}\right)^2 dz$$

Here A_0 denotes the area of the base of the original cone.

The total volume of the cone is $V = A_0 h/3$.

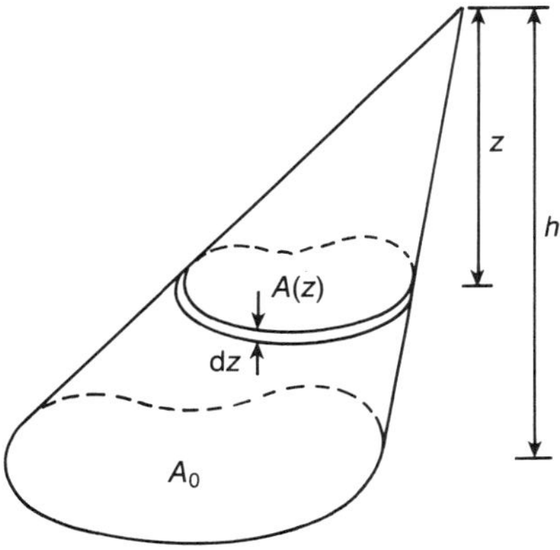

Figure 3.22

The z-coordinate of the centroid is obtained from Equations (3.32):

$$\bar{z} = \frac{1}{V} \int z_c \, dV = \frac{3}{A_0 h} \int_0^h z A_0 \left(\frac{z}{h}\right)^2 dz = \frac{3}{4} h$$

Obviously the result is independent of the area and the shape of the base. Nor is the exact location of the vertex important; only the height h is required. Thus the centre of mass of a completely arbitrary homogeneous cone of height h is located at a distance $h/4$ from the base.

In Worked Example 3.7 below we determine the location of the centre of mass 'sideways'.

REMARK

The result also applies to an arbitrary homogeneous *pyramid*.

If a body can be divided into elements in such a way that the centres of mass are all located on a straight line, then the centre of mass of the composite body must be located on this line. To show this we introduce a coordinate system with an x-axis along the line according to Figure 3.23. If there is a finite number of elements, we have for an arbitrary element, e.g. element number i, that $\bar{y}_i = 0$ and $\bar{z}_i = 0$.

For the composite body we then have

$$\bar{y} = \frac{1}{m} \sum_i m_i \bar{y}_i = 0 \qquad \bar{z} = \frac{1}{m} \sum_i m_i \bar{z}_i = 0$$

i.e. the centre of mass is located on the x-axis.

If the number of elements is infinite, we only have to replace $\bar{y}_i$ and $\bar{z}_i$ by y_c and z_c while the sums are replaced by integrals of the type $\int y_c \, dm$ where $y_c \equiv 0$.

In a similar manner it is possible to show that if a body can be divided into elements with their centres of mass in one plane, then the centre of mass of that body must be located in that plane. The proof is left as an exercise to the reader.

WORKED EXAMPLE 3.6

Locate the centroid of a triangular area.

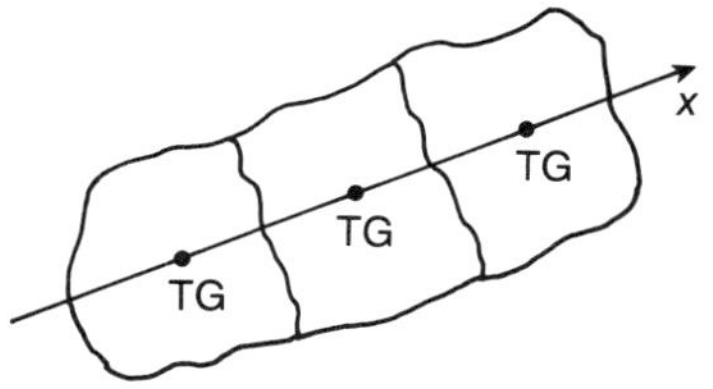

Figure 3.23

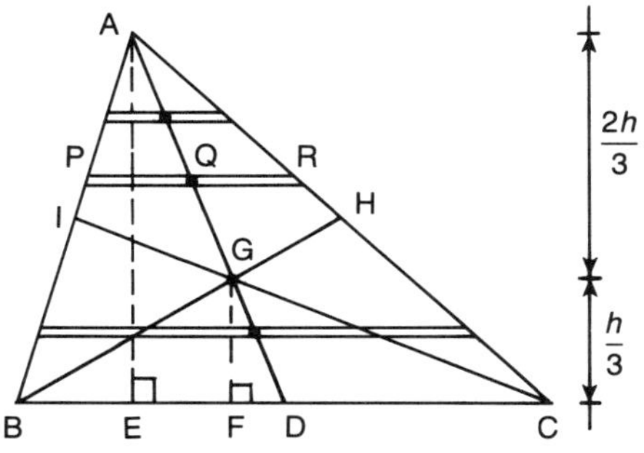

Figure 3.24

SOLUTION

The triangle is divided into elements as shown in Figure 3.24. These are approximately rectangles of infinitesimal height with two sides parallel to BC. One of these elements is denoted by PR. We note that triangle ABC is similar to triangle APR. D is the midpoint of BC, i.e. AD is a median. Triangles ABD and APQ are also similar to each other. We then have AP:AB = PR:BC = PQ:BD. Since BD = BC/2, we have PQ = PR/2. In other words, Q is the midpoint of PR, i.e. Q is the centroid of element PR. However, PR is quite arbitrarily chosen. Thus, the centroids of all elements must be located on the median AD. From the proof above we know that this must also be true for triangle ABC.

The reason may be repeated for each of the two other medians BH and CI. The centroid G of the triangle must then be located at the intersection of the medians. Furthermore, it is known from geometry that the medians intersect at the ratio 2:1, i.e. AG = 2 DG or AD = 3 DG. Utilizing the fact that triangles ADE and GDF are similar we conclude that the distance from the centroid to base BC is one-third of the corresponding altitude (AE) as shown in the figure.

WORKED EXAMPLE 3.7

Locate the centroid of an arbitrary cone.

SOLUTION

The cone is divided into thin slices parallel to the base as in Worked Example 3.5. All slices are similar to each other, and it is possible to show that their centres of mass are located on the line connecting the centre of mass of the base to the vertex, Figure 3.25. We refrain from proving this rigorously, however, and assume it will be intuitively clear to the reader.

From Worked Example 3.5 we know that the distance from the centroid to the base is one-quarter of the height of the cone. Therefore the location of the centroid is uniquely determined as shown.

REMARK

The result also applies to an arbitrary pyramid.

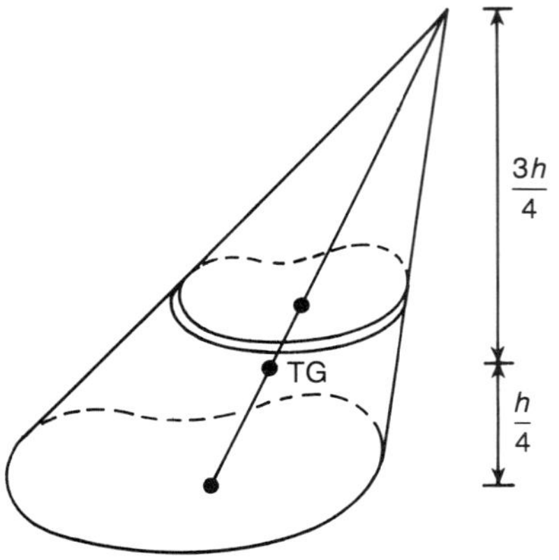

Figure 3.25

Some practical advice

1. If the shape of a body is 'geometrically simple' it is likely that the centre of mass may be found in tables or handbooks, for instance in Appendix 2 of this book.

2. If it is not possible to determine the location of the centre of mass from a table, it is recommended that the symmetry properties of the body, if there are any, are utilized to draw conclusions about the location of the centre of mass according to Section 3.2.1. Even if there is no plane of symmetry it may happen that a certain first moment is zero, which will simplify the problem. On some occasions it may be possible to divide the body into elements such that the centres of mass of all elements are located on a certain line or in a certain plane. In this way the number of coordinates to be determined may be reduced.

3. The next step is to try to divide the body into a finite number of elements such that the centre of mass of each element is known. We then use Equations (3.27)–(3.30) to find the centre of mass. We then take advantage of the fact that the first moment of mass of the composite body is equal to the sum of the first moments of the individual parts. In some cases the body can be considered as manufactured by cutting off parts

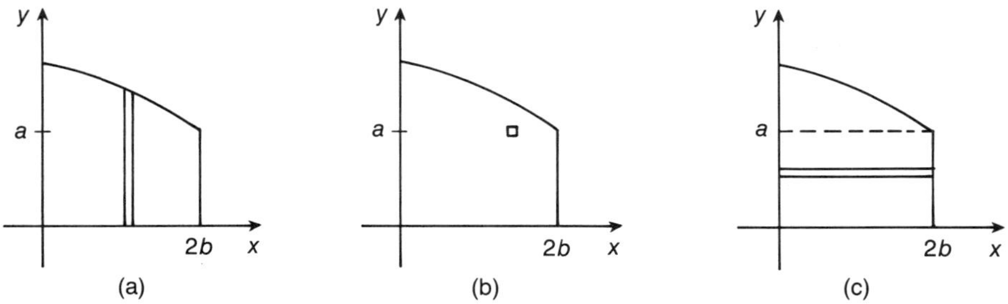

Figure 3.26

with known centres of mass from an original body, also with a known centre of mass. In such cases the first moments of mass of the portions removed must be subtracted.

4. If it is not possible to apply the method of step 3, it may be possible to divide the body into an infinite number of differential elements with known centres of mass. Some of Equations (3.31)–(3.33) can then be applied. Such elements should be chosen as 'large' as possible. As an example, in Figure 3.26 division (a) is preferable to division (b).

In order not to complicate the integrations more than necessary, division (a) is preferable to division (c). In the latter case the expressions for the area and the centre of mass of the element will be different depending on the integration variable being greater than or less than a. Consequently, twice as many integrals must be computed compared with division (a).

5. Finally, it is always a good idea to use common sense to check if the answer is reasonable. As an example, the coordinates of the centroid of the area of Figure 3.26 must satisfy the conditions $\bar{x} < b$ and $\bar{y} > a/2$.

PROBLEMS

○ **3.5** A bar of length L with end points A and B is non-uniform. The mass per unit length, denoted by λ, varies according to

$$\lambda = \lambda_0 \left(\frac{x}{L}\right)^3$$

where x is the distance from end point A and λ_0 is a constant. Determine the location of the centre of mass.

○ **3.6** A snowman is composed of three homogeneous spheres of radii a, $2a$ and $3a$. The mass of the hat is negligible. Compute the centre of mass of the snowman.

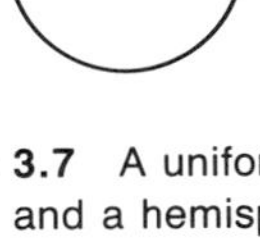

○ **3.7** A uniform body is composed of a cone and a hemisphere of radius a. Determine the height of the cone (h) so that the centre of mass of the composite body is located at the origin.

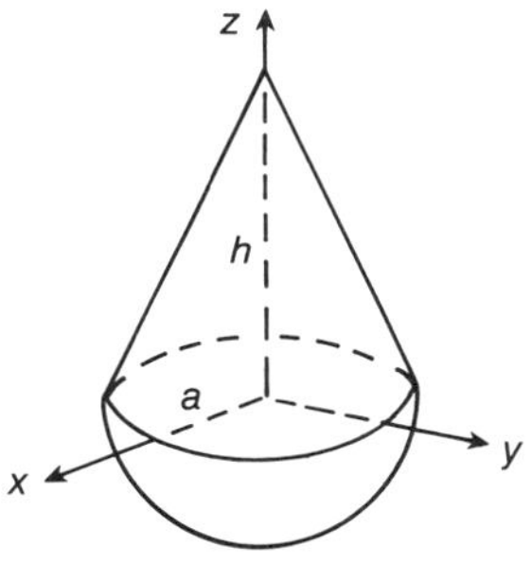

○ **3.8** A tin can has a circular cross-section and no lid. The height h is equal to the diameter. Where is the centre of mass located?

3.9 A body is generated by rotating the isosceles triangle ABC about the x-axis through 360°. Locate the centroid of the body.

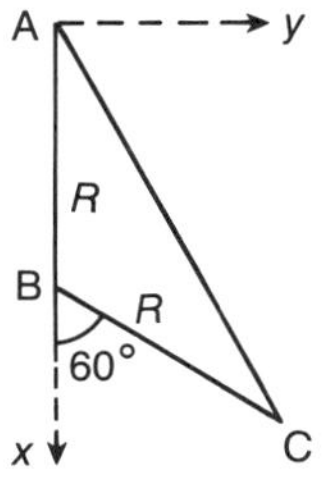

3.10 A hemispherical ceramics bowl is filled with water to the rim. The inner radius of the bowl is two-thirds of the outer radius. The distance from the centre of the sphere to the centre of mass of the whole system (bowl + water) is then two-fifths of the outer radius. What is the ratio of the density of the ceramic material to the density of water?

3.11 Locate the centroid of the shaded area.

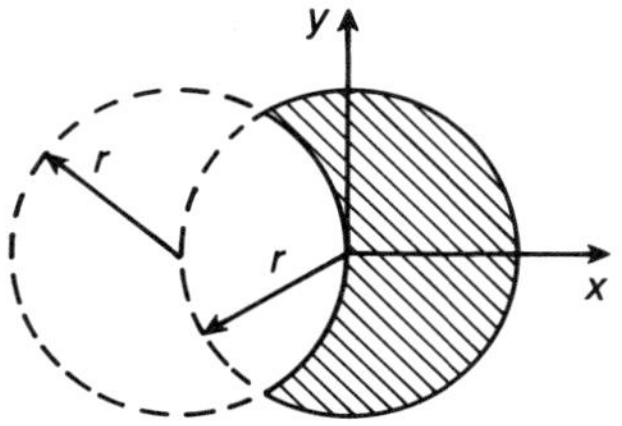

3.12 Locate the centroids of the shaded areas shown.

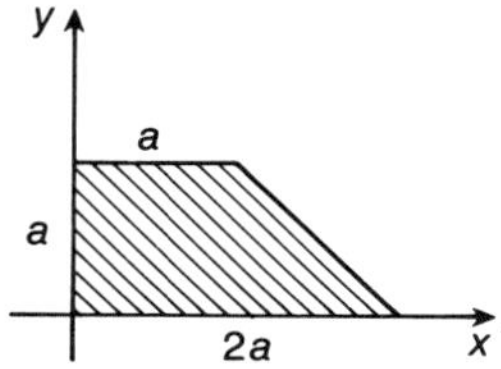

(a)

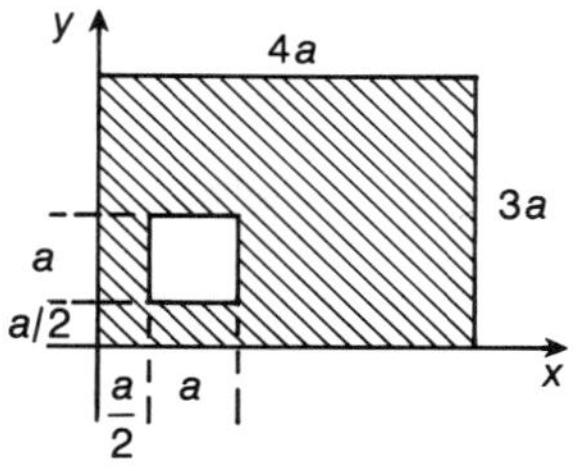

(b)

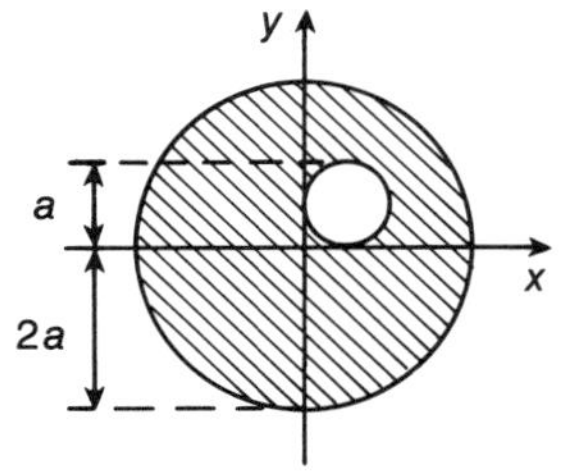

(c)

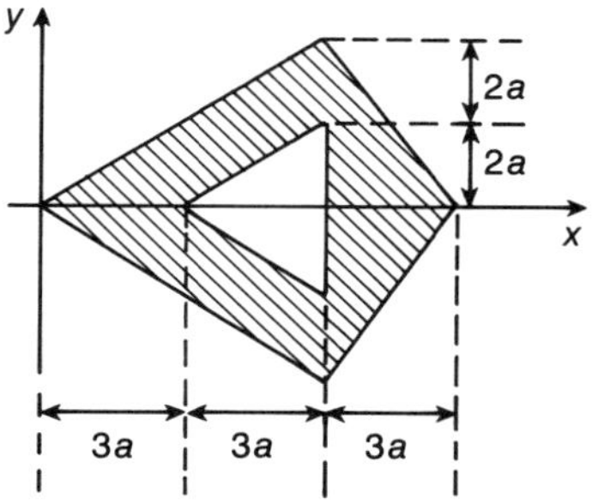

(d)

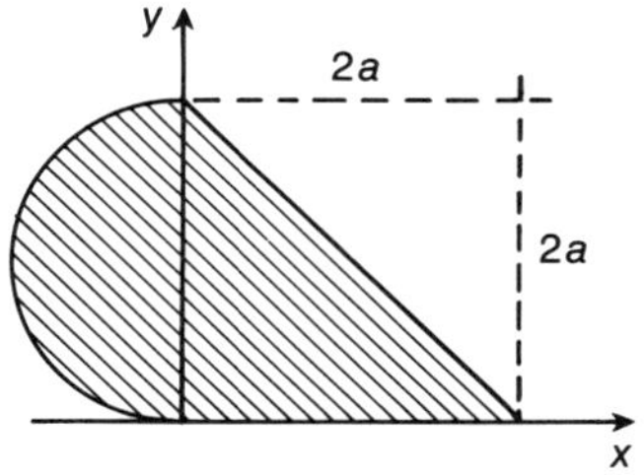

(e)

○ **3.13** Locate the centre of mass of the body composed of three bars of the same mass per unit length.

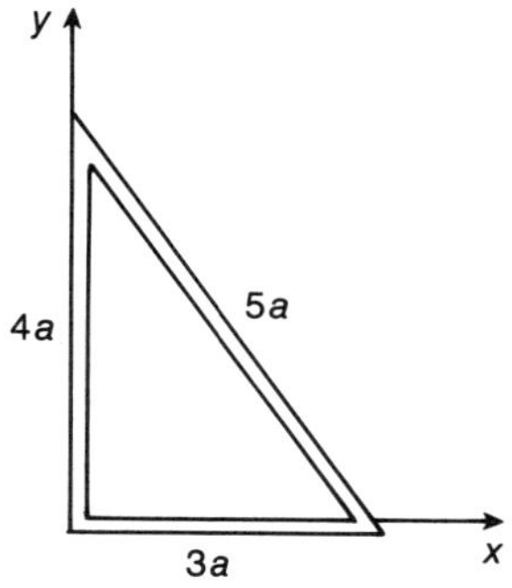

3.14 A thin pipe has been bent to the shape shown. Determine the position of the centre of mass.

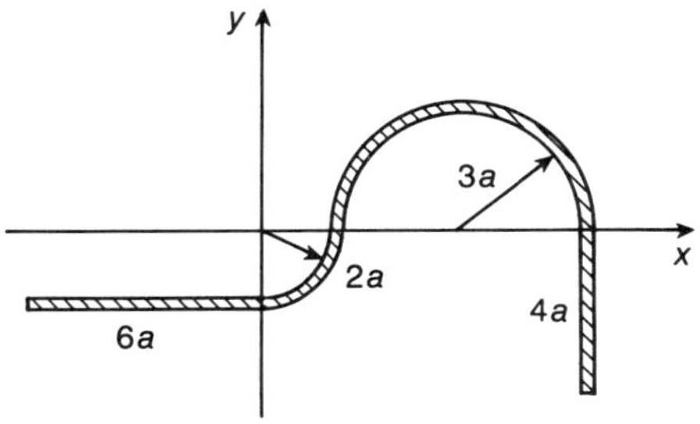

3.15 Determine the coordinates of the centroid of the shaded area. Check the answer by considering the special case $\alpha = 90°$.

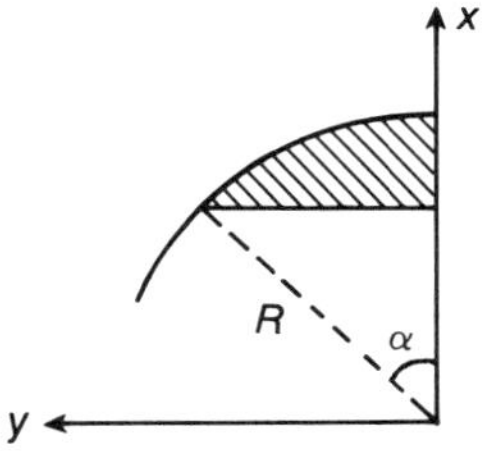

○ **3.16** A wire has been bent to the shape of a semicircular arc of radius r. The wire is joined to a thin semicircular disk of the same mass and the same radius as the wire such that the planes of the wire and the disk are perpendicular. Determine the coordinates of the centre of mass.

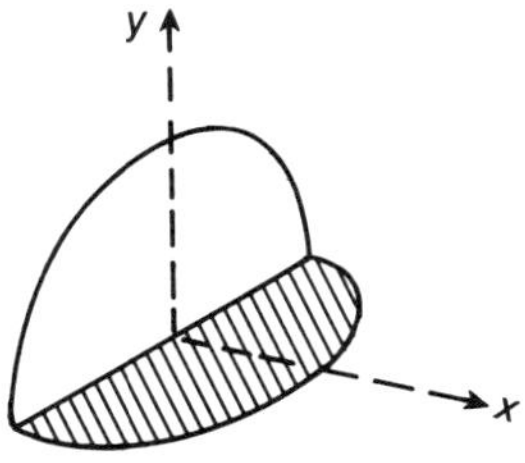

3.17 A barrel of semicircular cross-section is manufactured from sheet aluminium of constant thickness. There is no lid. Locate the centre of mass.

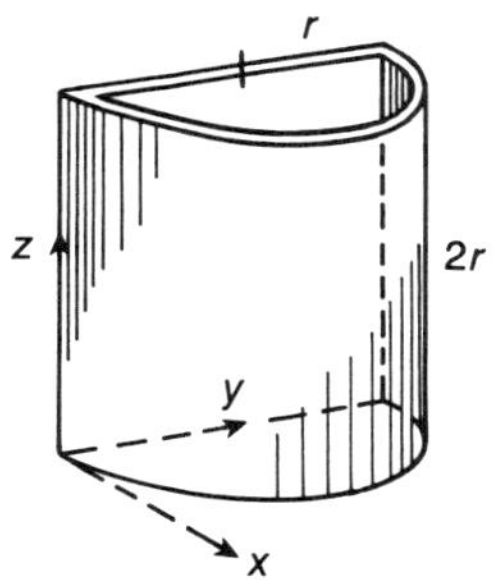

3.18 A scoop has a rectangular bottom and three walls. One of these is rectangular and the other two triangular. The same sheet-metal has been used for all parts. Locate the centre of mass.

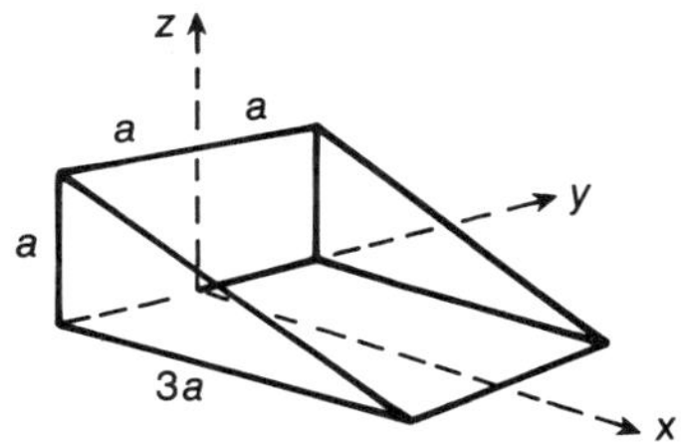

3.19 A uniform body is composed of two parts: a right triangular prism and a half, right circular cone as shown. Locate the centroid of the volume.

Hint: Use the result from Worked Example 3.7.

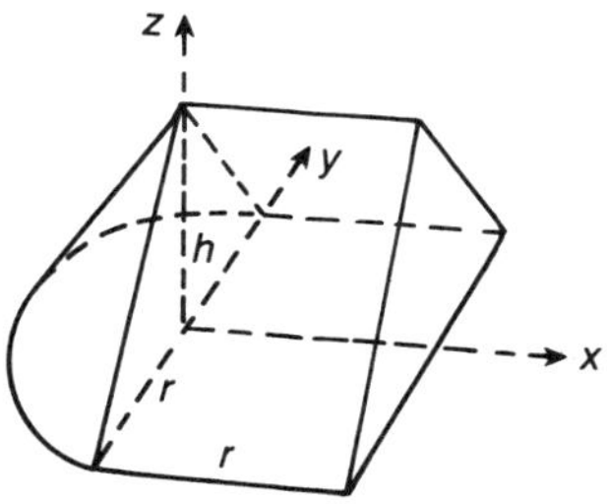

◎ **3.20** Return to Problem 3.19. Determine the coordinates of the centre of mass of that half of the body located on the positive side of the xz-plane.

3.21 A circular piece of uniform sheet-metal has been cut up along a radius. A portion of the piece in the shape of a quarter circular sector has been bent as shown. Compute the coordinates of the centre of mass of the body thus formed.

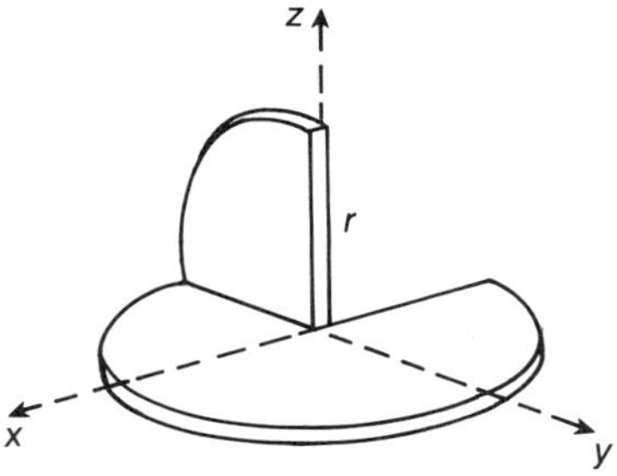

3.22 Determine the coordinates of the centroid of the area bounded by the curve $y = b(1 - x^2/a^2)$ and the coordinate axes.

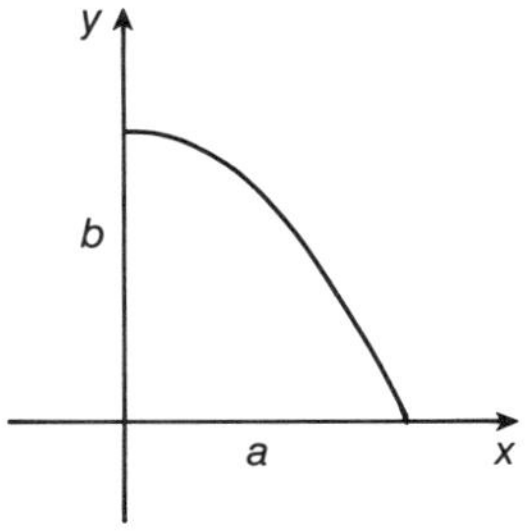

◎ **3.23** The curve $y = b \sin(\pi x/2a)$ passes through the origin and point B.

(a) The area OBC is bounded by the curve and two straight lines. Locate the centroid of the area.

(b) The area OBA is bounded by the curve and two other straight lines. Locate the centroid of the area.

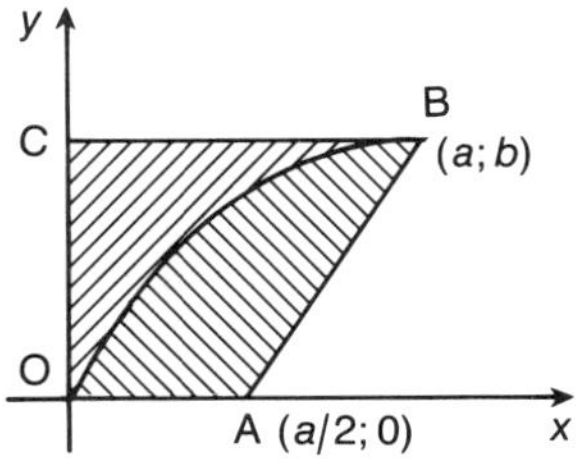

3.24 A piece of sheet-metal is bent to the shape of the lateral area of a right circular cone.

(a) Locate the centre of mass.

(b) The body is divided in half by a cut in the *yz*-plane. Where is the centre of mass of the left half located?

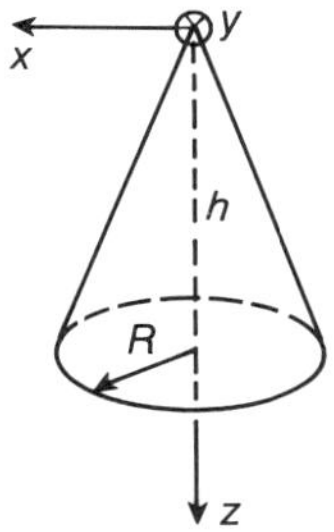

◎ **3.25** Locate the centre of mass of a spherical segment of one base (solid body) of radius R and height h. In particular, consider the special case $h = R$.

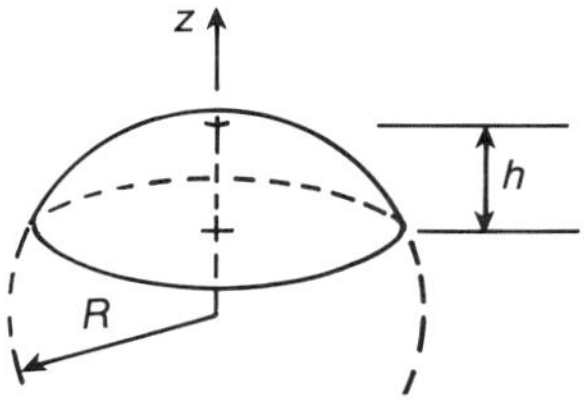

3.26 Two volumes are generated by rotating the areas shown in diagrams (a) and (b) through 360° about the *x*- and the *y*-axis, respectively. Determine the location of the centroids.

(a) Rotation about the *x*-axis

(b) Rotation about the *y*-axis

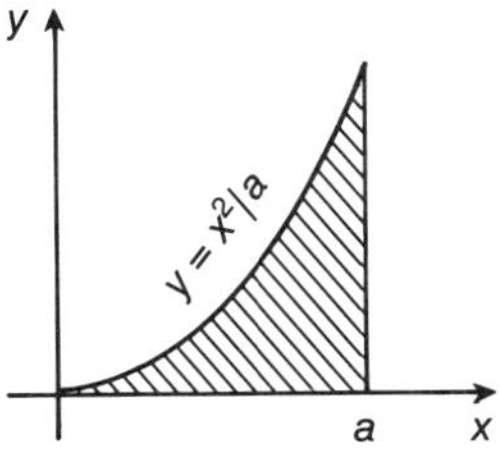

(a)

continued overleaf

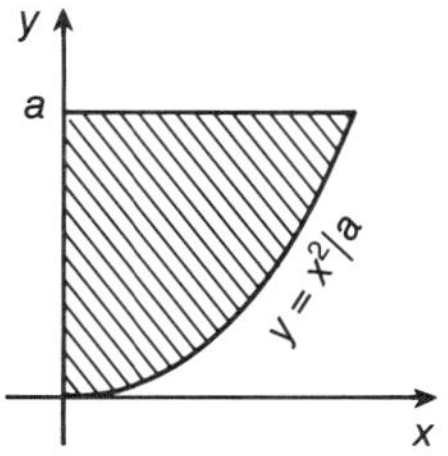

(b)

◎ **3.27** A body is generated by rotating the area of Problem 3.26(b) through 180° only, such that $x > 0$. Determine the location of the centroid.

3.28 A plate of mass 20 kg is suspended by two vertical cables as shown. Determine the tensions in the cables.

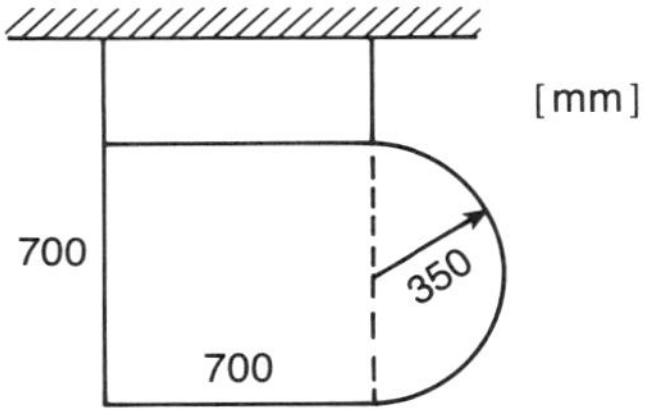

○ **3.29** A thin, uniform disk may rotate freely about a horizontal axis O as shown. Determine the length x such that the disk is in equilibrium in the position shown.

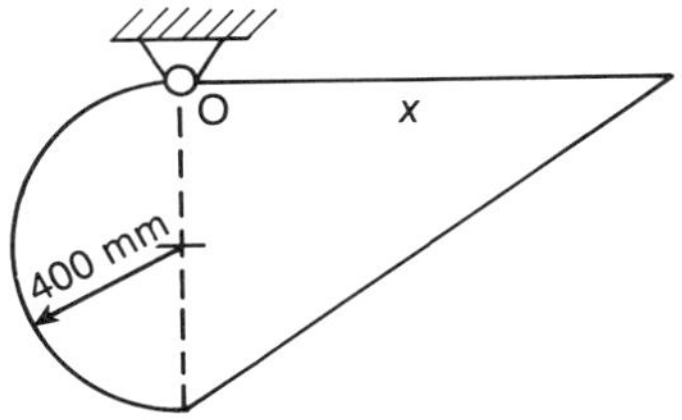

3.30 A body is composed of four identical cubes. If the body is suspended by a wire attached at a point P as shown, then face ABCD is horizontal at equilibrium. Determine the location of point P.

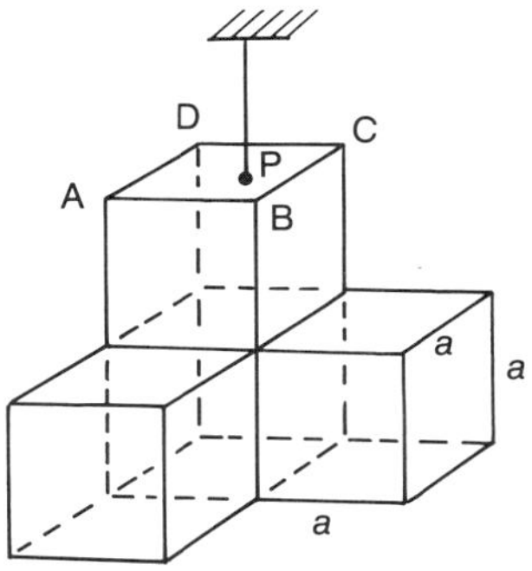

3.31 A wire has been bent to form a hook as shown. The hook is hung on a smooth, narrow bar inside the small circle. Determine the angle that the line connecting the centres of the circles makes with the vertical.

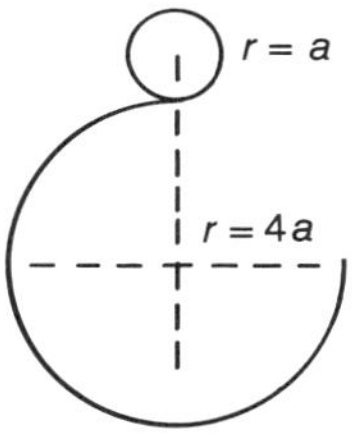

◎ **3.32** A semicircular arc is composed of two wires, AB and BC, in the shape of quarter circular arcs. The masses of AB and BC are $2m$ and m, respectively. The arc is supported by a smooth edge P. Determine the angle α that the diameter AC makes with the horizontal at equilibrium.

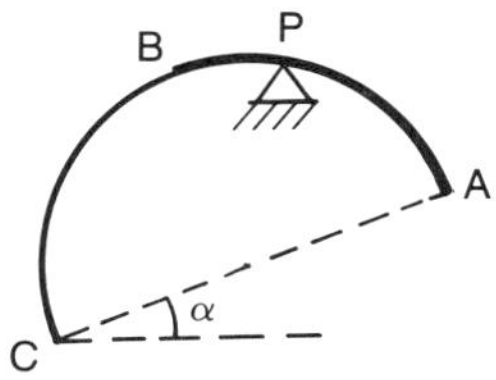

APPENDIX 1

Basic mathematics

A1.1 Formulas and tables

A1.1.1 Geometry

Elementary geometry

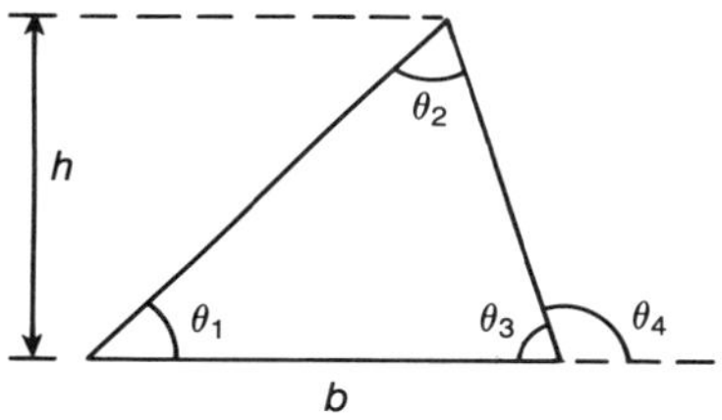

Triangle: Area $= \frac{1}{2} bh$

Angles: $\theta_4 = \theta_1 + \theta_2$

$\theta_1 + \theta_2 + \theta_3 = 180^\circ$

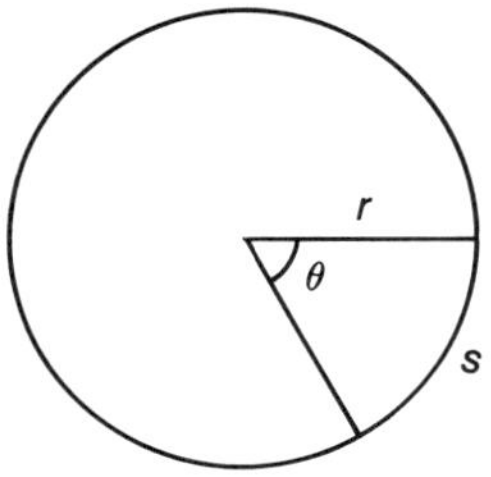

Circle: Circumference $= 2\pi r$ Area $= \pi r^2$

Arc length $s = r\theta$ Sector area $= \frac{1}{2} r^2\theta$

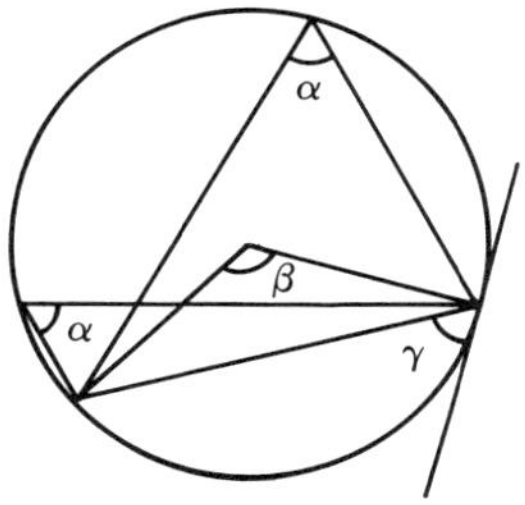

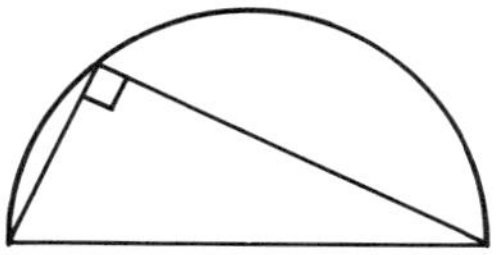

Central angle (β), periphery angle (α), angle between tangent and chord (γ):

$$\beta = 2\alpha \qquad \gamma = \alpha$$

In particular, a triangle inscribed within a semicircle is a right triangle.

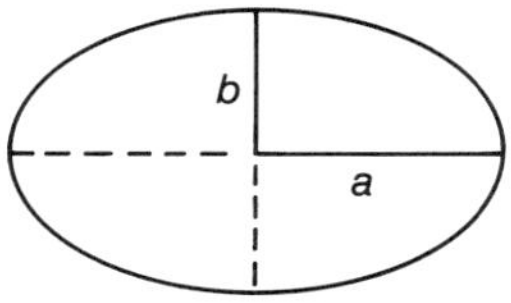

Ellipse: Area $= \pi ab$

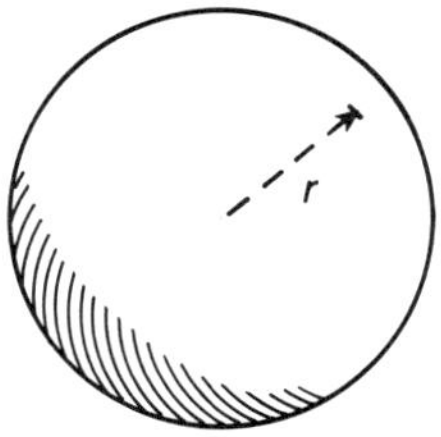

Sphere: Surface area $= 4\pi r^2$ Volume $= \dfrac{4}{3}\pi r^3$

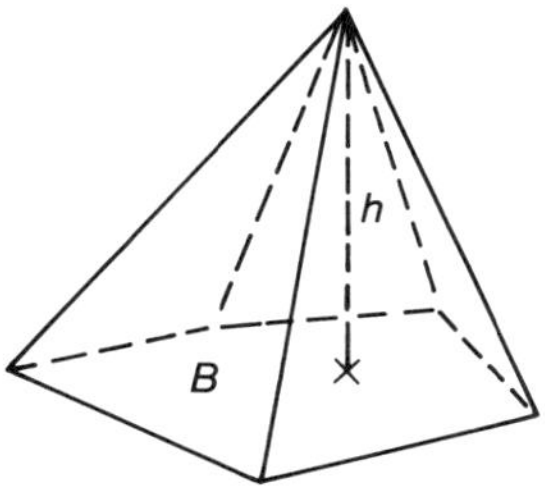

Pyramid or cone: Volume $= \frac{1}{3} Bh$ B is area of base

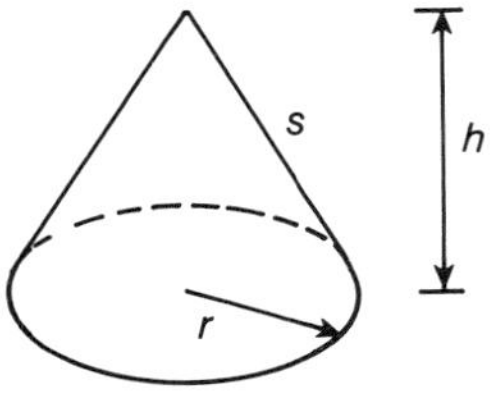

Right circular cone: Volume $= \frac{1}{3} \pi r^2 h$ Lateral area $= \pi rs$

Analytic geometry

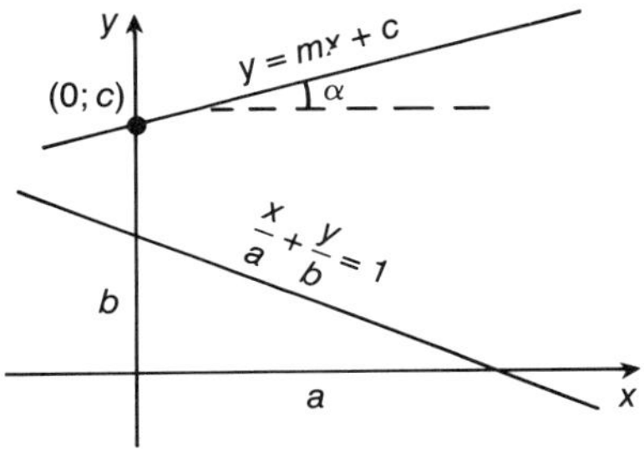

Straight line: $y = mx + c$ m (slope) $= \tan \alpha$ $\frac{x}{a} + \frac{y}{b} = 1$

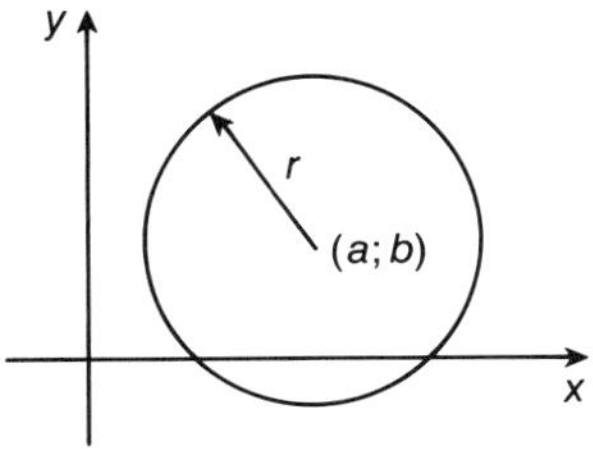

Circle: $(x-a)^2 + (y-b)^2 = r^2$

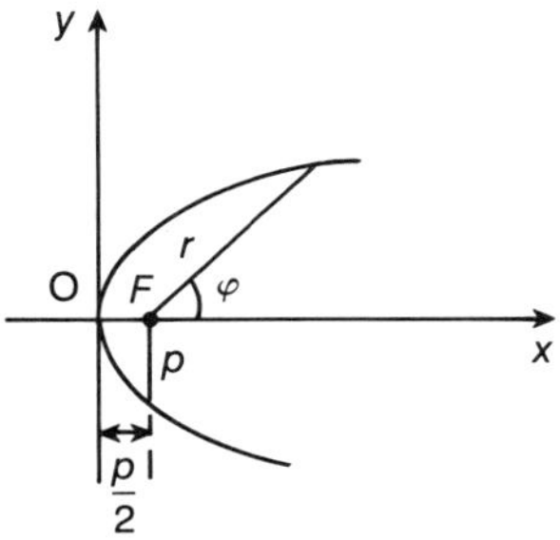

Parabola: $y^2 = 2px$ In polar coordinates: $r = \dfrac{p}{1 - \cos\varphi}$

F = focus, O = vertex

The parametric equations

$$\begin{cases} x = v_0 t \cos\alpha \\ y = v_0 t \sin\alpha - \dfrac{1}{2} g t^2 \end{cases}$$

represent the path of a projectile, where v_0 is the speed and α the angle of elevation at the start.

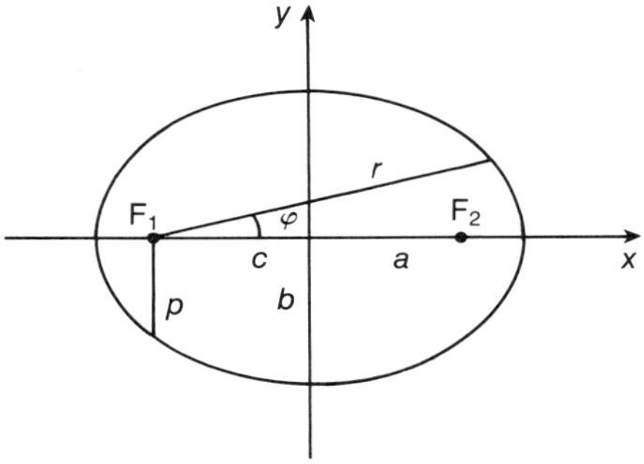

Ellipse: $\dfrac{x^2}{a^2} + \dfrac{y^2}{b^2} = 1$ Polar coordinates: $r = \dfrac{p}{1 - e\cos\varphi}$ $(e < 1)$

Eccentricity $e = \dfrac{c}{a}$ where $c = \sqrt{a^2 - b^2}$ $p = (1 - e^2)a$

F_1, F_2 = foci

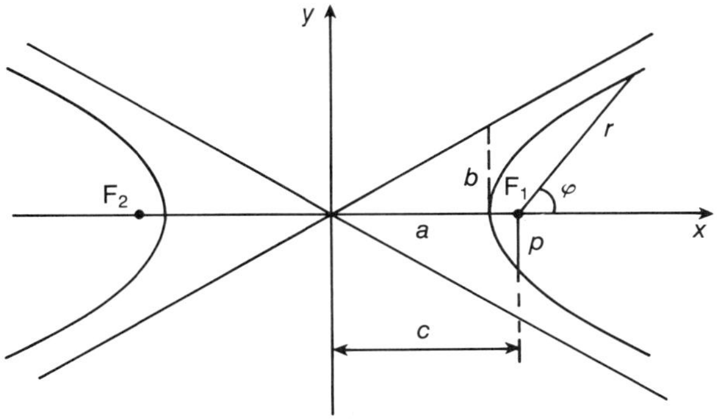

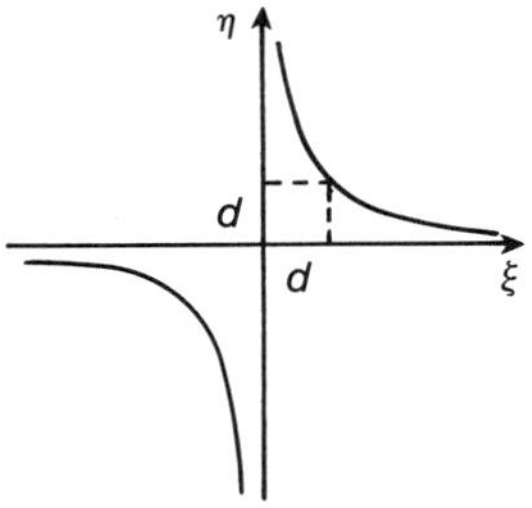

Hyperbola: $\frac{x^2}{a^2} - \frac{y^2}{b^2} = 1$ Polar coordinates: $r = \frac{p}{1 - e \cos \varphi}$ $(e > 1)$

Eccentricity $e = \frac{c}{a}$ where $c = \sqrt{a^2 + b^2}$ $p = (e^2 - 1)a$

F_1, F_2 = foci

In particular, if ξ- and η-axes are asymptotes, $\xi\eta = d^2$.

A1.1.2 Algebra

The quadratic equation

$x^2 + 2ax + b = 0$ $x = -a \pm \sqrt{a^2 - b}$

Logarithms

Definition: If $b^x = y$ then x is the logarithm of y to the base b: $\log_b y = x$.

Properties of logarithms:

$\log_b b = 1$

$\log_b (ac) = \log_b a + \log_b c$

$\log_b \frac{a}{c} = \log_b a - \log_b c$

$\log_b 1 = 0$

$\log_b \frac{1}{a} = -\log_b a$

$\log_b (a^n) = n \log_b a$

$\log_a y = \log_a b \cdot \log_b y$

Natural logarithms:

$b = \mathrm{e} = 2.718\ 28 \ldots$

$\log_e y = \ln y$

Matrices

$$\begin{pmatrix} a_{11} & a_{12} & a_{13} \\ a_{21} & a_{22} & a_{23} \\ a_{31} & a_{32} & a_{33} \end{pmatrix} \cdot \begin{pmatrix} b_{11} & b_{12} & b_{13} \\ b_{21} & b_{22} & b_{23} \\ b_{31} & b_{32} & b_{33} \end{pmatrix} = \begin{pmatrix} c_{11} & c_{12} & c_{13} \\ c_{21} & c_{22} & c_{23} \\ c_{31} & c_{32} & c_{33} \end{pmatrix} \text{ where } c_{ik} = \sum_m a_{im} b_{mk}$$

$$\begin{pmatrix} a_{11} & a_{12} & a_{13} \\ a_{21} & a_{22} & a_{23} \\ a_{31} & a_{32} & a_{33} \end{pmatrix} \cdot \begin{pmatrix} b_1 \\ b_2 \\ b_3 \end{pmatrix} = \begin{pmatrix} c_1 \\ c_2 \\ c_3 \end{pmatrix} \text{ where } c_i = \sum_m a_{im} b_m$$

Determinants

$$\begin{vmatrix} a_1 & a_2 \\ b_1 & b_2 \end{vmatrix} = a_1 b_2 - a_2 b_1$$

$$\begin{vmatrix} a_1 & a_2 & a_3 \\ b_1 & b_2 & b_3 \\ c_1 & c_2 & c_3 \end{vmatrix} = a_1 \begin{vmatrix} b_2 & b_3 \\ c_2 & c_3 \end{vmatrix} - a_2 \begin{vmatrix} b_1 & b_3 \\ c_1 & c_3 \end{vmatrix} + a_3 \begin{vmatrix} b_1 & b_2 \\ c_1 & c_2 \end{vmatrix}$$

A1.1.3 Trigonometric and hyperbolic functions

Trigonometric functions

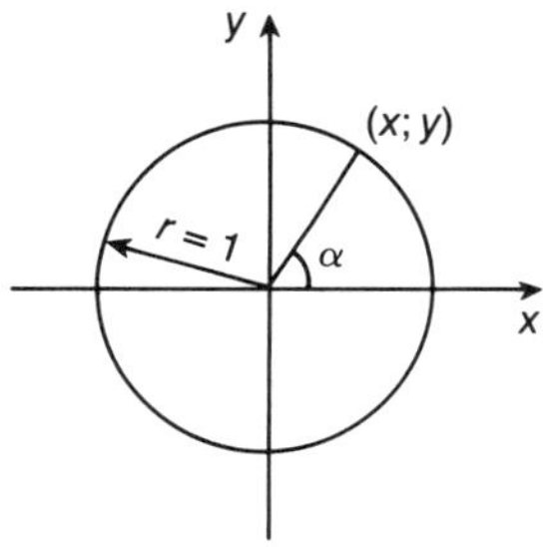

$$\sin \alpha = y \qquad \cos \alpha = x \qquad \tan \alpha = \frac{y}{x} \qquad \cot \alpha = \frac{x}{y} = \frac{1}{\tan \alpha}$$

$$\sec \alpha = \frac{1}{\cos \alpha} \qquad \csc \alpha = \frac{1}{\sin \alpha}$$

$$\sin^2 \alpha + \cos^2 \alpha = 1 \qquad 1 + \tan^2 \alpha = \sec^2 \alpha$$

$$\sin^2 \frac{\alpha}{2} = \frac{1}{2}(1 - \cos \alpha) \qquad \cos^2 \frac{\alpha}{2} = \frac{1}{2}(1 + \cos \alpha)$$

$$\sin(\alpha \pm \beta) = \sin \alpha \cos \beta \pm \cos \alpha \sin \beta$$

$$\cos(\alpha \pm \beta) = \cos \alpha \cos \beta \mp \sin \alpha \sin \beta$$

$$\sin 2\alpha = 2 \sin \alpha \cos \alpha$$

$$\cos 2\alpha = \cos^2 \alpha - \sin^2 \alpha = 1 - 2 \sin^2 \alpha = 2 \cos^2 \alpha - 1$$

Inverse functions: If $x = \sin y$ then $y = \arcsin x$ $\left(-\frac{\pi}{2} \leqslant y \leqslant \frac{\pi}{2}\right)$

$x = \cos y$ $\quad y = \arccos x$ $\quad (0 \leqslant y \leqslant \pi)$

$x = \tan y$ $\quad y = \arctan x$ $\quad \left(-\frac{\pi}{2} < y < \frac{\pi}{2}\right)$

$x = \cot y$ $\quad y = \operatorname{arccot} x$ $\quad \left(-\frac{\pi}{2} < y < \frac{\pi}{2}\right)$

Notations like $\sin^{-1} x$ are sometimes used for arc-functions. However, this is not recommended here; generally $\sin^n x$ means $(\sin x)^n$.

Hyperbolic functions

$$\sinh x = \frac{1}{2}(e^x - e^{-x})$$

$$\cosh x = \frac{1}{2}(e^x + e^{-x})$$

$$\tanh x = \frac{e^x - e^{-x}}{e^x + e^{-x}} = \frac{\sinh x}{\cosh x}$$

$$\cosh^2 x - \sinh^2 x = 1$$

Inverse functions: If $x = \sinh y$ then $y = \operatorname{arsinh} x$ etc. The inverse functions are sometimes written $\sinh^{-1} x$ etc. See remark above under trigonometric functions.

Complex functions ($i^2 = -1$)

$$e^{ix} = \cos x + i \sin x$$

$$\sin x = \frac{1}{2i}(e^{ix} - e^{-ix}) = \frac{1}{i}\sinh(ix)$$

$$\cos x = \frac{1}{2}(e^{ix} + e^{-ix}) = \cosh(ix)$$

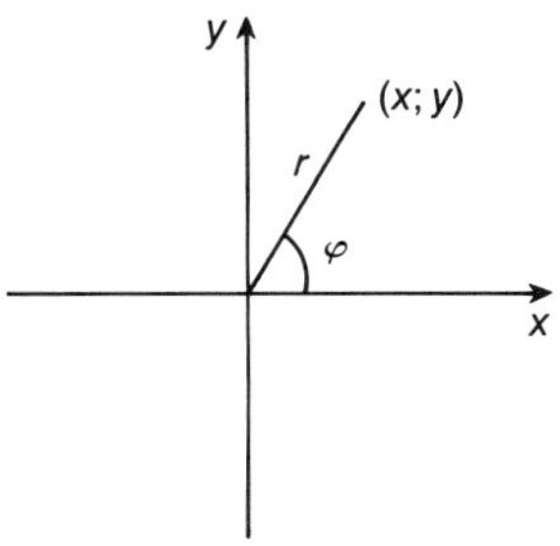

The complex number $z = x + iy$ can also be written $z = re^{i\varphi}$

Triangle laws

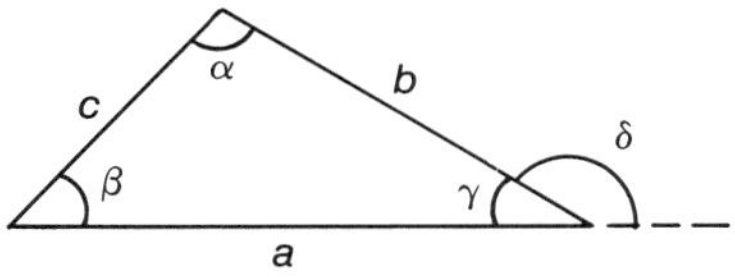

Sine law: $$\frac{a}{\sin\alpha}=\frac{b}{\sin\beta}=\frac{c}{\sin\gamma}$$

Cosine law: $$c^2=a^2+b^2-2ab\cos\gamma$$
$$c^2=a^2+b^2+2ab\cos\delta$$

A1.1.4 Series

Geometric progression

$$s=a+ak+ak^2+\ldots+ak^{n-1}=a\,\frac{k^n-1}{k-1}$$

for $n=\infty$, $|k|<1$: $s=\dfrac{a}{1-k}$

Series expansions

Taylor's formula ($f'=\dfrac{df}{dx}$, $f''=\dfrac{d^2f}{dx^2}$ etc.)

$$f(a+h)=f(a)+hf'(a)+\frac{h^2}{2!}f''(a)+\frac{h^3}{3!}f'''(a)+\ldots+\frac{h^n}{n!}f^{(n)}(a)+\ldots$$

Maclaurin's series

$$f(x)=f(0)+xf'(0)+\frac{x^2}{2!}f''(0)+\frac{x^3}{3!}f'''(0)+\ldots+\frac{x^n}{n!}f^{(n)}(0)+\ldots$$

$$\sin x=x-\frac{x^3}{3!}+\frac{x^5}{5!}-\frac{x^7}{7!}+\ldots$$

$$\cos x=1-\frac{x^2}{2!}+\frac{x^4}{4!}-\frac{x^6}{6!}+\ldots$$

$$\tan x=x+\frac{x^3}{3}+\frac{2x^5}{15}+\frac{17x^7}{315}+\ldots \qquad (|x|<\pi/2)$$

$$\ln(1+x)=x-\frac{x^2}{2}+\frac{x^3}{3}-\frac{x^4}{4}+\ldots \qquad (|x|<1)$$

$$e^x=1+x+\frac{x^2}{2!}+\frac{x^3}{3!}+\ldots$$

$$a^x = 1 + x \ln a + \frac{(x \ln a)^2}{2!} + \frac{(x \ln a)^3}{3!} + \ldots$$

$$\sinh x = \frac{1}{2}(e^x - e^{-x}) = x + \frac{x^3}{3!} + \frac{x^5}{5!} + \frac{x^7}{7!} + \ldots$$

$$\cosh x = \frac{1}{2}(e^x + e^{-x}) = 1 + \frac{x^2}{2!} + \frac{x^4}{4!} + \frac{x^6}{6!} + \ldots$$

A1.1.5 Derivatives

Chain rule:

If $f(u)$ is a function of u and u is a function of x, then $\frac{df}{dx} = \frac{df}{du}\frac{du}{dx}$

$$\frac{d}{dx} x^n = nx^{n-1}$$

$$\frac{d}{dx}(uv) = u\frac{dv}{dx} + v\frac{du}{dx} = uv' + u'v$$

$$\frac{d}{dx}\left(\frac{u}{v}\right) = \frac{1}{v^2}\left(v\frac{du}{dx} - u\frac{dv}{dx}\right) = \frac{vu' - uv'}{v^2}$$

$$\frac{d}{dx}\sin x = \cos x$$

$$\frac{d}{dx}\cos x = -\sin x$$

$$\frac{d}{dx}\tan x = \frac{1}{\cos^2 x} = \sec^2 x$$

$$\frac{d}{dx}\cot x = -\frac{1}{\sin^2 x} = -\csc^2 x$$

$$\frac{d}{dx}\arcsin x = \frac{1}{\sqrt{1-x^2}}$$ (where $\arcsin x$ lies between $-\frac{\pi}{2}$ and $+\frac{\pi}{2}$)

$$\frac{d}{dx}\arccos x = -\frac{1}{\sqrt{1-x^2}}$$ (where $\arccos x$ lies between 0 and π)

$$\frac{d}{dx}\arctan x = \frac{1}{1+x^2}$$ (where $\arctan x$ lies between $-\frac{\pi}{2}$ and $+\frac{\pi}{2}$)

$$\frac{d}{dx}\sinh x = \cosh x$$

$$\frac{d}{dx}\cosh x = \sinh x$$

$$\frac{d}{dx}\tanh x = \frac{1}{\cosh^2 x}$$

$$\frac{d}{dx}e^x = e^x$$

$$\frac{d}{dx}b^x = (\ln b)b^x$$

$$\frac{d}{dx}\ln x = \frac{1}{x}$$

A1.1.6 Integrals

Integration by parts:

$$\int g(x)f(x)\,dx = g(x)F(x) - \int \frac{dg(x)}{dx}F(x)\,dx \qquad \text{where } \frac{dF(x)}{dx} = f(x)$$

Note: In the following the constant of integration (C) is omitted but should be added to the result of every integration.

$$\int x^n\,dx = \frac{x^{n+1}}{n+1} \quad (n \neq -1)$$

$$\int \frac{dx}{x} = \ln x$$

$$\int \frac{dx}{a+bx} = \frac{1}{b}\ln(a+bx)$$

$$\int \frac{x\,dx}{a+bx} = \frac{1}{b^2}(a+bx-a\ln(a+bx))$$

$$\int \frac{dx}{x^2+a^2} = \frac{1}{a}\arctan\frac{x}{a}$$

$$\int \frac{dx}{x^2-a^2} = \frac{1}{2}\ln\frac{x-a}{x+a}$$

$$\int \frac{x\,dx}{a+bx^2} = \frac{1}{2b}\ln(a+bx^2)$$

$$\int \sqrt{a+bx}\,dx = \frac{2}{3b}\sqrt{(a+bx)^3}$$

$$\int x\sqrt{a+bx}\,dx = \frac{2}{15b^2}(3bx-2a)\sqrt{(a+bx)^3}$$

$$\int \frac{dx}{\sqrt{a+bx}} = \frac{2\sqrt{a+bx}}{b}$$

$$\int \frac{x\,dx}{\sqrt{a+bx}} = \frac{2}{3b^2}(bx-2a)\sqrt{a+bx}$$

$$\int \sqrt{a^2-x^2}\,dx = \frac{1}{2}\left(x\sqrt{a^2-x^2} + a^2 \arcsin\frac{x}{a}\right)$$

$$\int x\sqrt{a^2-x^2}\,dx = -\frac{1}{3}\sqrt{(a^2-x^2)^3}$$

$$\int \frac{dx}{\sqrt{a^2-x^2}} = \arcsin\frac{x}{a}$$

$$\int \frac{x\,dx}{\sqrt{a^2 \pm x^2}} = \pm\sqrt{a^2 \pm x^2}$$

$$\int \frac{x\,dx}{\sqrt{x^2-a^2}} = \sqrt{x^2-a^2}$$

$$\int \sqrt{x^2 \pm a^2}\,dx = \frac{1}{2}(x\sqrt{x^2 \pm a^2} \pm a^2 \ln(x + \sqrt{(x^2 \pm a^2)}))$$

$$\int x\sqrt{x^2 \pm a^2}\,dx = \frac{1}{3}\sqrt{(x^2 \pm a^2)^3}$$

$$\int \frac{dx}{\sqrt{x^2 \pm a^2}} = \ln(x + \sqrt{x^2 \pm a^2})$$

$$\int \sin x\,dx = -\cos x$$

$$\int \cos x\,dx = \sin x$$

$$\int \tan x\,dx = -\ln \cos x$$

$$\int \cot x\,dx = \int \frac{dx}{\tan x} = \ln \sin x$$

$$\int \sec x\,dx = \int \frac{dx}{\cos x} = \frac{1}{2}\ln\frac{1+\sin x}{1-\sin x}$$

$$\int \sin^2 x\,dx = \frac{x}{2} - \frac{\sin 2x}{4}$$

$$\int \cos^2 x\,dx = \frac{x}{2} + \frac{\sin 2x}{4}$$

$$\int \sin^3 x \, dx = -\frac{1}{3} \cos x \, (2 + \sin^2 x)$$

$$\int \cos^3 x \, dx = \frac{1}{3} \sin x \, (2 + \cos^2 x)$$

$$\int x \sin x \, dx = \sin x - x \cos x$$

$$\int x \cos x \, dx = \cos x + x \sin x$$

$$\int \arcsin x \, dx = x \arcsin x + \sqrt{1 - x^2}$$

$$\int \arccos x \, dx = x \arccos x - \sqrt{1 - x^2}$$

$$\int e^{cx} \, dx = \frac{1}{c} e^{cx}$$

$$\int x e^{cx} \, dx = \frac{1}{c^2} e^{cx}(cx - 1)$$

$$\int \ln x \, dx = x \ln x - x$$

$$\int b^{cx} \, dx = \frac{b^{cx}}{c \ln b}$$

$$\int \sinh x \, dx = \cosh x$$

$$\int \cosh x \, dx = \sinh x$$

$$\int \tanh x \, dx = \ln \cosh x$$

A1.1.7 Differential equations

Separable variables

$$\frac{dy}{dx} = f(x)g(y) \qquad \int \frac{dy}{g(y)} = \int f(x) \, dx$$

Linear, homogeneous equation with constant coefficients

$$\frac{d^2 y}{dx^2} + 2a \frac{dy}{dx} + by = 0 \quad \text{(A1.1)}$$

The following *real* functions $y(x)$ are general solutions:

$$a^2 > b: \; y = C_1 e^{\lambda_1 x} + C_2 e^{\lambda_2 x}$$

where λ_1 and λ_2 are the roots of the characteristic equation $\lambda^2 + 2a\lambda + b = 0$.

$$a^2 < b: \; y = e^{-ax}(A_1 \sin\sqrt{b - a^2}x + A_2 \cos\sqrt{b - a^2}x)$$

$$a^2 = b: \; y = (B_1 x + B_2)\, e^{-ax}$$

$$\frac{d^2 y}{dx^2} + 2a \frac{dy}{dx} + by = f(x) \quad \text{(A1.2)}$$

$$y = y_h + y_p$$

where y_h is the general solution of the corresponding homogeneous equation (A1.1) and y_p is a particular solution to equation (A1.2).

In particular:
If $f = c$ (constant): $y_p = c/b$.
If $f(x) = f_0 \sin \Omega x$ the constants c_1 and c_2 can be determined so that $y_p = c_1 \sin \Omega x + c_2 \cos \Omega x$ satisfies Equation (A1.2) (if $\Omega \neq \sqrt{b}$ or $a \neq 0$).
If $f(x) = de^{\alpha x}$ the constant c can be determined so that $y_p = ce^{\alpha x}$ satisfies Equation (A1.2).

A1.1.8 Vectors

Vector algebra

The vector $-\boldsymbol{B}$ is defined according to the above figure.

Addition

$\boldsymbol{A} + \boldsymbol{B} = \boldsymbol{C}$

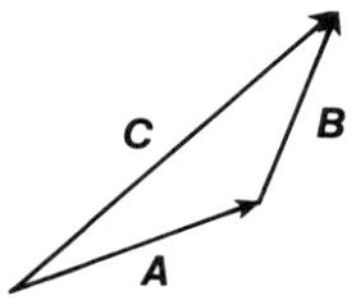

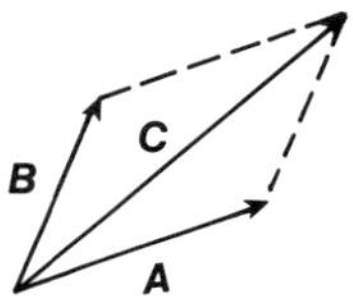

For *triangle* addition and *parallelogram* addition, see figure above.

Commutative law: $\mathbf{A}+\mathbf{B}=\mathbf{B}+\mathbf{A}$
Associative law: $(\mathbf{A}+\mathbf{B})+\mathbf{C}=\mathbf{A}+(\mathbf{B}+\mathbf{C})$
Subtraction: $\mathbf{A}-\mathbf{B}=\mathbf{A}+(-\mathbf{B})$

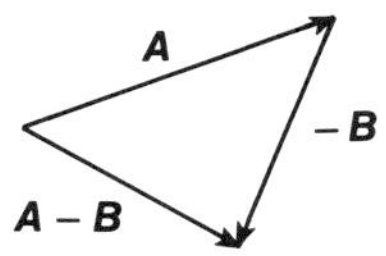

Unit vectors $\mathbf{e}_x$, $\mathbf{e}_y$, $\mathbf{e}_z$: $\mathbf{A}=A_x\mathbf{e}_x+A_y\mathbf{e}_y+A_z\mathbf{e}_z=(A_x,\ A_y,\ A_z)$
Vector magnitude: $|\mathbf{A}|=A=\sqrt{A_x^2+A_y^2+A_z^2}$

Direction cosines: $\cos\alpha=\frac{A_x}{A}$ $\cos\beta=\frac{A_y}{A}$ $\cos\gamma=\frac{A_z}{A}$

Scalar product

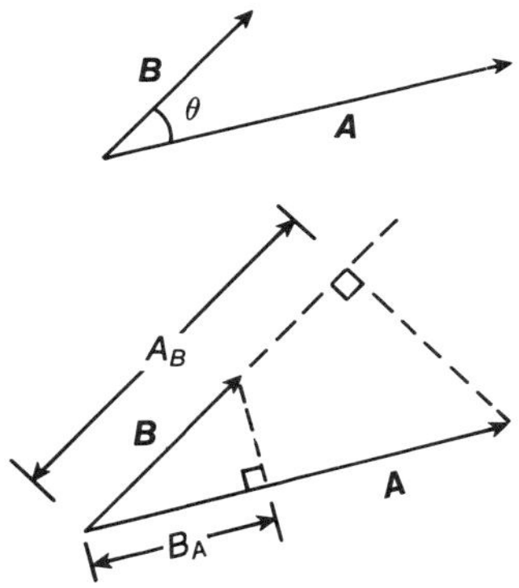

$$\mathbf{A}\cdot\mathbf{B}=AB\cos\theta=AB_A=A_BB=A_xB_x+A_yB_y+A_zB_z$$

Commutative law: $\mathbf{A}\cdot\mathbf{B}=\mathbf{B}\cdot\mathbf{A}$
Distributive law: $\mathbf{A}\cdot(\mathbf{B}+\mathbf{C})=\mathbf{A}\cdot\mathbf{B}+\mathbf{A}\cdot\mathbf{C}$
Scalar components: $A_x=\mathbf{A}\cdot\mathbf{e}_x$ $A_n=\mathbf{A}\cdot\mathbf{e}_n$

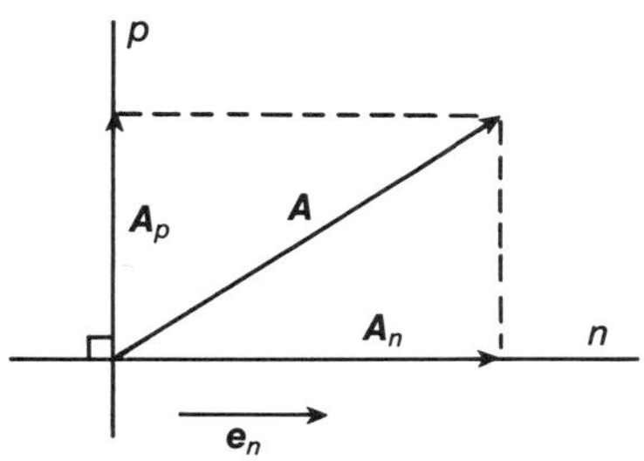

Vector components: $\boldsymbol{A}_n = (\boldsymbol{A} \cdot \mathbf{e}_n)\mathbf{e}_n \qquad \boldsymbol{A}_p = \boldsymbol{A} - \boldsymbol{A}_n$

Unit vector in the direction from P to Q: $\mathbf{e}_{PQ} = \dfrac{\overrightarrow{PQ}}{|\overrightarrow{PQ}|}$

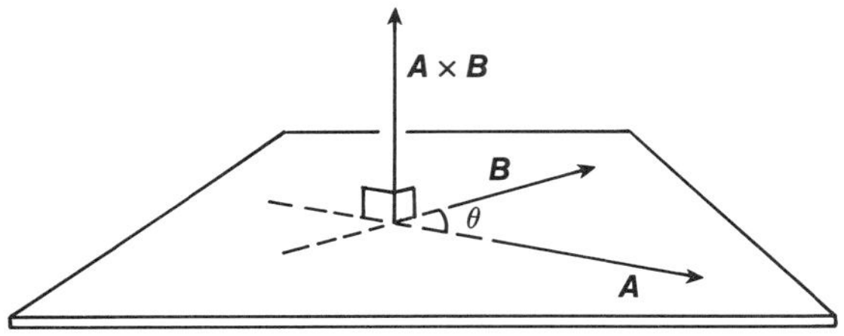

Vector product (cross-product)

Magnitude: $|\boldsymbol{A} \times \boldsymbol{B}| = AB \sin\theta$

Direction: Specified by the right-hand rule thus $\boldsymbol{A} \times \boldsymbol{B} = -\boldsymbol{B} \times \boldsymbol{A}$

Components: $(\boldsymbol{A} \times \boldsymbol{B})_x = A_y B_z - A_z B_y$, etc.

Determinant expansion:

$$\boldsymbol{A} \times \boldsymbol{B} = \begin{vmatrix} \mathbf{e}_x & \mathbf{e}_y & \mathbf{e}_z \\ A_x & A_y & A_z \\ B_x & B_y & B_z \end{vmatrix}$$

Triple scalar product

$\boldsymbol{A} \cdot (\boldsymbol{B} \times \boldsymbol{C}) = (\boldsymbol{B} \times \boldsymbol{C}) \cdot \boldsymbol{A}$

Permutation laws:

$$\begin{aligned} \boldsymbol{A} \cdot (\boldsymbol{B} \times \boldsymbol{C}) &= \boldsymbol{C} \cdot (\boldsymbol{A} \times \boldsymbol{B}) = \boldsymbol{B} \cdot (\boldsymbol{C} \times \boldsymbol{A}) \\ &= (\boldsymbol{A} \times \boldsymbol{B}) \cdot \boldsymbol{C} = (\boldsymbol{C} \times \boldsymbol{A}) \cdot \boldsymbol{B} = (\boldsymbol{B} \times \boldsymbol{C}) \cdot \boldsymbol{A} \end{aligned}$$

Determinant expansion:

$$\boldsymbol{A} \cdot (\boldsymbol{B} \times \boldsymbol{C}) = \begin{vmatrix} A_x & A_y & A_z \\ B_x & B_y & B_z \\ C_x & C_y & C_z \end{vmatrix}$$

Triple vector product

$\boldsymbol{A} \times (\boldsymbol{B} \times \boldsymbol{C}) = (\boldsymbol{C} \times \boldsymbol{B}) \times \boldsymbol{A}$

Equivalent form:

$$\boldsymbol{A} \times (\boldsymbol{B} \times \boldsymbol{C}) = (\boldsymbol{A} \cdot \boldsymbol{C})\boldsymbol{B} - (\boldsymbol{A} \cdot \boldsymbol{B})\boldsymbol{C}$$

Derivatives of vectors

The same rules as for pure scalar functions are applied for combinations of vector functions $\boldsymbol{F}(u)$ and $\boldsymbol{G}(u)$. A scalar function is denoted by $c(u)$.

Vector sum: $\dfrac{\mathrm{d}(\boldsymbol{F}+\boldsymbol{G})}{\mathrm{d}u}=\dfrac{\mathrm{d}\boldsymbol{F}}{\mathrm{d}u}+\dfrac{\mathrm{d}\boldsymbol{G}}{\mathrm{d}u}$

Scalar product: $\dfrac{\mathrm{d}(\boldsymbol{F}\cdot\boldsymbol{G})}{\mathrm{d}u}=\dfrac{\mathrm{d}\boldsymbol{F}}{\mathrm{d}u}\cdot\boldsymbol{G}+\boldsymbol{F}\cdot\dfrac{\mathrm{d}\boldsymbol{G}}{\mathrm{d}u}$ $\qquad$ $\dfrac{\mathrm{d}(\mathrm{c}\boldsymbol{F})}{\mathrm{d}u}=\dfrac{\mathrm{dc}}{\mathrm{d}u}\boldsymbol{F}+c\,\dfrac{\mathrm{d}\boldsymbol{F}}{\mathrm{d}u}$

In particular,

$$\frac{\mathrm{d}}{\mathrm{d}u}F^2=\frac{\mathrm{d}}{\mathrm{d}u}(\boldsymbol{F}\cdot\boldsymbol{F})=2\boldsymbol{F}\cdot\frac{\mathrm{d}\boldsymbol{F}}{\mathrm{d}u}$$

Cross-product: $\dfrac{\mathrm{d}(\boldsymbol{F}\times\boldsymbol{G})}{\mathrm{d}u}=\dfrac{\mathrm{d}\boldsymbol{F}}{\mathrm{d}u}\times\boldsymbol{G}+\boldsymbol{F}\times\dfrac{\mathrm{d}\boldsymbol{G}}{\mathrm{d}u}$

In *fixed* $\mathbf{e}_x\mathbf{e}_y\mathbf{e}_z$ basis

$$\frac{\mathrm{d}\boldsymbol{F}}{\mathrm{d}u}=\frac{\mathrm{d}F_x}{\mathrm{d}u}\mathbf{e}_x+\frac{\mathrm{d}F_y}{\mathrm{d}u}\mathbf{e}_y+\frac{\mathrm{d}F_z}{\mathrm{d}u}\mathbf{e}_z=\left(\frac{\mathrm{d}F_x}{\mathrm{d}u},\frac{\mathrm{d}F_y}{\mathrm{d}u},\frac{\mathrm{d}F_z}{\mathrm{d}u}\right)$$

APPENDIX 2

Centres of mass of uniform bodies

Table A2 **Centres of mass of uniform bodies**

Body	**Centre of mass**
Thin circular arc-shaped ring 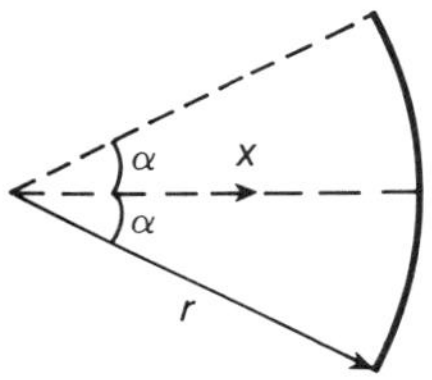	$\bar{x} = \dfrac{r \sin \alpha}{\alpha}$
Thin quartercircular ring 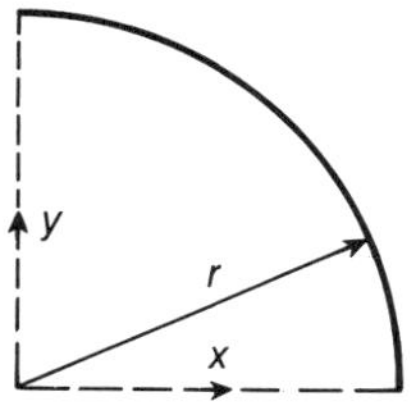	$\bar{x} = \bar{y} = \dfrac{2r}{\pi}$
Thin circular sector-shaped disk 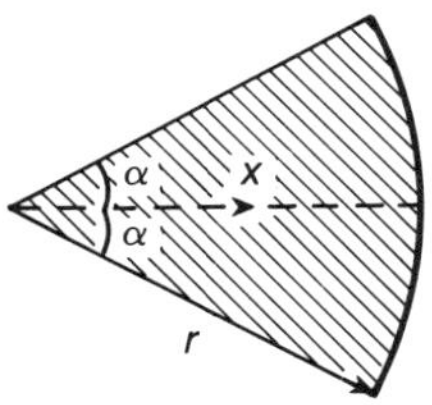	$\bar{x} = \dfrac{2r \sin \alpha}{3\alpha}$

Table A2 **(continued)**

Body	**Centre of mass**
Thin quartercircular disk	
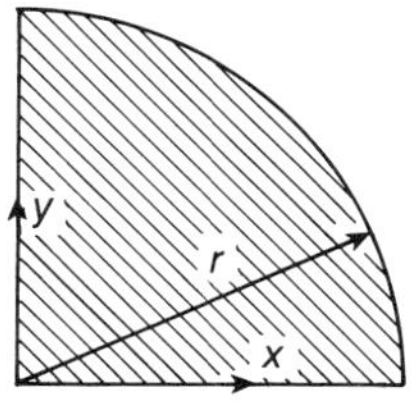	$\bar{x} = \bar{y} = \dfrac{4r}{3\pi}$
Thin triangular disk	
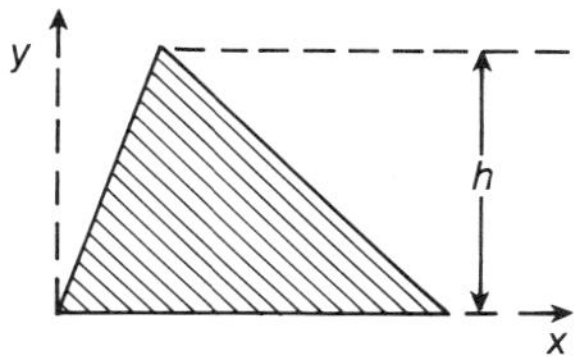	$\bar{y} = \dfrac{h}{3}$
Semicircular cylinder	
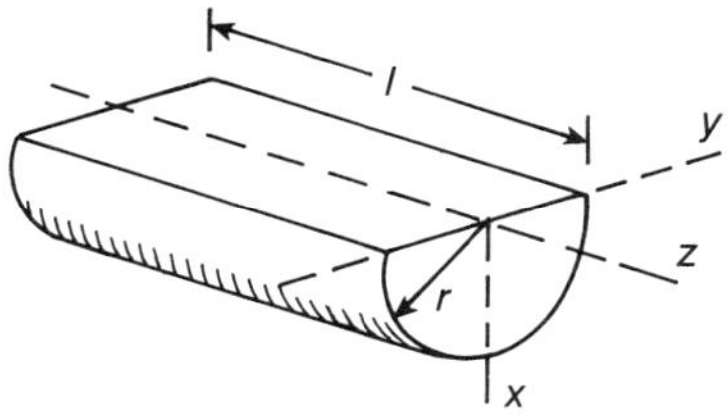	$\bar{x} = \dfrac{4r}{3\pi}$
Thin semicircular cylindrical shell	
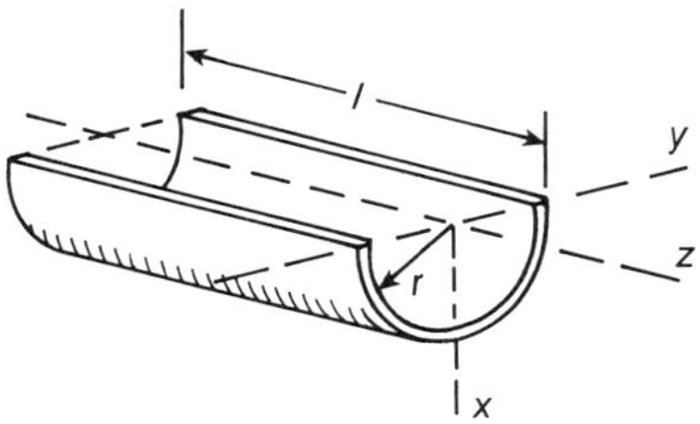	$\bar{x} = \dfrac{2r}{\pi}$

Table A2 **(continued)**

Body	Centre of mass
Right circular cone	
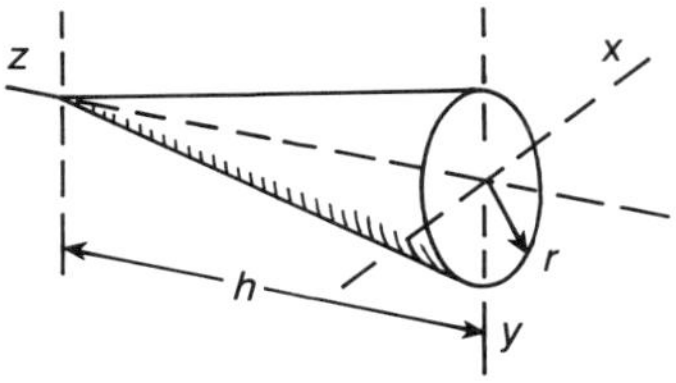	$\bar{z} = \frac{h}{4}$
Semicircular cone	
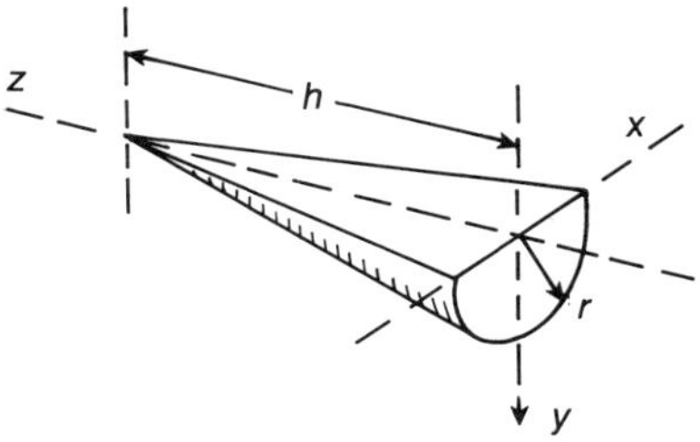	$\bar{z} = \frac{h}{4}$ $\bar{y} = \frac{r}{\pi}$
Circular conical shell	
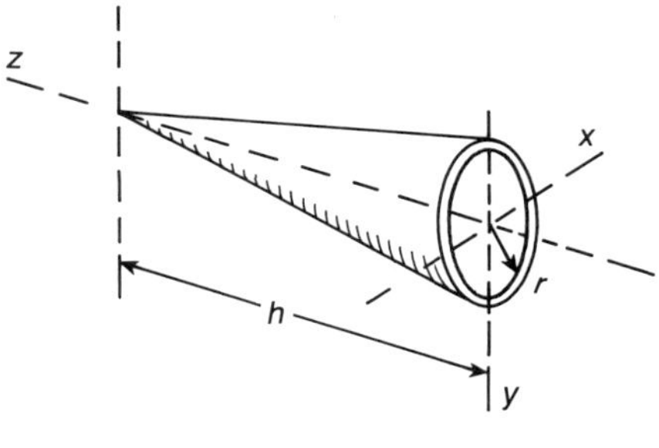	$\bar{z} = \frac{h}{3}$
Semicircular conical shell	
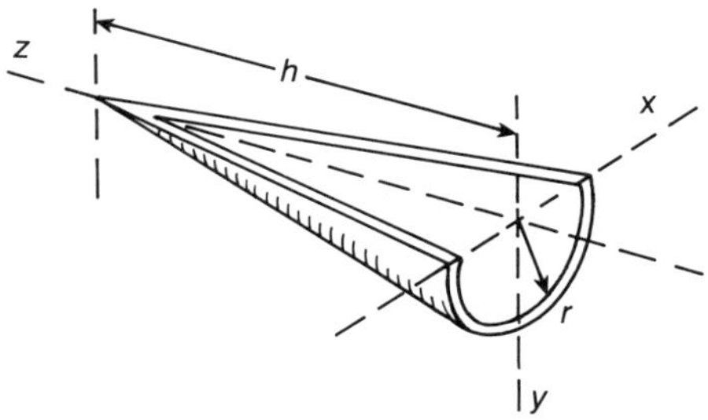	$\bar{z} = \frac{h}{3}$ $\bar{y} = \frac{4r}{3\pi}$

Table A2 **(continued)**

Body	**Centre of mass**
Hemisphere 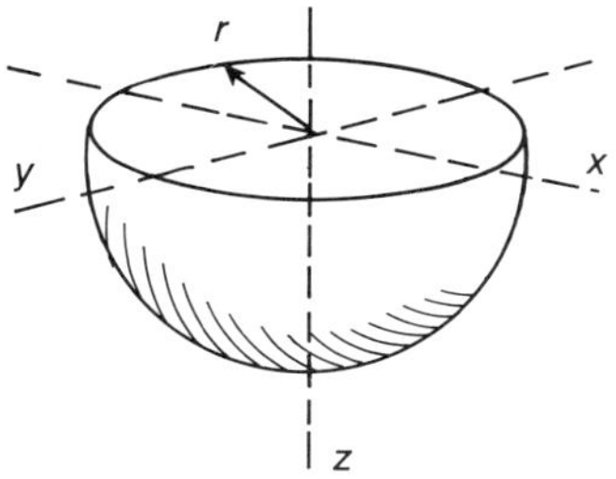	$\bar{z} = \dfrac{3r}{8}$
Thin hemispherical shell 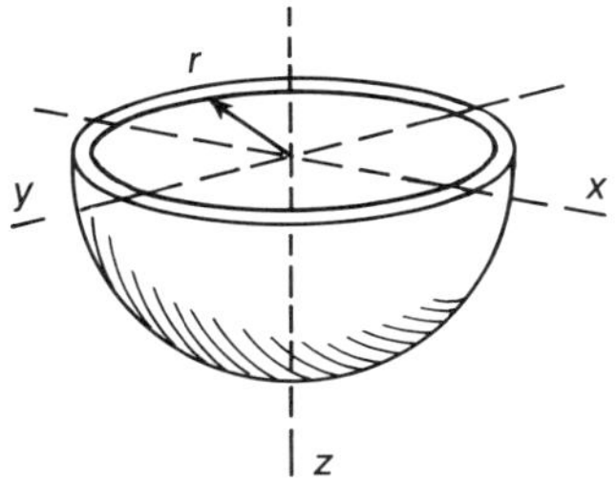	$\bar{z} = \dfrac{r}{2}$

Answers to problems

Chapter 1

Section 1.2

1.1 $P\sqrt{3}$
1.2 8.3 N and 17.8 N
1.3 4.71 N 15.8°
1.4 1.09 kN and 1.35 kN in the directions of AB and BC, respectively
1.5 (a) 8.8 N or 4.2 N
(b) 27.3° or 12.7°
1.6 (a) $F_1 = 0.98$ kN $F_2 = 0.67$ kN
(b) $F_2 = 0.51$ kN $\alpha = 70^\circ$ $F_1 = 1.41$ kN
1.7 (a) $\boldsymbol{F}_A = (144, -111)$ N
(b) $\boldsymbol{F}_B = (-144, 111)$ N
1.8 (a) $F_x = 32.1$ N $F_y - 38.3$ N
(b) $F_{AB} = 47.0$ N $F_\perp = 17.1$ N
1.9 $F_{AB} = 0.59F$
1.10 $FL \sin \alpha$
1.11 $M_A = 0$, $M_B = 0.366Fa$ counterclockwise, $M_C = Fa$ clockwise
1.12 $F(b \sin \alpha + h \cos \alpha)$ clockwise
1.13 $M_C = 3.60$ N m clockwise
1.14 (a) $M_O = 39.1$ N m counterclockwise
(b) $M_B = 71.0$ N m counterclockwise
(c) $M_C = 10.1$ N m counterclockwise
(d) $M_D = 5.8$ N m counterclockwise
(e) $M_E = 35.8$ N m clockwise
1.15 (a) 14.8 N m counterclockwise
(b) $|M_O|_{max} = 16.3$ N m for $\alpha = -80^\circ, 100^\circ$
$|M_O|_{min} = 0$ for $\alpha = -170^\circ, 10^\circ$
1.16 (a) $M_O = 157$ N m counterclockwise
(b) $M_P = 188$ N m counterclockwise
1.17 $\sum F_x = 199$ N $\sum F_y = 14.3$ N
$\left|\sum F\right| = 200$ N
1.18 (a) $F(1, 3)$
(b) $2Fa$ counterclockwise
(c) $7Fa$ counterclockwise
(d) $8Fa$ counterclockwise
(e) Fa clockwise
1.19 (a) $F(1.12, 1.26)$
(b) $6.29Fa$ clockwise
(c) $1.81Fa$ clockwise
1.20 (a) 5 N m
(b) 2.5 N m
(c) 2.5 N m
(d) 5 N m
(e) 15 N m
(f) 5 N m
(g) 5 N m – all counterclockwise
1.21 6.26 N
1.22 (a) (153, 388) N
(b) 342 N m
1.23 If $x = 0$ at the left end of the beam:
(a) 5 kN at $x = 3.4$ m
(b) 3 kNi$x = 3$ m
(c) 3 kN at $x = 2.5$ m
(d) 1 kNi$x = 1$ m
1.24 (a) 50 N downwards
(b) 2.5 N m clockwise
1.25 $R = 0.35$ kN; makes an angle of 4.3° with the 200 N force
$M = 4.0$ N m clockwise

1.26 $R = 40$ N downwards
$M = 85$ N m counterclockwise

1.27 The resultant force is $(-13, 194)$ N in both cases.
(a) $M = 1.74$ kN m clockwise
(b) line of action $194x + 13y + 1740 = 0$

1.28 $(55.1, 14.8)$ N, line of action
$y = 0.268x - 0.035$

1.29 200 N, 169 N m counterclockwise

1.30 200 N, 350 N m counterclockwise

1.31 6.81°

1.32 400 N downwards along a line 30 mm to the right of the original 400 N force

Section 1.3

1.33 (a) $F = 69.3$ N $F_x = 11.9$ N $F_y = 32.6$ N
(b) 80.2° and 62.0°

1.34 $F_{AB} = \dfrac{P}{\sqrt{2}}$ $F_{DC} = \dfrac{P}{\sqrt{2}}$ $F_{EG} = \dfrac{P}{2}$

$F_{HF} = -\dfrac{P}{2}$ $F_{BD} = -\dfrac{P}{2}$ $F_{EB} = F_{BE} = 0$

1.35 (a) $\left(\dfrac{4P}{\sqrt{29}}, \dfrac{-3P}{\sqrt{29}}, \dfrac{-2P}{\sqrt{29}}\right)$

(b) $\left(\dfrac{-7P}{\sqrt{58}}, \dfrac{-P}{\sqrt{58}}, \dfrac{-2P}{\sqrt{29}}\right)$

1.36 $-\dfrac{P}{\sqrt{87}}$

1.37 (a) 320 N
(b) 716 N
(c) -523 N
(d) 0

1.38 88.7 N and 74.4 N

1.39 $\boldsymbol{F} = (498\boldsymbol{e}_x + 98\boldsymbol{e}_y + 1006\boldsymbol{e}_z)$ mN

1.40 (a) $\boldsymbol{F} = (-82, 53, -107)$ N
(b) $\boldsymbol{F} = (116\boldsymbol{e}_{EC} + 10\boldsymbol{e}_{BC} + 82\boldsymbol{e}_{CD})$ N

1.41 (a) $(1, -2, 0)Pa$
(b) Pa
(c) $-2Pa$
(d) 0
(e) $\dfrac{-3Pa}{\sqrt{6}}$

1.42 (a) $(-2, -6, 2)$ N m
(b) $(-6, -4, -1)$ N m
(c) $\boldsymbol{M}_B = 0$

1.43 (a) -6 N m
(b) -1 N m
(c) 0

1.44 $\boldsymbol{M}_A = (-1, -1, 0)aQ$
$\boldsymbol{M}_B = (-1, -1, 0)aQ$
$\boldsymbol{M}_C = (1, -1, 0)aQ$
$\boldsymbol{M}_O = (-1, 1, 0)aQ$
where
$aQ = 0.2 \cdot 78 \cdot 9.81$ N m ≈ 153 N m

1.45 (a) $(1, -1, -1)Pa$
(b) $(1, -2, 0)Pa$
(c) 0 (**P** is parallel to the position vector)
(d) $-\dfrac{Pa}{\sqrt{3}}$
(e) 0 (**P** is parallel to the axis)
(f) $-\dfrac{Pa}{\sqrt{2}}$

1.46 $M_x = \dfrac{Fa}{\sqrt{2}}$ $M_y = \dfrac{Fa}{\sqrt{2}}$ $M_z = -\dfrac{Fa}{\sqrt{2}}$

1.47 $\dfrac{Fa}{\sqrt{6}}$

1.48 (a) $(0, 0, Fa/\sqrt{2})$
(b) $Fa/\sqrt{6}$

1.49 272 kN m $\cdot\, \boldsymbol{e}_x$

1.50 (a) $(256, 128, 93)$ kN m
(b) $(57, 114, 83)$ kN m
(c) 0, 93 and 83 kN m, respectively

1.51 (a) 29.9 N m
(b) $(19.5, 0, -29.9)$ N m

1.52 $P \cdot 0.57$ m

1.53 $0.5mga \sin\alpha \sin\beta$

1.54 $(0, 2P, P)$

1.55 (a) $(2Pa, 3Pa, -2Pa)$
(b) $3Pa$
(c) $(2Pa, 6Pa, -8Pa)$
(d) $-\dfrac{26Pa}{3}$
(e) $\sum \boldsymbol{M}_E = \sum \boldsymbol{M}_O + \vec{EO} \times \sum \boldsymbol{F}$
$= (2Pa, 3Pa, -2Pa)$

1.56 (a) $(10.0, 1.70, 4.43)$ N
(b) $(0, -3.54, -22.6)$ N m
(c) $(-5.31, -3.54, -10.6)$ N m
(d) 3.80 N m
(e) 10.6 N m
(f) 0

1.57 $(9, 12, 18)$ N m

1.58 $P = \dfrac{Q}{\sqrt{3}}$ $F = \dfrac{2Q}{\sqrt{3}}$

$\sum \boldsymbol{M}_O = \sum \boldsymbol{M}_A = (1/\sqrt{3}, 1 - 1/2\sqrt{3}, 0)Qa$

1.59 (a) $(0, 0, Fa)$
(b) $(0, 0, Fa)$
(c) $(Fa, 0, 0)$
(d) $(0, Fa, Fa)$
(e) $(0, -Fa, Fa)$

1.60 $\frac{Pa}{\sqrt{2}}(-1, -1, -1)$

1.61 −120 N m and −24 N m

1.62 $\boldsymbol{R} = 20(-3 - 2\sqrt{2}, 0, 2\sqrt{2})$ N
(a) $(0, 0, -48)$ N m
(b) $32\sqrt{2}\,(1, 0, 1)$ N m
(c) $32\sqrt{2}\,(1, -1, 1)$ N m
(d) $16(2\sqrt{2}, 3, 2\sqrt{2})$ N m

1.63 $\boldsymbol{R} = 20(-3 - 2\sqrt{2}, 0, 2\sqrt{2})$ N
A couple (0, 0, 50) N m is added to each of the answers to Problem 1.62.

1.64 $-19\mathbf{e}_z$ N applied at (2.74; 3) m

1.65 $-240\mathbf{e}_z$ N applied at (102; 527) mm

1.66 (a) $(12, 9.7, -30.3)$ N
$(-8.8, 10.5, -5.6)$ N m
(b) $(12, 9.7, -30.3)$ N
$(-10.4, 12.4, -5.6)$ N m

1.67 $\boldsymbol{R} = 140(1, 0, -\sqrt{3})$ N
(a) $(-38.8, -33.8, -22.4)$ N m
$M_\perp = 44.8$ N m
(b) $(0, -43.6, 0)$ N m $M_\perp = 43.6$ N m

1.68 $(0, -P, P)$ and $(0, -3Pa, 0)$

Chapter 2

Section 2.2

2.1 681 N upwards to the right making an angle of 7.5° with the vertical

2.2 (a) 19.5°
(b) 1.06 mg
(c) 0.35 mg

2.3 (a) $N_A \approx 154$ N $N_B \approx 785$ N
$F_B \approx 154$ N
(b) $N_A \approx 244$ N $N_B \approx 1324$ N
$F_B \approx 244$ N

2.4 0.30 m

2.5 From left to right:
(a) 6.98 N 6.17 N
(b) 10.2 N 7.65 N 16.1 N

2.6 (a) $N_A = mg \tan \alpha$ $N_B = \frac{mg}{\cos \alpha}$
(b) 30°

2.7 (a) 500 N upwards 2.0 kN m counterclockwise
(b) Left: zero Right: 500 N upwards
(c) Left: 1.5 kN downwards
Right: 2.0 kN upwards
(d) Left: 167 N upwards, 770 N to the right
Right: 674 N normal to the incline

2.8 A: $(r_1 + r_2)M_1/r_1L$ upwards
B: $(r_1 + r_2)M_1/r_1L$ downwards

2.9 7.28 kN

2.10 With the denotations shown here:
(a) $S = 455$ N $H = 439$ N
(b) $S = 118$ N $H = 114$ N
$V = 87$ N $M = 175$ N m
(c) $S = 118$ N $H = 107$ N $V = 68$ N
$M = 256$ N m

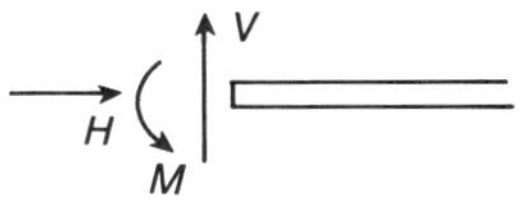

2.11 (a) $\frac{F}{3k}$
(b) $\frac{F}{3k}$
(c) $\frac{3F}{2k}$

2.12 (a) 5.53 N (tensile force)
(b) 4.47 N to the right, 10 N downwards

2.13 (a) 76.0°
(b) $S_1 = 1.62$ kN $S_2 = 2.50$ kN

2.14 (a) 59.0 kN
(b) 86.4 kN

2.15 415 N and 13.0 kN respectively

2.16 67 N upwards, 3.04 kN to the left, 0.53 kN m counterclockwise

2.17 (a) 981 N (b) 981 N
(c) 654 N (d) 654 N
(e) 490 N (f) 654 N
(g) 490 N

2.18 (a) $F = 114$ N
(b) $M = 34.3$ N m
(c) $F = 73.4$ N

2.19 With the denotations shown here
(a) $V_A = V_B = \frac{M}{2a}$
(b) $M_A = M$

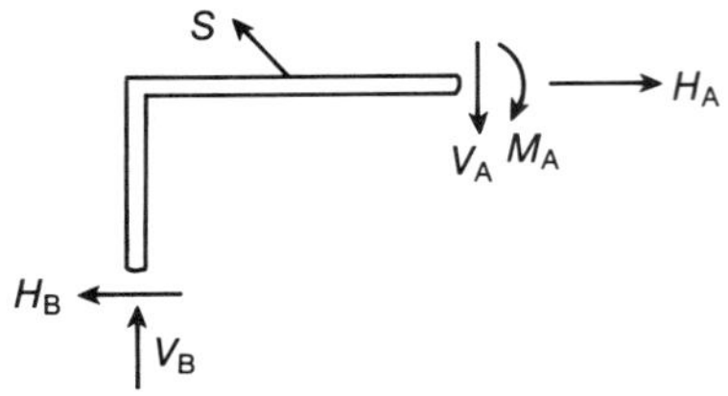

(c) $H_A = H_B = \dfrac{M}{a}$

(d) $V_A = H_A = \dfrac{M}{a}$ $\quad S = \dfrac{M\sqrt{2}}{a}$

2.20 With the denotations of Problem 2.19 above:

(a) $V_A = V_B = \dfrac{F}{4}$ $\quad H_A = -F$

(b) $H_A = -F$ $\quad M_A = \dfrac{Fa}{2}$

(c) $H_A = -\dfrac{F}{2}$ $\quad H_B = \dfrac{F}{2}$

(d) $V_A = \dfrac{F}{2}$ $\quad H_A = -\dfrac{F}{2}$ $\quad S = \dfrac{F}{\sqrt{2}}$

2.21 (a) At A: 245 N upwards
At B: 245 N to the right
(b) At A: 245 N upwards, 282 N to the left
At B: 282 N to the right

2.22 (a) 63.4° (b) 65.1°
(c) 29.0° (d) 52.5°
(e) 55.7°

2.23 $S_{AC} = 17.3$ N $\quad S_{BC} = 86.7$ N

2.24 $N_A = N_B = N_C = \dfrac{Q}{\sqrt{2}}$ $\quad N_D = Q\sqrt{2}$

2.25 72 N

2.26 On the disk:

At A: $\dfrac{3}{4}mg$ upwards; $\dfrac{3}{4}mg$ to the right

At B: $\dfrac{1}{4}mg$ upwards; $\dfrac{3}{4}mg$ to the left

On the bar:

At C: $\dfrac{5}{4}mg$ upwards; $\dfrac{3}{4}mg$ to the left

2.27 (a) At A: 200 N downwards; 400 N to the left
At B: 400 N upwards; 400 N to the right
(b) Same forces as in case (a)
(c) At A: 64 N upwards; 48 N to the right
At B: 136 N upwards; 48 N to the left
(d) At A: 100 N upwards; 200 N to the left
At B: 100 N upwards; 200 N to the right
(e) At A: 100 N upwards;
100 N m counterclockwise
At B: 100 N upwards

2.28

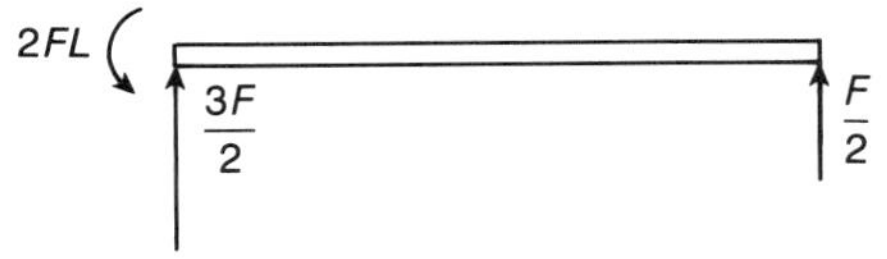

(a)

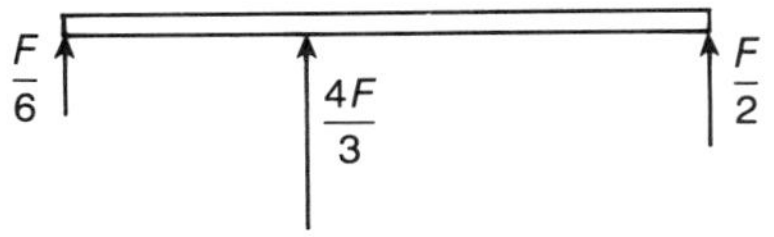

(b)

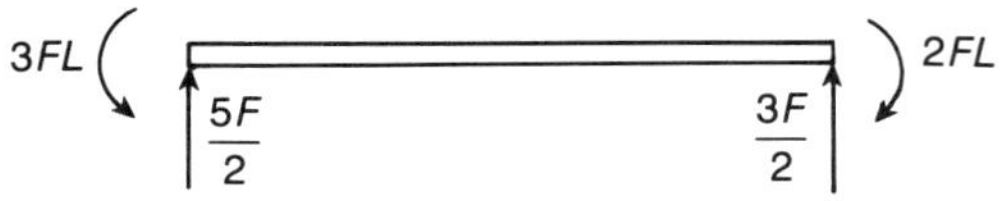

(c)

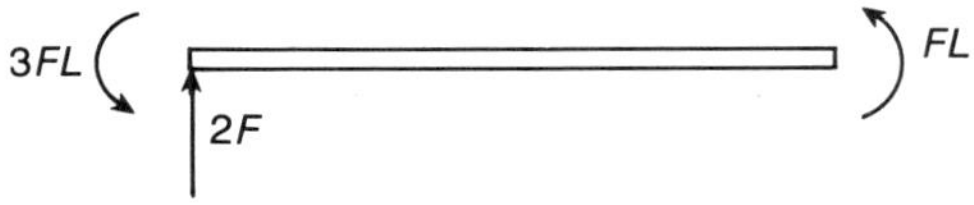

(d)

2.29 (a) At A: *mg* downwards; $(2\sqrt{3} - 1)\,mg$ to the left
At B: 4 *mg*
(b) At A: 2 *mg* downwards;
$4\,mg\sqrt{3}$ to the left
At B: 8 *mg*

(c) At A: $3\,mg/2$ downwards; $5\,mg\,\sqrt{3}/2$ to the left
At B: $6\,mg$

2.30 342 N m

2.31 (a) $S_1 \approx 249$ N $S_2 = S_3 \approx 497$ N
(b) 438 N

2.32 (a) 230 N (b) 169 N

2.33 At A: $0.024F$ upwards; $0.59F$ to the right
At C: $0.33F$ upwards; $1.18F$ to the right

2.34 (a) 803 N
(b) At A: 3.62 kN to the right
At B: 4.19 kN to the left; 4.98 kN upwards

2.35 At A: $mg/2$ upwards; mg to the right
At B: $3mg/2$ upwards; mg to the left

2.36 331 N m

2.37 10 $R_A \approx 10.9P$

2.38 4.42 kN

2.39 2.28 kN

2.40 (a) 11.9 kN (compression)
(b) At A: 3.75 kN along the link
At B: 3.57 kN to the left; 6.72 kN upwards

2.41 (a) 2.50 kN on the rear wheels, 4.26 kN on the front wheels
(b) No, not separately
(c) Same normal forces as in case (a)
Friction forces:
1.19 kN on the front wheels, no forces on the rear wheels
(d) Normal forces:
2.01 kN on the rear wheels, 4.76 kN on the front wheels
Same friction forces as in case (c)

2.42 At A: 14.3 kN vertically; 5.91 kN horizontally
At C: 18.7 kN along BC
At D and E: 1.63 kN along DE
At F: 3.89 kN vertically; 1.56 kN horizontally

2.43 (a) 857 N m counterclockwise
(b) 657 N m clockwise

2.44 $6\,aq$ to the right; $\frac{5}{2}aq$ downwards

2.45 $Q_1 \dfrac{2r - r_1 - r_2}{r}$

2.46 (a) $x_1 = a$ $x_2 = \dfrac{a}{2}$ $x_3 = \dfrac{a}{3}$
(b) $\dfrac{a}{n}$
(c) Infinitely wide (theoretically)

Section 2.3

2.47 $S_{AB} = mg\sqrt{2}$ $S_{AC} = \dfrac{3mg\sqrt{5}}{4}$ $S_{AD} = \dfrac{mg\sqrt{21}}{4}$

2.48 Location: $y = 4$ m. Forces 164 N and 67.1 N

2.49 $\dfrac{Q \sin \varphi}{\sqrt{3 - 2\cos \varphi}}$ normal to AB

2.50 $\dfrac{8\sqrt{2}}{5} P \approx 2.26P$

2.51 $S = \dfrac{mg(a + r)}{\sqrt{a^2 - r^2 + 2ar}}$ $N = \dfrac{mgr}{\sqrt{a^2 - r^2 + 2ar}}$

2.52 All tensions are $Q/2$

2.53 $(-12, 0, -10)$ N, $(0, 2.48, -3)$ N m

2.54 $P = \dfrac{Q}{4}$ $R_{Ax} = \dfrac{Q}{4}$ $R_{Bx} = 0$

$R_{Az} = R_{Bz} = \dfrac{Q\sqrt{3}}{4}$

where $Q = 2.65$ kN

2.55 $R_{Ax} = 33.3$ N $R_{Ay} = -75$ N
$R_{Bx} = -33.3$ N $R_{Bz} = -50$ N
$R_{Cy} = 75$ N $R_{Cz} = 50$ N

2.56 Tension $= Q \approx 14 \cdot 9.81$ N

A: $(1, -1, 0)\,\dfrac{Q}{2}$ D: $(-3, 1, -6)\,\dfrac{Q}{2}$

2.57 $\dfrac{a}{3}$

2.59 (a) $S_{CE} = 0.62$ kN $S_{CD} = 1.22$ kN
(b) 0.23 kN

2.60 D is located on the bisector of the angle between the two vertical planes containing the cables. The tension is 1.27 kN

2.61 $S_{CD} = \dfrac{3\sqrt{29}Q}{16}$ $S_{BE} = \dfrac{3\sqrt{14}Q}{16}$

$R_{Ax} = -\dfrac{9Q}{8}$ $R_{Ay} = -\dfrac{3Q}{16}$ $R_{Az} = -\dfrac{Q}{8}$

where $Q = 120 \cdot 9.81$ N

2.62 $\dfrac{Q}{\sqrt{3}}$ in the shortest cord,

$\dfrac{Q}{4}\sqrt{\dfrac{5}{3}}$ in the other ones

2.63 $\dfrac{90}{1+\cos\alpha}$ N in OB and OC,

$\dfrac{180\cos\alpha}{1+\cos\alpha}$ N in OA

2.64 $\dfrac{Q}{2}$, $\dfrac{Q}{4}$ and $\dfrac{Q}{4}$ where

$Q \approx 3 \cdot 9.81$ N

2.65 From the left: 76.5 N 27.5 N
43.2 N

2.66 $S_A = \dfrac{167\cos\alpha}{1+\cos\alpha}$ N

$S_B = S_C = \dfrac{83}{1+\cos\alpha}$ N

2.67 $\bar{x} = 1.08$ m $\bar{y} = -0.76$ m

2.68 $S_A = \dfrac{11P}{8}$ $S_B = \dfrac{43P}{8}$ $S_D = \dfrac{21P}{4}$

2.69 237 N each

2.70 $P = 160$ N E: $(-325, 0, -160)$ N
F: $(325, 300, -22)$ N

2.71 $0.225Q$ upwards

2.72 $\dfrac{a}{\sqrt{3}}$

2.73 A: $0.5Q$ upwards
B: $0.286Q$ downwards $0.286Q$ along $\overrightarrow{FB}$

2.74 $F = 1019\cos\alpha$ N
$R_{Ax} = -181\cos\alpha$ N $R_{Ay} = 137$ N
$R_{Cx} = 1199\cos\alpha$ N $R_{Cy} = 137$ N

2.75 Force at E: $8Q/15$ normal to EB
A: $(Q/2, 0)$ B: $(11Q/150, -8Q/25)$

2.76 A: $(-Q/8, Q)$ B: $(7Q/8, 0)$

2.77 $F = 410$ N $R_{Ax} = 130$ N $R_{Ay} = 160$ N
$R_{Cx} = -650$ N $R_{Cy} = 160$ N

2.78 $R_{Ax} = -31$ N $R_{Ay} = 306$ N
$R_{Bx} = 280$ N $R_{By} = 701$ N

2.79 $T = 1.26$ kN $R_A = 1.08$ kN $R_B = 1.36$ kN

Section 2.4

2.80 (a)

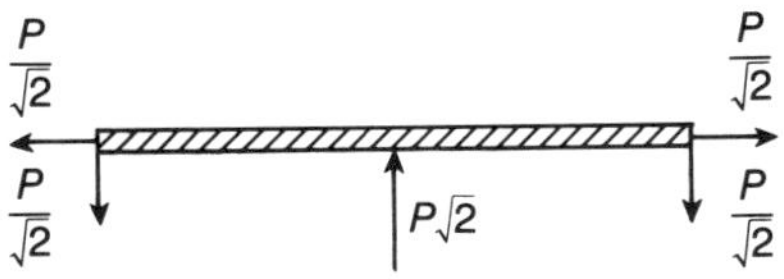

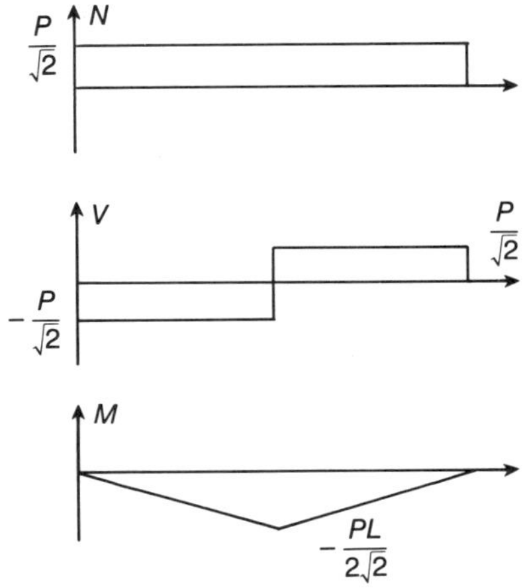

(b)

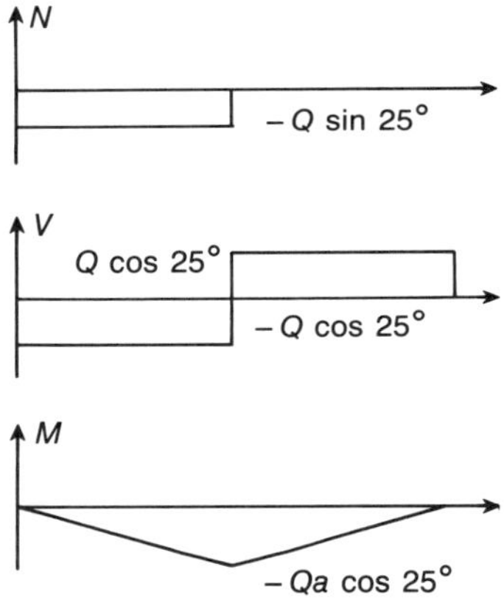

2.81

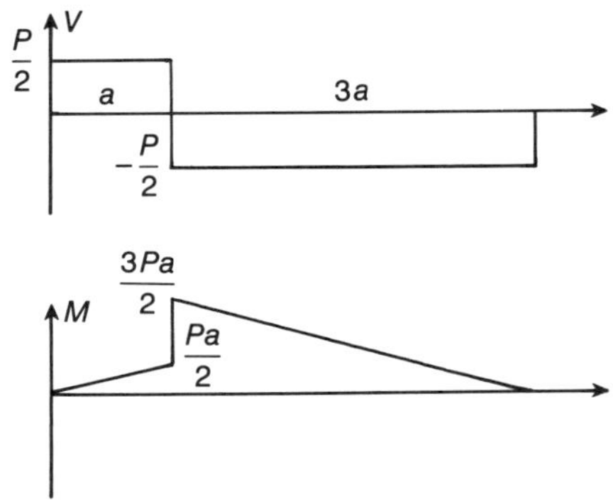

2.82

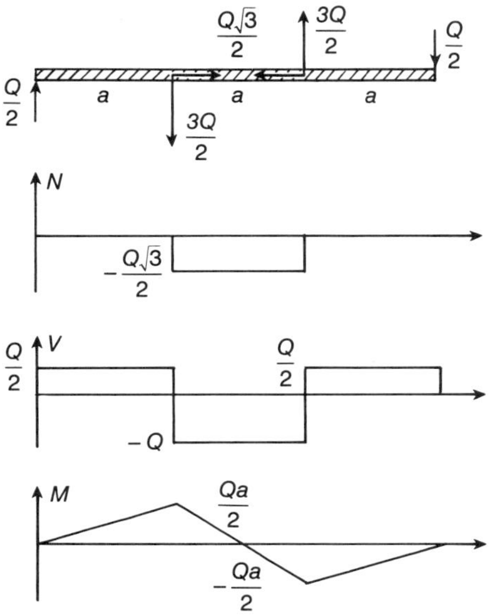

2.83

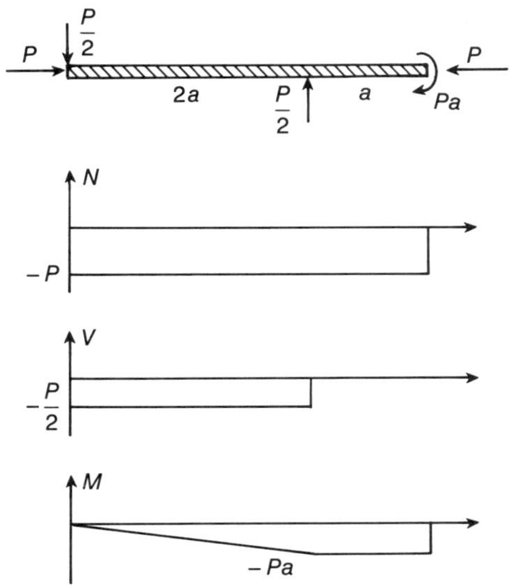

2.84 For the beam to the right:

$$V = N = -\frac{1}{\sqrt{2}}\left(\frac{P}{2} + mg\,\frac{x}{L}\right)$$

$$M = \frac{L-x}{2L\sqrt{2}}\,(PL + mg(L+x))$$

2.85 $N = -mg\left[\dfrac{\cos\alpha}{2\tan\alpha} + \left(1 - \dfrac{x}{L}\right)\sin\alpha\right]$

$$V = mg\cos\alpha\left(\frac{1}{2} - \frac{x}{L}\right)$$

$$M = \frac{mg\cos\alpha}{2L}\,x(L-x)$$

2.86 BD: $N = \dfrac{P}{\sqrt{3}}$ $V = 0$ $M = 0$

AC ($x = 0$ at A):

$x < a$: $N = -\dfrac{P\sqrt{3}}{4}$ $V = \dfrac{P}{4}$ $M = \dfrac{Px}{4}$

$x > a$: $N = -\dfrac{5P}{4\sqrt{3}}$ $V = -\dfrac{P}{4}$

$$M = \frac{P}{4}\,(2a - x)$$

2.87 $y = c\sqrt{x}$ where c is a constant

2.88 $N_{AB} = 2F$ $N_{BC} = -F\sqrt{3}$ $N_{AC} = \dfrac{F\sqrt{3}}{2}$

2.89 $N_{AB} = N_{BD} = N_{DF} = -\dfrac{2}{\sqrt{3}}$ kN

$N_{AC} = \dfrac{1}{\sqrt{3}}$ kN

$N_{BC} = N_{CE} = N_{EG} = \dfrac{2}{\sqrt{3}}$ kN

$N_{CD} = N_{DE} = N_{EF} = 0$ $N_{FG} = -\dfrac{4}{\sqrt{3}}$ kN

2.90 $N_{AB} = 0$ $N_{AC} \approx -2.77$ kN

2.91 $\dfrac{7}{8}F$

2.92 (a) $N_{AB} = N_{AC} = 623$ N $N_{AD} = -1246$ N
(b) $N_{AB} = 456$ N $N_{AC} = 559$ N
$N_{AD} = -1246$ N

2.93 (a) $N_{OA} = N_{OB} = N_{OC} = -45.7$ N
(b) $N_{OA} = N_{OB} = -0.650P$
$N_{OC} = 1.30P$

2.94 $N_{AB} = N_{AC} = \dfrac{2P}{\sqrt{3}}$ $N_{BC} = -\dfrac{P}{\sqrt{3}}$

$N_{BE} = N_{BF} = N_{CF} = 0$

$N_{AE} = N_{AF} = -2P\sqrt{\dfrac{2}{3}}$ $N_{AD} = \dfrac{4P}{\sqrt{3}}$

Section 2.5

2.95 (a) 0.27 (b) 0.69
(c) 0.15 (d) 0.13

2.96 (a) 0.57 (b) 0.97

2.97 $\dfrac{\sin\alpha\cos\alpha}{2-\sin^2\alpha}$

2.98 Only in case (a)

2.99 (a) $\alpha < 53.1^\circ$
(b) $\alpha < 110.6^\circ$

2.100 0.46

2.101 (a) $\dfrac{\sin\alpha}{1+\cos\alpha}$

(b) $\dfrac{1}{2}\tan\alpha$

2.102 (a) 0.75
(b) 0.72

2.103 $\dfrac{3}{\sqrt{55}} \approx 0.40$

2.104 $\mu\left(a-\dfrac{d}{2}\right)$

2.105 $\tan\dfrac{\alpha}{2}$

2.106 $\tan\dfrac{\alpha}{2}$

2.107 38.6°

2.108 2.34 m

2.109 89.1 N

2.110 0.125 m

2.111 $\dfrac{(\mu^2+\mu)mgr}{\mu^2+1}$

2.112 $0.249 < \dfrac{Q_1}{Q_2} < 6.21$

2.113 (a) $0.233Q$ (b) $0.805Q$

2.114 0.57 kg $< m <$ 1.43 kg

2.115 (a) $\varphi > 56.3^\circ$
(b) $56.3^\circ < \varphi < 71.6^\circ$

2.116 $50.2^\circ < \varphi < 73.3^\circ$

2.117 (a) $\dfrac{\mu bQ}{2(a+b)}$

(b) $F_A = \dfrac{\mu Q}{2}$

$F_B = \dfrac{\mu aQ}{2(a+b)}$

(in the opposite sense)
Impending motion at A

2.118 (a) $\tan\alpha$ (b) $\dfrac{\cos\alpha}{1+\sin\alpha}$

(impending motion at A)

2.119 2.13 m (impending motion at B)

2.120 0.16 (impending motion to the left)

2.121 $18.4^\circ < \alpha < 45^\circ$

2.122 $\dfrac{b}{2\mu}$

2.123 (a) 31.0°
(b) 36.9°

2.124 0.70

2.125 $\dfrac{\sqrt{P^2+(mg\sin\alpha)^2}}{mg\cos\alpha}$

2.126 10.6 N

2.127 $\sqrt{2}-1 \approx 0.41$

2.128 $\sqrt{3}-\sqrt{2} \approx 0.32$

2.129 (a) 876 N (b) 16.8 kN
(c) 322 kN

2.130 (a) 45.7 N (b) 2.38 N
(c) 124 mN

2.131 5.9 kg $< m <$ 1.7 tonnes

2.132 79.7 N $< P <$ 125 N

2.133 11.5 N (without the drum: 219 N)

2.134 $\dfrac{M(a-R+be^{1.5\mu\pi})}{cR(e^{1.5\mu\pi}-1)}$

2.135 31.0 kg

Chapter 3

Section 3.1

3.1 (a) 1.10 Å from the Cl-nucleus (the isotope ^{35}Cl is assumed)
(b) On the line of symmetry at a distance of 0.065 Å from the O-nucleus

3.3 (– 0.375; 0.083; 0.083) m

3.4 $\left(\dfrac{5a}{8}; -\dfrac{3a}{8}; -\dfrac{7a}{8}\right)$

Section 3.2

3.5 At a distance of $4L/5$ from A
3.6 $13a/3$ from the ground
3.7 $h = a\sqrt{3}$
3.8 $2h/5$ from the bottom
3.9 At B
3.10 $\frac{16}{7}$
3.11 $\bar{x} = 0.32r \quad \bar{y} = 0$
3.12 (a) $\left(\frac{7a}{9}; \frac{4a}{9}\right)$
(b) $\left(\frac{23a}{11}; \frac{17a}{11}\right)$
(c) $\left(-\frac{a}{30}; -\frac{a}{30}\right)$
(d) $(5a; 0)$
(e) $\left(\frac{4a}{3(4+\pi)}; \frac{(8+3\pi)a}{3(4+\pi)}\right) \approx (0.187a; 0.813a)$
3.13 $\left(a; \frac{3a}{2}\right)$
3.14 $\left(\frac{(18+15\pi)a}{10+4\pi}; -\frac{3a}{5+2\pi}\right) \approx (2.89a; -0.27a)$
3.15 $\bar{x} = \frac{2\sin^3\alpha}{3(\alpha - \sin\alpha\cos\alpha)}R$
$\bar{y} = \frac{\cos^3\alpha - 3\cos\alpha + 2}{3(\alpha - \sin\alpha\cos\alpha)}R$
3.16 $\left(\frac{2r}{3\pi}; \frac{r}{\pi}\right)$
3.17 $\left(\frac{28r}{3(5\pi+8)}; r; \frac{4(\pi+2)r}{5\pi+8}\right) \approx (0.39\,r; r; 0.87r)$
3.18 $\left(\frac{12a}{11}; 0; \frac{2a}{11}\right)$
3.19 $\bar{x} = 0.22r \quad \bar{y} = 0 \quad \bar{z} = 0.30h$
3.20 $\bar{x} = 0.22r \quad \bar{y} = 0.33r \quad \bar{z} = 0.30h$
3.21 $\left(\frac{r}{3\pi}; 0; \frac{r}{3\pi}\right)$
3.22 $\bar{x} = \frac{3a}{8} \quad \bar{y} = \frac{2b}{5}$
3.23 (a) $\bar{x} = 0.261a \quad \bar{y} = 0.688b$
(b) $\bar{x} = 0.509a \quad \bar{y} = 0.431b$
3.24 (a) $\left(0; 0; \frac{2h}{3}\right)$ (b) $\left(\frac{4R}{3\pi}; 0; \frac{2h}{3}\right)$
3.25 $\frac{3(2R-h)^2}{4(3R-h)}$ from the centre of the sphere
3.26 (a) $\bar{x} = \frac{5}{6}a$ (b) $\bar{y} = \frac{2}{3}a$
3.27 $\left(\frac{16a}{15\pi}; \frac{2a}{3}; 0\right)$
3.28 Left: 58.7 N Right: 137.5 N
3.29 566 mm
3.30 $3a/4$ from edges AD and CD
3.31 8.26°
3.32 18.4°

Glossary

Acceleration A vector quantity defined as $\boldsymbol{a} = \mathrm{d}\boldsymbol{v}/\mathrm{d}t$ where $\boldsymbol{v}$ is the velocity vector of a moving point; *see also* Normal and tangential components; *see also* Polar coordinates, plane.
Acceleration of a freely falling body *see* Gravity.
Action, law of action and reaction The forces exerted by two bodies on each other are equal in magnitude and opposite in direction. ('Action equals reaction.') *see also* Newton's laws.
Actuator A power mechanism used to effect motion of a *robot,* i.e. a motor which converts electrical, hydraulic or pneumatic energy to effect motion of the robot.
Amplitude *see* Harmonic oscillator.
Angle of elevation *see* Projectile motion.
Angular acceleration

Particle motion *see* Normal and tangential components.
Rotation of a rigid body *see* Rotation of a rigid body.

Angular frequency *see* Harmonic oscillator; Vibrations.
Angular impulse If $\boldsymbol{M}_\mathrm{O}$ is a moment acting on a body during the time interval $t_1 \rightarrow t_2$ the angular impulse

$$\int_{t_1}^{t_2} \boldsymbol{M}_\mathrm{O}\, \mathrm{d}t$$

is added to the body.

If $\Sigma \boldsymbol{M}_\mathrm{O}$ is the sum of moments acting on the body about a fixed point O the *angular momentum* $\boldsymbol{L}_\mathrm{O}$ of the body is changed according to

$$\int_{t_1}^{t_2} \sum \boldsymbol{M}_\mathrm{O}\, \mathrm{d}t = \boldsymbol{L}_\mathrm{O}(t_2) - \boldsymbol{L}_\mathrm{O}(t_1)$$

This equation is called the *angular impulse–momentum* principle.
Angular impulse–momentum principle *see* Angular impulse.
Angular momentum A vector quantity ($\boldsymbol{L}_\mathrm{O}$) defined as the moment about a certain point (O) of the *linear momentum* $m\boldsymbol{v}$ of a moving particle P. Thus, $\boldsymbol{L}_\mathrm{O} = \overrightarrow{\mathrm{OP}} \times m\boldsymbol{v}$.
Angular momentum, principle of For the total *angular momentum* ($\boldsymbol{L}_\mathrm{O}$) of a body or an arbitrary system of particles there is a relation to the total moment ($\Sigma \boldsymbol{M}_\mathrm{O}$) of the external forces: $\Sigma \boldsymbol{M}_\mathrm{O} = \dot{\boldsymbol{L}}_\mathrm{O}$. Here, O is a fixed point or the centre of mass.
Angular velocity

Particle motion *see* Normal and tangential components.
Rotation of a rigid body *see* Rotation of a rigid body.

Approach vector *see* Tool

Areal velocity A moving point P, usually representing a particle, is assumed to move in a plane containing a fixed point O. If the quantity A represents the area swept by the *radius vector* $\overrightarrow{OP}$ the time derivative $\dot{A}$ is the areal velocity of P with respect to O. There is a close relation between A and the *angular momentum* $\boldsymbol{L}_O$, i.e. $2m\dot{A} = |\boldsymbol{L}_O|$ where m is the particle mass. *See also* Central force motion; Planetary motion.

Axisymmetry *see* Rotational symmetry.

Balancing a rotating body A body, e.g. a flywheel, which rotates about a fixed axis (the z-axis) can cause undesirable bearing reactions, forces and couples that change directions permanently; the body is not balanced for rotation about the actual axis.

Balancing the body means that the mass distribution of the body is adjusted so that its centre of mass will be located on the z-axis ('static balancing') *and* the z-axis will be a *principal axis of inertia* ('dynamic balancing'). The latter condition means that $I_{xz} = I_{yz} = 0$.

Balancing may be performed by mounting balance weights on the body. The required balance weights are normally small and may for instance be made out of lead.

Ball-and-socket joint A joint connecting two bodies that can support an arbitrary force (three components) between the bodies but no couple; *see also* Constraints.

Base units *see* Fundamental quantities.

Basis vectors Form the basis of a vector space – a set of linearly independent vectors such that every vector of the space is equal to a linear combination of the basis vectors. In mechanics an *orthonormal* basis is usually used. This means that the vectors of the basis are normalized and mutually orthogonal. For basis vectors $(\mathbf{e}_1, \mathbf{e}_2, \mathbf{e}_3)$ of a three-dimensional space this means that $\mathbf{e}_i \cdot \mathbf{e}_i = 1$ and $\mathbf{e}_i \cdot \mathbf{e}_k = 0$ for $i \neq k$.

Beams Important bodies in technical applications. A beam is a long and narrow body, normally straight, whose length is much greater than the typical dimension of the cross-section. Beams are used to carry loads, usually transverse loads, in various applications in engineering. The components of the section force are the *axial force* (parallel to the beam) and the *shear force* (perpendicular to the beam). The components of the section couple are the *bending moment* and the *torque* or *torsional moment*. *See also* Section force system.

Bearing reactions When a body is rotating about a fixed axis it is subjected to a reaction from the bearings of the axis. If the axis is supported by two bearings the reactions as a rule consist of two constraining forces $\boldsymbol{R}_1$ and $\boldsymbol{R}_2$, one at each bearing. If the support consists of one bearing the system of reaction forces is best described by its resultant $\boldsymbol{R}$ and a couple-moment $\boldsymbol{M}$. Like all constraining forces the bearing reactions are not known in advance. They may be determined by use of the *law of motion of the centre of mass*, which contains the components of the forces ($\boldsymbol{R}_1$, $\boldsymbol{R}_2$, $\boldsymbol{R}$), and the *principle of angular momentum*, where components of the couple-moment $\boldsymbol{M}$ and one of the forces $\boldsymbol{R}_1$ and $\boldsymbol{R}_2$ enter into the equations.

Bending moment *see* Beams.

Brahe, Tycho (1546–1601) Danish astronomer. He was the last of the great naked-eye astronomers. His enormous collection of observational data was taken over by his assistant *Johannes Kepler* and became of great importance for the formulation of Kepler's laws of planetary motion. *see* Planetary motion.

Cartesian coordinates The axes (x- y- and z-axes) are mutually perpendicular. They are often assumed to be fixed in space. The unit vectors $\mathbf{e}_x$, $\mathbf{e}_y$ and $\mathbf{e}_z$ are then constant.

Central force field A force field in which a particle is subjected to a force $\boldsymbol{F}$ such that the line of action is constantly passing through a certain fixed point O, the centre of force. Furthermore, the magnitude of the force is a function of the distance r to O, i.e. $\boldsymbol{F} = F(r)\mathbf{e}_r$. All central forces are *conservative*.

Central force motion The motion of a particle which is free to move under the influence of a central force. The planetary motion around the sun and the motion of satellites around the earth are well-known examples of central force motion. In central force motion the following quantities are constant: the mechanical energy $(T+V)$, the angular momentum $(\boldsymbol{L}_O)$, the areal velocity $(r^2\dot{\varphi}/2)$ with respect to the force centre O. *See also* Planetary motion.
Centre of mass *see* Mass, centre of.
Centrifugal force *see* Fictitious force.
Centripetal acceleration *see* Normal and tangential components.
Circular motion *see* Normal and tangential acceleration.
Classical mechanics The study of *mechanics* without considering the effects of *relativistic mechanics* or *quantum physics, see also* Mechanics.
Component of a vector One of a set of two or more vectors, the sum of which is equivalent to the given vector, is called a *vector component*. A *scalar component* of a vector in a given direction is the projection of the vector on a line in the given direction.
Conservative force *see* Potential energy.
Constraining forces *see* Constraints.
Constraints Most practical situations, in statics as well as in dynamics, concern bodies whose abilities to move are restricted in some way owing to contact with other bodies. Such a body is subjected to a constraint. A force which maintains a constraint is called a *constraining force* or a *reaction*. In general the system of forces at a point of contact between two bodies is represented by one force vector and one couple vector, six scalar components in all. This is the case when a body is completely fixed at a point. One or more of the components can be zero. Examples of this are the contact between two smooth surfaces – only a normal force – and the *ball-and-socket joint* – an arbitrary force vector.
Coulomb's law Two particles at rest with electric charges Q_1 and Q_2 exert forces on each other according to Coulomb's law

$$F=\frac{Q_1Q_2}{4\pi\varepsilon_0 r^2}$$

Here r is the distance between the particles and $\varepsilon_0 \approx 8.854\cdot 10^{-12}\ \mathrm{F\,m^{-1}}$ is called the permittivity of free space. The force is in the direction joining the particles. It is a repulsive force if Q_1 and Q_2 have the same sign, and an attractive force if they have the opposite sign.
Couple, couple-moment A force system consisting of two parallel forces of the same magnitude but opposite directions is called a couple (force couple). The moment of a couple about an arbitrary point (O) is independent of the position of the point O and can be written $\boldsymbol{M}=\boldsymbol{r}\times\boldsymbol{F}$. Here $-\boldsymbol{F}$ and $\boldsymbol{F}$ are the force vectors and $\boldsymbol{r}$ is the relative position vector between the points of application in the given order.
Damping factor *see* Vibrations.
Degrees of freedom The number of independent variables required to specify the configuration of a mechanical system is referred to as the number of degrees of freedom.
Density In classical mechanics it is assumed that matter is continuously distributed. The density function ρ is defined according to

$$\rho(\mathrm{P})=\lim_{\Delta V\to 0}\frac{\Delta m}{\Delta V}$$

Here Δm is the mass of a material system of volume ΔV including point P.
Dimension The unit of every physical quantity can be expressed as a combination of the *base units*, the units of the seven *fundamental quantities*. For the quantities of mechanics three base units

are sufficient, namely the units of mass, length and time. The dimension of a quantity expresses in what way the base units constitute the unit of the actual quantity.

The dimensions of mass, length and time are denoted by M, L and T. If a quantity is a product/quotient of two other quantities its dimension is a product/quotient of the dimensions of the factors. For a velocity v we then have dim $v = LT^{-1}$ (or L^1T^{-1}). If F is a force, we have dim $F = MLT^{-2}$ etc. Quantities that are added or subtracted must have the same dimensions.

Dynamics Refers to the study of mechanical systems in motion. It thus includes *kinematics* and *kinetics*. Courses and textbooks in *mechanics* are often divided into *statics* and *dynamics*.

Efficiency The ratio between the power output and the power input of a machine, e.g. an electric motor, is called the efficiency (η). Thus, $\eta = P_{out}/P_{in}$; *see also* Power.

Einstein, Albert (1879–1955) Albert Einstein is considered to be the greatest scientific genius of our time. In the first decade of this century he advanced physics with his theory of the photoelectric effect, special relativity (1905) and the theory of Brownian motion.

In the theory of general relativity Einstein proposed in 1916 the fundamental hypothesis that the principle of equivalence between gravitation and acceleration of the reference frame applies to all physical phenomena. Typical consequences are the deflection of light rays by a gravitational field and the redshift of spectral lines in radiation from the stars. These phenomena have been observed and have been proved to be in agreement with the principle of equivalence. Nor has the principle been refuted by any experiments. *See also* Relativity, theory of.

Elastic force A body in contact with an elastic medium is subjected to a force that depends on the deformation of the medium. The simplest example is a body which is attached to a perfectly elastic spring. If such a spring is stretched or compressed a distance δ from its unstretched length the force on the body is $F = k\delta$. The constant k is called the *spring constant* ('stiffness'). *See also* Hooke's law.

End-effector *see* Robot; Tool.

Equilibrium A material system is said to be in equilibrium if every portion of the system is at continuous rest. Thereby equilibrium and rest relative to an *inertial frame* is understood. The study of systems in equilibrium is undertaken in *statics*. *See also* Stability.

Equilibrium, conditions of A material system will remain in equilibrium if the system of *external forces* acting on each portion of the system is equivalent to zero, i.e. $\Sigma \boldsymbol{F} = 0$, $\Sigma \boldsymbol{M} = 0$. These are the *equations of equilibrium*.

Equilibrium, equations of *see* Equilibrium, conditions of.

Euler, Leonhard (1707–83) Swiss mathematician who made fundamental contributions to the development within various areas of pure and applied mathematics and also within physics and astronomy. He derived the equations for the motion of a rigid body about a fixed point.

Euler's angles Three angles which determine the orientation in space of a rigid body. The angles are denoted by φ (precession angle), θ (nutation angle) and ψ (spin angle).

Fictitious force When describing the motion of bodies with respect to a frame of reference which is not an inertial frame, it is sometimes desired to obtain a 'law of motion' of the type $\boldsymbol{F}' = m\boldsymbol{a}_{rel}$ where $\boldsymbol{a}_{rel}$ is the relative acceleration. $\boldsymbol{F}'$ cannot be expected to be the sum of real forces $\boldsymbol{F}$ but can always be written in the form $\boldsymbol{F}' = \boldsymbol{F} + \boldsymbol{F}_{fic}$. The quantity $\boldsymbol{F}_{fic}$ is called the fictitious force.

If the frame of reference has a translatory motion with acceleration $\boldsymbol{a}_A$ the fictitious force on a particle with mass m is $-m\boldsymbol{a}_A$.

Another example is the *centrifugal force*, a fictitious force that belongs to rotating frames of reference. Consider a particle at a distance x from the axis of rotation. If the mass of the particle is m the centrifugal force is directed radially outwards from the axis and has a magnitude of $mx\omega^2$ where ω is the angular velocity of the frame.

Force In general physics all forces are traced back to the four fundamental forces, which originate from *gravitation*, *electromagnetism* and the *weak* and *strong interaction*. In classical mechanics *force* is a quantity that describes the influence on a material system by its surroundings, e.g. by other material systems. If the motion of a particle is related to an *inertial frame of reference* Newton's second law, $\Sigma \boldsymbol{F} = m\boldsymbol{a}$ will be obeyed. Here $\boldsymbol{a}$ is the absolute acceleration and m the inertial mass of the particle. This relation establishes the Newtonian concept of force. Quantities which violate this equation if brought into its left member are not real forces. Examples of such quantities are *centrifugal forces* and other *fictitious forces*.

Force, external, internal When considering a body or a more general material system the term 'external forces' is used for forces exerted by bodies outside the system. Internal forces are all forces exerted by various portions of the system on each other. *See also* Free-body diagram.

Force, fundamental operations Aims at the following two operations leaving the resultant sum of forces and sum of moments of a force system unchanged:

1. Two forces with a common point of application may be added according to the *parallelogram law*. In the opposite way a force can be resolved into components along two arbitrary directions.
2. A force may be displaced in an arbitrary way along its line of action.

Force, line of action The line of action is parallel to the force and passes through its point of application.

Force, moment of A vector quantity defined by $\boldsymbol{M}_A = \boldsymbol{r} \times \boldsymbol{F}$. Here A is the moment point and $\boldsymbol{r}$ is the position vector from A to the point of application of the force $\boldsymbol{F}$. In everyday life the concept of moment is usually associated with the tendency of a force to rotate a certain body. The moment about an axis (λ) passing through A can be obtained as the scalar product $M_\lambda = \mathbf{e}_\lambda \cdot \boldsymbol{M}_A$ where $\mathbf{e}_\lambda$ is a unit vector along the axis. If the force is perpendicular to the axis the moment can be written $M_\lambda = Fd$ where d is the distance between the axis and the line of action of the force. The term *torque* is frequently used for the moment of a force.

Force resultant, sum of forces The resultant ($\Sigma \boldsymbol{F}$) of a force system is obtained by adding the individual force vectors without taking into account that they may have different points of application. Thus, we have

$$\Sigma \boldsymbol{F} = \sum_{i=1}^{n} \boldsymbol{F}_i$$

where $\boldsymbol{F}_i$ are the individual forces.

Force system, reduction of Every force system can be reduced by *fundamental operations* to a resultant $\boldsymbol{R}$ with an arbitrary point of application (A) and a couple moment ($\boldsymbol{M}$). The resultant $\boldsymbol{R} = \Sigma \boldsymbol{F}$ is independent of the location of the reduction point A while $\boldsymbol{M} = \Sigma \boldsymbol{M}_A$ will vary depending on the choice of A. In the special case of a *parallel force system* it can be reduced to a single resultant applied at an arbitrary point on a specific line, the line of action of $\boldsymbol{R}$.

If a force system can be reduced to a system with $\boldsymbol{R} = 0$ and $\boldsymbol{M} = 0$ it is a *zero system of force*. Such systems are of fundamental importance in *statics*. *See also* Equilibrium, conditions of.

Frame of reference Every measurement of a position, a velocity or an acceleration is assumed to be made with respect to a certain rigid system, a frame of reference.

A frame of reference where Newton's laws may be applied is called an *inertial frame of reference*. These frames are all of equal standing and may be defined without any internal contradiction.

Once a certain inertial frame has been chosen, it is customary to refer to this as the *resting frame*. A transformation between two inertial frames is called a *Galilean transformation*. It leaves the quantity acceleration unchanged.

Free-body diagram Once a material system has been chosen to which the laws of statics or dynamics are to be applied it is advantageous to isolate the system in a diagram. In this free-body diagram the limits of the system are clearly identified. The *external forces* – and no other forces – are also introduced into the diagram.

Freely falling body *see* Gravity.
Frequency *see* Harmonic oscillator.
Friction, static, kinetic, coefficient of The contact force between two bodies is resolved into one component (F, the *friction force*) in the common tangential plane and one component (N, the *normal force*) normal to the tangential plane. An approximate model for the relation between $\mathbf{F}$ and N is commonly used: in statics $|\mathbf{F}| \leqslant \mu_s N$, in dynamics $|\mathbf{F}| = \mu_k N$. Here μ_s is the *coefficient of static friction* and μ_k the *coefficient of kinetic friction*. As a general rule $\mu_k \leqslant \mu_s$. This type of friction is known as dry friction or Coulomb friction.

In many technical applications lubricants of various types are used in order to reduce the effects of friction. This type of friction is called fluid friction, and the theory presented above is not applicable.
Fundamental operations *see* Force, fundamental operations.
Fundamental quantities The SI system contains units of seven fundamental quantities: length, mass, time, electric current, thermodynamic temperature, amount of substance, luminous intensity.

The corresponding *base units* are, respectively: 1 metre (1 m), 1 kilogram (1 kg), 1 second (1 s), 1 ampere (1 A), 1 kelvin (1 K), 1 mole (1 mol), 1 candela (1 cd).

In mechanics the three fundamental quantities of length, mass and time are of special interest. Their units form the base of the units for mechanical quantities. *See also* Dimension; SI system.
Galilean transformation *see* Frame of reference.
Galilei, Galileo (1564–1642) Italian astronomer, physicist and mathematician. He was one of the pioneers in utilizing experiments performed in a systematic manner in dynamics. He investigated the laws of freely falling bodies and motion down an incline. Galileo constructed telescopes and discovered four of the moons of Jupiter. Through his conception of the universe he came into conflict with the Church.
Geostationary satellite A satellite which is orbiting the earth so that it is constantly located above the same point on the ground. The path is a circle in the equatorial plane of the earth. The altitude of the satellite above the earth is $35.8 \cdot 10^3$ km $\approx 5.6R$ where R is the radius of the earth.
Gravitation, law of *see* Gravity.
Gravitational field *see* Gravity.
Gravitational force *see* Gravity.
Gravity Force of attraction between two material objects. According to the *law of gravitation* two particles with masses m_1 and m_2 attract each other with the *gravitational force* Gm_1m_2/r^2. Here r is the distance between the particles and G is the *universal gravitational constant* with the value $6.673 \cdot 10^{-11}$ N m^2/kg^2.

The gravitational force on an object at the surface of the earth is often called *weight* in everyday life. Its magnitude is mg where m is the mass of the object and $g \approx 9.81$ N/kg (=m/s^2) is called the strength of the *gravitational field* or the *acceleration of a freely falling body* at the surface of the earth. An approximate formula for the variation of g with latitude angle (λ) is $g = 9.8322(1 - 0.005\,30 \cos^2\lambda)$ N/kg. The relatively small variation depends upon the rotational motion and the lack of spherical symmetry of the earth. The approximate average value of $g \approx GM/R^2$, where M and R are the mass and the average radius of the earth respectively, can often be used in calculations.

On the moon the value of g is 1.62 N/kg.
Gravity, centre of A system of parallel gravitational forces on a material system can be reduced to a resultant acting along a line passing through a certain point, the centre of gravity. In a homogeneous gravitational field this point coincides with the *centre of mass* of the system.
Gyroscope By a gyroscope we normally mean an axisymmetric body with a very large angular velocity, the *spin* velocity, about the axis of symmetry. The body may be free, e.g. a bullet,

supported by a horizontal surface, e.g. a spinning top, or suspended with one point fixed. The last case, which has several interesting technical applications, is achieved by various kinds of suspension depending on the actual situation. The importance of the gyroscope is mainly due to its ability to keep its direction in space in the absence of external moments.

The terms *gyro* and *spinning top* can be used to describe two types of body in gyroscopic motion. The following physical properties are common to them:

> The body is axisymmetric, which means that two of the principal moments of inertia at a point on the axis of symmetry are equal. The body is rotating ('spinning') at a very large angular velocity about its axis of symmetry. In addition 'small' angular velocity components may be superposed on the spinning motion.

One such motion is *steady precession*: the body is spinning about the axis of symmetry while this axis is rotating in a regular sense about a fixed line in space. An example is the *spinning top*, which is supported at a fixed point and performs a steady precession motion, where its spin axis rotates about a vertical line through the point of suspension.

Harmonic oscillator A system that is moving according to the equation $x(t)=A\sin(\omega t+\delta)$. It performs a 'harmonic vibration'. The following symbols and designations are used for some important quantities:

$\lvert A\rvert$	amplitude
ω	angular frequency
$f=\omega/2\pi$	frequency
$\omega t+\delta$	phase
δ	phase angle
$\tau=1/f=2\pi/\omega$	period

See also Vibrations.

Homogeneous coordinates Used in robotics. For a certain point of a *robotic arm* they are obtained from the three ordinary coordinates by adding an extra coordinate which by convention is set equal to one. Let the homogeneous coordinates be represented by the column matrix $\mathbf{r}_0$ in the $x_0y_0z_0$-system and the column matrix $\mathbf{r}_n$ in the $x_ny_nz_n$-system. Thus,

$$\mathbf{r}_0=\begin{pmatrix}x_0\\y_0\\z_0\\1\end{pmatrix}\quad\text{and}\quad\mathbf{r}_n=\begin{pmatrix}x_n\\y_n\\z_n\\1\end{pmatrix}$$

Then

$$\mathbf{r}_0=\mathbf{T}\mathbf{r}_n\quad\text{where }\mathbf{T}=\begin{pmatrix}R_{11}&R_{12}&R_{13}&p_x\\R_{21}&R_{22}&R_{23}&p_y\\R_{31}&R_{32}&R_{33}&p_z\\0&0&0&1\end{pmatrix}$$

This is called a *homogeneous transformation*. Here the 4×4 matrix $\mathbf{T}$ is the *homogeneous transformation matrix*. If the coordinates $(x_n; y_n; z_n)$ are the *tool* coordinates and $(x_0; y_0; z_0)$ the base coordinates, the transformation matrix $\mathbf{T}$ completely specifies the *tool configuration*, i.e. the position and orientation of the tool. The 3×3 submatrix $\mathbf{R}$ in the upper left corner of $\mathbf{T}$ defines the orientation of the tool and the 3×1 column matrix in the upper right corner defines the position of the tool; $(p_x; p_y; p_z)$ are the coordinates of the tool in the base coordinate system. *See also* Robot; Tool.

Homogeneous transformation *see* Homogeneous coordinates.
Homogeneous transformation matrix *see* Homogeneous coordinates.
Hooke's law When a solid is deformed, it resists the deformation with a force proportional to the amount of deformation, provided the deformation is not too large. This statement is based on the assumption that the undeformed solid is at a potential energy minimum and that the potential energy may be expanded in a Taylor series in the coordinate describing the deformation. If a solid is deformed beyond a certain point, called the elastic limit, it will remain permanently deformed. *See also* Elastic force.
Impact The main characteristics of impact between two bodies are

- The time of interaction is very short but the velocity changes of the bodies are appreciable.
- The impact force between the bodies is very large. Hence, external forces that are not of impact character may be neglected during the time of impact. Such forces are the gravitational force (mg) and forces exerted by elastic connections, e.g. spring forces.

If such forces are the only external ones the total *linear momentum* of the system is conserved during the short time of impact. *See also* Impact, coefficient of restitution.
Impact, central, eccentric If the centres of mass of two bodies are located on the *line of impact* (the normal to the common tangential plane at the contact point) the impact is called central impact. If this condition is not fulfilled the impact is eccentric.
Impact, coefficient of restitution From experiments it is known that the ratio of the relative velocities along the line of impact before and after impact is relatively independent of the individual velocities before collision. This relation may be expressed as

$$\frac{|\,\textit{relative velocity after impact}\,|}{|\,\textit{relative velocity before impact}\,|} = e$$

The quantity e is called the coefficient of restitution. It is treated as a property of the materials involved. The limits of e are given by $0 \leqslant e \leqslant 1$. The limiting cases are designated as follows:

$e = 0$ *inelastic* or *perfectly plastic impact*
$e = 1$ (*perfectly*) *elastic impact*

Impact, direct, oblique If the relative velocity of the contact points of the bodies is directed along the *line of impact* immediately prior to impact, the impact is called *direct impact*. If this condition is not fulfilled the impact is *oblique*.
Impact, elastic, inelastic *see* Impact, coefficient of restitution.
Impact, line of *see* Impact, central, eccentric.
Impulse If F is a force acting on a body during the time interval $t_1 \rightarrow t_2$ the impulse

$$\int_{t_1}^{t_2} F \, \mathrm{d}t$$

is added to the body.

If ΣF is the sum of forces acting on the body the *linear momentum* p of the body is changed according to

$$\int_{t_1}^{t_2} \sum F \, \mathrm{d}t = p(t_2) - p(t_1)$$

This equation is called the *impulse–momentum* principle.
Impulse–momentum principle *see* Impulse.
Inertia, law of *see* Newton's laws.

Inertia matrix A symmetric matrix $\mathbf{I}_0$ containing (*mass*) *moments of inertia* and (*mass*) *products of inertia* of a body according to

$$\mathbf{I}_0 = \begin{pmatrix} I_{xx} & -I_{xy} & -I_{xz} \\ -I_{yx} & I_{yy} & -I_{yz} \\ -I_{zx} & -I_{zy} & I_{zz} \end{pmatrix}$$

If the body is rotating about an axis through O, the *angular momentum* in matrix form can be written $\mathbf{L}_0 = \mathbf{I}_0\omega$ where ω is the *angular velocity* matrix. If the body is rotating about the fixed point O the *kinetic energy* is the scalar quantity given by $T = \frac{1}{2}\omega^{\mathrm{T}}\mathbf{I}_0\omega$. If the coordinate axes are *principal axes of inertia* $(\xi\eta\zeta)$ at the point O the inertia matrix is diagonal

$$\mathbf{I}_0 = \begin{pmatrix} I_{\xi\xi} & 0 & 0 \\ 0 & I_{\eta\eta} & 0 \\ 0 & 0 & I_{\zeta\zeta} \end{pmatrix}$$

See also Moment of mass.

Inertial frame of reference *see* Frame of reference.

Instantaneous centre of rotation If a rigid body is assumed to be in plane motion with a non-zero angular velocity ω, it is always possible to find a point C of the body whose velocity is zero at the instant of consideration. The expression 'point C of the body' does not necessarily refer to a material point of the body. Momentarily the body is rotating about the point C. Hence the velocity of an arbitrary point P can be written $v_{\mathrm{P}} = \boldsymbol{\omega} \times \overrightarrow{\mathrm{CP}}$.

Jacobian *see* Manipulator Jacobian.

Joints, method of *see* Truss.

Kepler, Johannes (1571–1630) German astronomer and mathematician. After a systematic study of the observations performed by the Danish astronomer *Tycho Brahe* he formulated his famous laws of planetary motion. *see* Planetary motion.

Kepler's laws *see* Planetary motion.

Kinematics In kinematics the motion of particles and extended bodies is described from a geometrical point of view. The cause of the motion, the forces that are acting, is not considered here. *See also* Mechanics; Kinetics.

Kinetic energy The kinetic energy of a particle with mass m and velocity v is $T = \frac{1}{2}mv^2$. For an arbitrary system of particles the kinetic energy can be written as

$$T = \sum \frac{1}{2} m_i v_i^2 = \frac{1}{2} m\bar{v}^2 + \bar{T} \text{ (König's theorem)}$$

Here $\bar{v}$ is the magnitude of the velocity of the centre of mass and $\bar{T}$ the contribution from the motion relative to the centre of mass of the individual particles.

For a rigid body $\bar{T}$ can be written $\frac{1}{2}\bar{I}_\omega\omega^2$ where $\bar{I}_\omega$ is the *moment of inertia* with respect to an axis passing through the centre of mass and parallel to the vector $\boldsymbol{\omega}$. More generally $\bar{T} = \frac{1}{2}\boldsymbol{\omega}^{\mathrm{T}}\bar{\mathbf{I}}\boldsymbol{\omega}$ where $\bar{\mathbf{I}}$ is the *inertia matrix* of the body in a centre of mass coordinate system and $\boldsymbol{\omega}^T = (\omega_x\ \omega_y\ \omega_z)$ is the transposed $\boldsymbol{\omega}$-matrix.

For a rigid body rotating about a fixed point O the kinetic energy is $T = \frac{1}{2}I_\lambda\omega^2 = \frac{1}{2}\boldsymbol{\omega}^T\mathbf{I}_0\boldsymbol{\omega}$. Here I_λ is the moment of inertia about the instantaneous axis of rotation and $\mathbf{I}_0$ is the inertia matrix in the $Oxyz$ system. *See also* Work–energy principle.

Kinetics The term 'kinetics' is used for the study of motion of material systems under the influence of forces. The interaction between the motion and the forces is the main subject of study in kinetics. *See also* Mechanics; Kinematics.

König's theorem *see* Kinetic energy.

Line of action *see* Force, line of action.

Linear momentum For a particle of mass m and velocity v its linear momentum p is defined as $p = mv$. For an arbitrary particle system the total linear momentum is $p = m\bar{v}$ where m is the total mass and $\bar{v}$ the velocity of the centre of mass of the system. *Newton's laws* yield $F = \dot{p}$ for a particle and $\Sigma F = \dot{p}$ for a particle system. This is the *principle of linear momentum*.
Linear momentum, principle of *see* Linear momentum.
Link *see* Robot.
Lorentz force A particle of electric charge Q, which is travelling with a velocity v in an electromagnetic field of electric field strength E and magnetic flux density B, is subjected to the force $F = Q(E + v \times B)$. The contribution $Q(v \times B)$ is called the Lorentz force.
Magnification factor *see* Vibrations.
Manipulator, arm *see* Robot.
Manipulator Jacobian Used in robotics to describe the relation between the motion of the *tool* and the motion of the *links*. Let the velocity $v = \dot{x}_0 \mathbf{e}_{x_0} + \dot{y}_0 \mathbf{e}_{y_0} + \dot{z}_0 \mathbf{e}_{z_0}$ and the angular velocity $\boldsymbol{\omega} = \omega_x \mathbf{e}_{x_0} + \omega_y \mathbf{e}_{y_0} + \omega_z \mathbf{e}_{z_0}$ describe the motion of the tool and the n variables $q_1, q_2, \ldots, q_n$ specify the positions of the links. Then the desired transformation is

$$\begin{pmatrix} \dot{x}_0 \\ \dot{y}_0 \\ \dot{z}_0 \\ \omega_x \\ \omega_y \\ \omega_z \end{pmatrix} = \mathbf{J} \begin{pmatrix} \dot{q}_1 \\ \dot{q}_2 \\ \cdot \\ \cdot \\ \cdot \\ \dot{q}_n \end{pmatrix}$$

where $\mathbf{J}$ is a $6 \times n$ matrix whose elements are functions of $q_1, q_2, \ldots, q_n$.
Mass There are two fundamentally different possibilities for measuring the mass of a body:

1. By weighing the body, for instance at the surface of the earth. Using the law of gravitation a value of the *gravitational mass* is obtained.
2. By subjecting the body to a known force and measuring the acceleration. Using Newton's second law, $F = ma$, the inertial mass is obtained.

The inertial mass turns out to be proportional to the gravitational mass. With a suitable choice of units we may say that the inertial and gravitational masses of a body are equal. This experience is a great discovery. It is one of the cornerstones of the general *theory of relativity* formulated by Einstein. *See also* Fundamental quantities.
Mass, centre of The centre of mass of a system of particles is a point whose coordinates $(\bar{x}; \bar{y}; \bar{z})$ are weighted averages of the coordinates of the particles according to

$$\bar{x} = \frac{\sum m_i x_i}{\sum m_i} \quad \bar{y} = \frac{\sum m_i y_i}{\sum m_i} \quad \bar{z} = \frac{\sum m_i z_i}{\sum m_i}$$

For a continuous mass distribution all sums are replaced by integrals. Here m_i is the mass of an individual particle and $(x_i; y_i; z_i)$ the coordinates.

In a homogeneous gravitational field the centre of mass coincides with the *centre of gravity* of the system. *See also* Moment of mass.
Mass, centre of, law of motion The centre of mass of a particle system with total mass m has an acceleration $\bar{a}$ given by $\Sigma F = m\bar{a}$ where ΣF is the sum of external forces acting on the system.
Mechanics Mechanics is the study of material systems at rest or in motion and the corresponding forces acting. Mechanics is usually divided into two main areas:

Statics, which refers to the study of systems at rest.

Dynamics, which refers to the study of systems in motion. This includes *kinematics* where the pure motion is studied and *kinetics* where the forces acting and their interaction with the motion are also taken into account.

Within engineering the term 'mechanics' usually refers to what is called *classical mechanics*. It may be applied to systems with an enormous variation in size, from the stars and other celestial bodies down to objects approaching a molecule in size. When we proceed down further and reach the atomic level, *quantum mechanics* (*physics*) must be applied. When the velocity v of a body is of the same order of magnitude as the speed of light c it is necessary to use *relativistic mechanics*.
Member *see* Truss.
Moment of inertia *see* Moment of mass.
Moment of mass; first moments, second moments The first moment of mass or *static moment of mass* with respect to the yz-plane is denoted by S_{yz} and defined by $S_{yz} = \Sigma m_i x_i$. The first moment of mass with respect to any plane is defined in an analogous way.

The coordinates of *centre of mass* can be written as

$$\bar{x} = S_{yz}/m \quad \bar{y} = S_{xz}/m \quad \bar{z} = S_{xy}/m$$

The second moments of mass are quadratic functions of the coordinates x_i, y_i and z_i. The following quantities are called (*mass*) *moments of inertia*:

$$I_x = \sum m_i(y_i^2 + z_i^2) \quad I_y = \sum m_i(x_i^2 + z_i^2) \quad I_z = \sum m_i(x_i^2 + y_i^2)$$

Here I_x is the moment of inertia with respect to the x-axis etc. Also, the denotations I_{xx}, I_{yy} and I_{zz} are used for I_x, I_y and I_z.

The moment of inertia with respect to any axis (λ) can be written $I_\lambda = \Sigma m_i a_i^2$ where a_i is the distance of the mass element to the actual axis. The moment of inertia can always be written in the form mk^2 where k is a quantity of dimension length. It is known as the *radius of gyration*. If $I_x = mk_x^2$ the 'radius of gyration with respect to the x-axis' is k_x. The quantities

$$I_{xy} = \sum m_i x_i y_i \quad I_{xz} = \sum m_i x_i z_i \quad I_{yz} = \sum m_i y_i z_i$$

are also second moments of mass. They are called (*mass*) *products of inertia*.

For a continuous mass distribution all sums are replaced by integrals. *See also* Mass, centre of; Gravity, centre of; Steiner's theorem.
Momentum *see* Linear momentum.
Natural coordinates *see* Normal and tangential components.
Natural frequency *see* Vibrations.
Newton, Isaac (1642–1727) Professor in Cambridge, UK. One of the greatest scientists of all time. He made pioneering mathematical discoveries in differential and integral calculus. In physics he studied light and proposed theories on the nature of light. He split white light into its components using prisms and also constructed telescopes.

Most of all Newton is known for his contributions to mechanics. Based on his studies of the dynamics of celestial bodies he summarized the results of his predecessors into three *laws of motion* and demonstrated their general validity. The *law of gravitation* was formulated in connection with this work.

Newton's greatest and most famous work, *Principia* (*Philosophiae naturalis principia mathematica*), was first published in 1687. *See also* Newton's laws; Gravity.

Newton's laws In his greatest and most famous work, *Principia* (1687), Isaac Newton published the fundamental laws of mechanics. The following three statements are known as 'Newton's laws' and are summarized here with the use of modern terminology:

1. The *law of inertia*:
   A particle remains in its state of rest or motion along a straight line at a constant velocity unless it is forced by acting forces to change this state.
2. The *law of motion*:
   A particle that is subjected to a force $\boldsymbol{F}$ is given an acceleration $\boldsymbol{a}$ so that $\boldsymbol{F} = m\boldsymbol{a}$, where the constant m is the *inertial mass* of the body.
3. The *law of action and reaction*:
   The forces exerted by two bodies on each other are equal in magnitude and opposite in direction.

Normal and tangential components The path of a moving particle P is assumed to lie in a plane. The arc length s and the radius of curvature ρ of the path curve at P are sometimes called the *natural coordinates* of the path. The unit vectors used as a basis are $\mathbf{e}_s$ along the tangential direction and $\mathbf{e}_n$ along the normal direction towards the concave side of the path. The components of the velocity and acceleration vectors are

$$v_s = v = \dot{s} \quad v_n = 0 \quad a_s = \dot{v} = \ddot{s} \quad a_n = \frac{v^2}{\rho} = \frac{\dot{s}^2}{\rho}$$

a_s is the *tangential acceleration* and a_n is the *normal acceleration*, sometimes called *centripetal acceleration*.

In *circular motion* the arc length s is normally replaced by the central angle φ using $s = r\varphi$ where r is the radius. The time derivative $\dot{\varphi}$, which is called *angular velocity*, is often denoted by ω. The components of the velocity and acceleration vectors can be written as

$$v_s = v = r\dot{\varphi} = r\omega \quad a_s = \ddot{s} = r\ddot{\varphi} = r\dot{\omega}$$

$$v_n = 0 \qquad a_n = \frac{v^2}{\rho} = \frac{\dot{s}^2}{\rho} = r\dot{\varphi}^2 = r\omega^2$$

Here $\ddot{\varphi} = \dot{\omega}$ is the *angular acceleration*.

Normal force *see* Friction.

Normal vector *see* Tool.

Parallel force system *see* Force system, reduction of.

Parallelogram law Two forces with a common point of application (A) may be added according to the *parallelogram law* as follows. If a parallelogram is constructed with the two force vectors as sides the vector along the diagonal through A represents the sum of the forces.

Particle A body which is so small that its extension in space may be neglected, i.e. its position is specified by a point. Its state of rotation is of no interest in classical mechanics. A velocity vector, given as a function of time, contains all relevant information about the motion.

Pendulum

Compound pendulum: A rigid body which may rotate freely about a horizontal axis which does not pass through the centre of gravity. Normally the term is used of a body that oscillates about its stable position of equilibrium.
Simple pendulum: If the mass of a compound pendulum is concentrated at a point at a distance l from the axis it is known as a simple pendulum. It may, for instance, consist of a light cord from which a point mass is suspended.

If a pendulum vibrates with a small amplitude it moves as a *harmonic oscillator*. The period is $\tau = 2\pi(I/mg\bar{r})^{1/2}$ for a compound pendulum of mass *m*, *moment of inertia* *I* with respect to the axis and with its centre of mass at a distance $\bar{r}$ from the axis. For a simple pendulum the period is $\tau = 2\pi(l/g)^{1/2}$.

Period *see* Harmonic oscillator.

Phase, phase angle *see* Harmonic oscillator.

Plane motion of a rigid body Plane motion is characterized by the fact that all points of a body are moving in fixed planes which are all parallel to a certain plane of reference.

Planetary motion To a good approximation it can be said that a planet moves under the influence of a central force, the *gravitational force* from the sun. *Kepler's laws* for planetary motion, which Kepler formulated after combining astronomical data, can also be verified by Newtonian mechanics. The laws can be written as follows:

1. A planet is moving in an elliptic orbit such that the sun is located at one of the foci.
2. A planet is moving so that equal areas are swept in equal times by the radius vector, i.e. the *areal velocity* is constant for each planet.
3. The cubes of the major axes of the orbits are proportional to the squares of the periods.

See also Central force motion.

Polar coordinates, plane The position of a point P in a plane can be located by its distance *r* to the pole O and the angle φ that the line OP makes with a certain fixed line (the axis) through O. The following unit vectors are used as a base: $\mathbf{e}_r$ parallel to $\overrightarrow{OP}$ and $\mathbf{e}_\varphi$ perpendicular to $\mathbf{e}_r$ in the direction corresponding to increasing values of φ. The components of the velocity and acceleration vectors are

$$v_r = \dot{r} \qquad a_r = \ddot{r} - r\dot{\varphi}^2$$

$$v_\varphi = r\dot{\varphi} \qquad a_\varphi = r\ddot{\varphi} + 2\dot{r}\dot{\varphi} = \frac{1}{r}\frac{\mathrm{d}}{\mathrm{d}t}(r^2\dot{\varphi})$$

Potential energy If a force $\boldsymbol{F}$ is a function of position, such that the work integral

$$\int_{s_P}^{s_Q} F_s \, \mathrm{d}s$$

between two arbitrary points P and Q is independent of the path between the points, the force (or 'force field') is said to be *conservative*. For a conservative force there exists a potential energy function *V* such that the work $W^{PQ} = -\Delta V = -(V_Q - V_P)$.

More direct relations between the force components and *V* are

$$F_x = -\frac{\mathrm{d}V}{\mathrm{d}x} \quad F_y = -\frac{\mathrm{d}V}{\mathrm{d}y} \quad F_z = -\frac{\mathrm{d}V}{\mathrm{d}z}$$

Power If the work done by a force during a time interval Δt is ΔW the average power during the time is defined as $\bar{P} = \Delta W/\Delta t$.

The instantaneous power is $P = \mathrm{d}W/\mathrm{d}t = \boldsymbol{F} \cdot \boldsymbol{v}$ where $\boldsymbol{v}$ is the velocity of the point of application of $\boldsymbol{F}$. Further, $W = \int P \, \mathrm{d}t$.

If *P* is the total power of the forces acting on a particle system and *T* is the kinetic energy of the system the general relation $P = \mathrm{d}T/\mathrm{d}t$ holds.

A couple *M* acting on a rotating body develops the power $P = M\omega$. Here ω is the angular velocity of the body and *M* the moment about the axis of rotation.

Precession, steady *see* Gyroscope.

Principal axes of inertia For a given body and a given point O it is always possible to find at least one set of orthogonal axes ($\xi\eta\zeta$) so that the corresponding *inertia matrix* is diagonal, i.e. the *products of inertia* $I_{\xi\eta}$, $I_{\xi\zeta}$ and $I_{\eta\zeta}$ are zero. The corresponding *moments of inertia* $I_{\xi\xi}$, $I_{\eta\eta}$ and $I_{\zeta\zeta}$ are called *principal moments of inertia*. *See also* Inertia matrix.
Principal moments of inertia *see* Principal axes of inertia.
Principle of equivalence of classical mechanics If mechanical experiments are carried out in a certain frame of reference it is impossible to decide if the frame of reference is resting in a gravitational field or if it is accelerated. There is no possibility of distinguishing between gravitation and acceleration in this way. In this sense gravitation and acceleration are equivalent. *See also* Relativity, theory of; Einstein, Albert.
Product of inertia *see* Moment of mass.
Projectile motion Usually refers to the case of a body that is moving in a homogeneous gravitational field where the air resistance is negligible. If the body is moving in the xy-plane, where the y-axis is vertical, the equation of the path – the projectile parabola – can be written as

$$y = -\frac{g}{2v_0^2 \cos^2\alpha} x^2 + \tan \alpha \cdot x$$

Here $v_0 = (v_0 \cos \alpha,\ v_0 \sin \alpha)$ is the initial velocity at the origin. The angle α is the *angle of elevation*.

If a target is located above a certain curve in the xy-plane it can never be hit by a projectile launched from the origin at a speed v_0. It does not matter how the angle of elevation α is chosen. The actual curve is a parabola, the *safety parabola* with the equation

$$y = \frac{v_0^2}{2g} - \frac{gx^2}{2v_0^2}$$

Quantum mechanics, physics When the objects treated approach a molecule in size the need for a physical model that is better than *classical mechanics* increases considerably. Quantum mechanics was developed during the first part of the twentieth century and has been established as the model for describing atomic and molecular physics. Important achievements were originally made by Planck, Einstein and Bohr among others. A characteristic quantity in quantum physics is Planck's constant $h = 6.626 \cdot 10^{-34}$ kg m/s. Quantities of the same dimension (MLT^{-1}), e.g. a linear momentum, may be compared with h. Obviously, there are no quantities occurring in traditional engineering problems that are comparable with h. Thus, Newton's model is good enough.
Radius of gyration *see* Moment of mass.
Reaction, law of action and reaction *see* Action; Newton's laws.
Reduction of force system *see* Force system.
Relativistic mechanics *see* Relativity, theory of.
Relativity, theory of The special theory of relativity (Einstein, 1905) must be used in the study of bodies whose velocity v is of the same order of magnitude as the velocity of light c. Relativistic mechanics includes Newton's mechanics as a limiting case and turns into it when the ratio v^2/c^2 may be neglected compared with 1.

The theory of general relativity (Einstein, 1916) is a fundamental, general theory of gravitation. One of its fundamental principles is that the *principle of equivalence* applies to all physical phenomena. There is no possibility of distinguishing between gravitation and acceleration with *any* physical experiment. Typical consequences are that light rays are deflected by a gravitational field and that a redshift appears in the spectral lines of the radiation from the stars. *See also* Principle of equivalence of classical mechanics.
Resisting force Also referred to as *viscous drag*, a force F on a moving body which increases when the speed of the body increases. The models most widely used are the linear approximation,

$F = cv$, and the quadratic approximation, $F = kv^2$. A combination of these models, $F = bv + dv^2$, is always better than any of the two basic approximations. The constants involved (c, k, b, d) depend on the medium and the shape and size of the moving body.

A falling body of mass m is subjected to a resisting force. The velocity approaches a *terminal velocity* v_g characterized by the fact that the acceleration – and the force sum – have become zero. Thus $mg - F = 0$. Therefore, in the linear case $v_g = mg/c$ and in the quadratic case $v_g = (mg/k)^{1/2}$.

Resonance *see* Vibrations.

Resting frame *see* Frame of reference.

Restitution, coefficient of *see* Impact, coefficient of restitution.

Resultant, force resultant *see* Force system.

Rigid body The characteristic property of a rigid body is that the deformations are negligibly small, i.e. the distance between two arbitrary points of the body is constant.

Robot A robot or *robotic arm*, also called a *manipulator* or *manipulator arm*, is a mechanical device which can be programmed to perform various tasks. In industry robots are used for welding, component mounting, spray painting, etc.

A robotic arm is often modelled as a chain of rigid *links* connected by *joints*. At the end of the arm an *end-effector* or *tool* is mounted.

Robotic arm *see* Robot.

Rotation matrix If two coordinate systems share the same origin O but the axes do not coincide the transformation from one system to the other is described by a rotation matrix **R**. If the coordinates of a certain point Q are $(x_0;\ y_0;\ z_0)$ and $(x_1;\ y_1;\ z_1)$, respectively, the relation between these sets of coordinates is

$$\begin{pmatrix} x_0 \\ y_0 \\ z_0 \end{pmatrix} = \begin{pmatrix} R_{11} & R_{12} & R_{13} \\ R_{21} & R_{22} & R_{23} \\ R_{31} & R_{32} & R_{33} \end{pmatrix} \begin{pmatrix} x_1 \\ y_1 \\ z_1 \end{pmatrix} \quad \text{where } \mathbf{R} = \begin{pmatrix} R_{11} & R_{12} & R_{13} \\ R_{21} & R_{22} & R_{23} \\ R_{31} & R_{32} & R_{33} \end{pmatrix}$$

represents the transformation from $Ox_1y_1z_1$ to $Ox_0y_0z_0$.

The inverse transformation, from $Ox_0y_0z_0$ to $Ox_1y_1z_1$, is described by the transposed matrix $\mathbf{R}^T$ because $\mathbf{R}$ is orthogonal; $\mathbf{R}^T = \mathbf{R}^{-1}$. *See also* Homogeneous coordinates.

Rotation of a rigid body

> Rotation about a fixed axis:
> Every point P in the body moves in a circle with its centre O on the axis. The position of P can be described by the angle φ between the line OP and a fixed reference radius of the circle. The *angular velocity* ω is defined as the time derivative of φ: $\omega = \dot{\varphi}$. The angular velocity vector $\boldsymbol{\omega}$ is directed along the axis according to the right-hand rule. The magnitude is $|\boldsymbol{\omega}| = |\dot{\varphi}|$. The *angular acceleration* $\dot{\boldsymbol{\omega}}$ has the magnitude $|\ddot{\varphi}|$ and is directed along this axis.
> Rotation of a body about a fixed point O, *spherical motion*:
> If one point O of the body is fixed it can be proved that the motion at every instant may be described as a rotation about an axis through O. The *angular velocity* vector $\boldsymbol{\omega}$ describes the instantaneous motion of the body in the same way as in the case of rotation about a fixed axis; $\boldsymbol{\omega}$ is directed along the axis and the velocity $\boldsymbol{v}$ of an arbitrary point P of the body is given by $\boldsymbol{v} = \boldsymbol{\omega} \times \boldsymbol{r}$ where $\boldsymbol{r} = \overrightarrow{OP}$. The direction, and not only the magnitude, of $\boldsymbol{\omega}$ can change with time in this general case. The *angular acceleration* $\dot{\boldsymbol{\omega}}$ is the time derivative of the vector $\boldsymbol{\omega}$.

In general the direction of $\dot{\boldsymbol{\omega}}$ differs from the direction of $\boldsymbol{\omega}$.

Rotational symmetry Also called *axisymmetry*. If the mass distribution of a body is only a function of the distance to a certain straight line, the axis of symmetry, the body has rotational symmetry. A body is also said to have *dynamic rotational symmetry* if two *principal moments of inertia* are equal. Thus, if $I_{xx} = I_{yy}$ the z-axis is the symmetry axis.
Safety parabola *see* Projectile motion.
Section force system The system of internal forces at a cross-section of a body, e.g. a beam. The section force system is usually described in terms of *axial force, shear force, torsional moment* and *bending moment*. *See also* Beams.
Shear force *see* Beams.
SI system Système International d'Unités; international system of units, ratified 1960. The SI system contains seven *fundamental quantities* and the corresponding seven *base units*. *See also* Fundamental quantities.
Sliding vector *see* Tool.
Speed Magnitude of *velocity*, $|\boldsymbol{v}|$.
Spin *see* Gyroscope.
Spinning top *see* Gyroscope.
Spring constant Also called stiffness. Denoted by the constant k in the relation $F = k\delta$ between the force F and the deformation δ of an elastic body, e.g. a spring, *see also* Elastic force; Hooke's law.
Stability If a mechanical system in a state of equilibrium remains in the neighbourhood of the equilibrium position after a small disturbance and moves with small velocities, then the position of equilibrium is *stable*.

As an example, consider a pendulum at its bottom position (stable) and at its top position (unstable). An intermediate situation can be exemplified by a ball on a horizontal surface; all positions are equilibrium positions (neutral equilibrium).

For a system where work can only be done by conservative forces it is possible to show that, for the equilibrium position to be stable, the potential energy must have a relative minimum. The equilibrium is neutral if the potential energy is constant.

Also, a dynamic state can be characterized in terms of stability. Consider, for example, a steadily spinning top (stable) and a bicycle that is sent away straight ahead in an upright position (unstable).
Static moment of mass *see* Moment of mass.
Statical determinacy, indeterminacy A problem in statics where all the unknowns may be determined from the equations of equilibrium is called statically determinate. If this is not the case the problem is said to be statically indeterminate. For example: a rigid body is hanging in three vertical cables not in the same plane; the forces in each cable can be determined. A fourth cable is added; the force in each individual cable cannot then be determined.
Statics That part of *mechanics* where systems at rest are studied, *see also* Equilibrium.
Steiner's theorem Also called the parallel-axis theorem. Consider an axis passing through the centre of mass of a certain body and another, parallel, axis. The two corresponding *mass moments of inertia* are $\bar{I}$ and I respectively. The following relation holds: $I = \bar{I} + md^2$ where m is the mass of the body and d the distance between the axes.
Stevin, Simon (1548–1620) Dutch mathematician. He was aware of the parallelogram law and used it without explicitly stating it.
Tangential acceleration *see* Normal and tangential components.
Terminal velocity *see* Resisting force.
Theory of relativity *see* Relativity, theory of.
Tool The last link of a robotic arm. Its position and orientation are essential for the task to be performed. The z_n-axis of the tool coordinate system ($x_n y_n z_n$) is usually assigned along the

direction the tool is pointing. The corresponding unit vector $\boldsymbol{a}$ is called the *approach vector*. If the tool is designed for gripping objects the y_n-axis is defined along the 'open–close axis' of the tool. The corresponding unit vector $\boldsymbol{s}$ is the sliding vector. The unit vector $\boldsymbol{n}$ along the x_n-axis is the *normal vector*.

These vectors together with the coordinates of the tool in the base coordinate system specify the *tool configuration*. The *homogeneous transformation matrix* **T** contains information on these quantities. *See also* Homogeneous coordinates; Robot.

Tool configuration *see* Tool; Homogeneous coordinates.

Torque *see* Force, moment of.

Torsional moment *see* Beams.

Translation A type of motion characterized by the fact that the points of a body at every instant all have the same velocity. This means that every line which is fixed with respect to the body must remain parallel to its original direction. To describe the motion of the body it is sufficient to consider one point, e.g. the centre of mass, and its velocity $\boldsymbol{v}(t)$.

Truss A construction that is used in civil engineering applications, e.g. bridges and roofs, to support various types of load. A truss consists of straight *members* that are connected by smooth joints. In most cases gravitational forces on the members may be neglected compared with the applied loading. This is usually assumed to be applied at the joints only. To determine the necessary dimensions of the members and to choose a suitable material the forces in the members are to be determined. Therefore a systematic method, the *method of joints*, is often used; each joint is isolated and the equilibrium of that joint is studied.

Varignon's theorem The *moment of a force* about a point does not change if the force is first resolved into components according to the *parallelogram law* and the moments of the components are then added. It follows that the theorem also applies to the *moment about an axis* as this can always be regarded as a scalar component of a moment vector about a point.

Velocity A vector quantity defined as $\boldsymbol{v} = d\boldsymbol{r}/dt$ where $\boldsymbol{r}$ is the position vector of the moving point. *Speed* $= |\boldsymbol{v}|$, *see* Normal and tangential components; Polar coordinates.

Vibrations

Free, undamped vibrations *see* Harmonic oscillator.

Free, linearly damped vibrations:

Vibrations according to the equation $x(t) = Ae^{-\zeta\omega t}\sin(\omega_d t + \delta)$. Here ω is the *natural frequency*, ζ the *damping factor* and ω_d the *angular frequency*. The equation of motion is $\ddot{x} + 2\zeta\omega\dot{x} + \omega^2 x = 0$.

Forced, linearly damped vibrations under harmonic excitation:

The equation of motion is $\ddot{x} + 2\zeta\omega\dot{x} + \omega^2 x = \xi_0 \sin \Omega t$. The solution is $x(t) = x_h(t) + x_p(t)$ where x_h is the function of x as above and x_p is the particular solution $x_p = C\sin(\Omega t - \psi)$, the steady-state function. Here $C = \lambda\xi_0$ where λ is the *magnification factor*.

For small values of ζ the amplitude C becomes very large for $\Omega \approx \omega$. This is the case of *resonance*.

Viscous drag *see* Resisting force.

Weight The gravitational force with which a body is affected by the earth, *see* also Gravity.

Work The work of force $\boldsymbol{F}$ during displacement $d\boldsymbol{r}$ is $dW = \boldsymbol{F}\cdot d\boldsymbol{r}$. Accordingly, the work performed during a displacement along a curve from a point P to a point Q is

$$W_{PQ} = \int_{s_P}^{s_Q} F_s \, ds$$

where s is the arc length from a reference point on the curve.

The work done by a couple or a single force during a rotation around a fixed axis from a position P to a position Q is

$$W_{PQ} = \int_{\varphi_P}^{\varphi_Q} M \, d\varphi$$

Here φ is the rotation angle from a reference position and M is the moment about the axis of the couple or the force.

Work–energy principle The change in kinetic energy is equal to the total work done by the forces acting. This can be written as $W = \Delta T = T_2 - T_1$. Here W is the total work done by the active forces and T_1 and T_2 the kinetic energies at the beginning and the end of the process, respectively.

The work done by *conservative forces* is $-\Delta V$ where V is the potential energy function. If the work done by non-conservative forces is denoted by $W^{(nc)}$ the work–energy relation can be written as

$$W^{(nc)} = \Delta(T + V) = T_2 + V_2 - (T_1 + V_1)$$

If $W^{(nc)} = 0$ the total mechanical energy is conserved; $T + V$ is a constant.

Zero system of force *see* Force system, reduction of.

Bibliography

Classical mechanics

Danby, J.M.A. (1992) *Fundamentals of Celestial Mechanics* Willmann-Bell, Inc., Richmond, VA

Desloge, E.A. (1982) *Classical Mechanics* John Wiley & Sons, New York

Ginsberg, J.H. (1988) *Advanced Engineering Dynamics* Harper & Row Publishers, New York

Goldstein, H. (1980) *Classical Mechanics* Addison-Wesley, Reading, MA

Haug, E.J. (1989) *Computer Aided Kinematics and Dynamics of Mechanical Systems* Allyn and Bacon, Needham Heights, MA

Mathews, J.H. (1987) *Numerical Methods for Computer Science, Engineering and Mathematics* Prentice Hall, Englewood Cliffs, NJ

Meirovitch, L. (1970) *Methods of Analytical Dynamics* McGraw-Hill, New York

Meirovitch, L. (1985) *Introduction to Dynamics and Control* John Wiley & Sons, New York

Newton, Sir Isaac *Principia* (translated into English and revised) University of California Press, Berkeley (1962)

Symon, K.R. (1971) *Mechanics* Addison-Wesley, Reading, MA

Synge, J.L. and Griffith, B.A. (1959) *Principles of Mechanics* McGraw-Hill, New York

Whittaker, E.T. (1937) *Analytical Dynamics* Cambridge University Press, London

Wittenburg, J. (1977) *Dynamics of Systems of Rigid Bodies* B.G. Teubner, Stuttgart

Solid mechanics

Gere, J.M. and Timoshenko, S.P. (1991) *Mechanics of Materials* Chapman & Hall, London

Timoshenko, S. (1956) *Strength of Materials* D. Van Nostrand Company, New York

Vibrations

Hagedorn, P. (1988) *Non-linear Oscillations* Clarendon Press, Oxford

Meirovitch, L. (1967) *Analytical Methods in Vibrations* The Macmillan Company, New York

Moon, F.C. (1987) *Chaotic Vibrations* John Wiley & Sons, New York

Nayfeh, A.H. and Mook, I.T. (1979) *Nonlinear Oscillations* John Wiley & Sons, New York

Robotics

Asada, H. and Slotine, J.J.E. (1985) *Robot Analysis and Control* John Wiley & Sons, Chichester
Craig, J.J. (1989) *Introduction to Robotics: Mechanics and Control* Addison-Wesley, Reading, MA
McKerrow, P.J. (1991) *Introduction to Robotics* Addison-Wesley, Sidney
Schilling, R.J. (1990) *Fundamentals of Robotics: Analysis and Control* Prentice Hall, Englewood Cliffs, NJ

General physics

Alonso, M. and Finn, E.J. (1992) *Physics* Addison-Wesley, Reading, MA
Einstein, A. (1955) *The Meaning of Relativity* Princeton University Press, Princeton, NJ
Einstein, A. and Infeld, L. (1938) *The Evolution of Physics* Cambridge University Press, London
Feynman, R.P., Leighton, R.B. and Sands, M. (1963) *The Feynman Lectures on Physics* Addison-Wesley, Reading, MA
Merzbacher, E. (1970) *Quantum Mechanics* John Wiley & Sons, New York

Index